The New Naturalist Libra.

A SURVEY OF BRITISH N:

PLANT DISEASE – A NAT

The aim of this series is to interest the general reader in the wildlife of Britain by recapturing the enquiring spirit of the old naturalists. The editors believe that the natural pride of the British public in the native flora and fauna, to which must be added concern for their conservation, is best fostered by maintaining a high standard of accuracy combined with clarity of exposition in presenting the results of modern scientific research.

The New Naturalist

PLANT DISEASE
A NATURAL HISTORY

David Ingram & Noel Robertson

With 8 colour plates and over 100 black
and white photographs and line drawings

HarperCollins*Publishers*

HarperCollins*Publishers*
77–85 Fulham Palace Road
Hammersmith
London W6 8JB

The HarperCollins website address is:
www.**fire**and**water**.com

Collins is a registered trademark of HarperCollins*Publishers* Ltd.

First published 1999

ISBN 0 00 220074 0 (Hardback)
ISBN 0 00 220075 9 (Paperback)

Printed and bound in Great Britain by The Bath Press
Colour reproduction by Colourscan, Singapore

Contents

List of Plates

Plate 1 *Phytophthora infestans,* **cause of late blight of potato**
Leaf lesion with a necrotic centre; note the white 'haze' of sporangiophores and sporangia on the green tissue around the margin of the lesion.
Infected tuber; note the foxy red discolouration of the infected cells.
Potato plot with a primary focus of infection in the top left hand corner.
The same plot ten days later; spores have spread from the primary focus to infect most of the plants.

Plate 2 Zoosporic pathogens
Pythium ultimum causing damping-off in a pot of lettuce (*Lactuca sativa*) seedlings.
Phytophthora sp. (probably *cactorum*) causing partial death of Japanese maple (*Acer palmatum*) at Benmore, Argyll.
Bremia lactucae causing downy mildew of lettuce; note that the infected tissues between the veins bear white spores and are quite green.
Albugo candida causing white blister on shepherd's purse (*Capsella bursa-pastoris*) at Gullane, East Lothian; note the masses of white spores and the distortion of the infected flower stalks.

Plate 3 Pathogens spread by conidia or ascospores
Botryotinia fuckeliana (= *Botrytis cinerea*) causing grey mould on *Kalanchoe blossfeldiana*; note the grey 'haze' of conidiophores and conidia on the rotted tissues.
Coniothyrium hellebori causing target spot on a leaf of Christmas rose (*Helleborus niger*).
Taphrina deformans causing 'peach' leaf curl on almond (*Prunus amygdalus*); note the distortion and red coloration of the infected leaves.
An ergot of *Claviceps purpurea*, collected from sea lyme grass (*Elymus arenarius*), germinating to produce drum-stick perithecial stromas.

Plate 4 Pathogens spread by conidia or ascospores continued
Epichloe typhina causing choke of fescue (*Festuca* sp.) near Kilmartin, Argyll.
Venturia inaequalis causing scab of crab apple (*Malus* sp.) in Aberlady, East Lothian.
Early signs of disease on one of the elms (*Ulmus* sp.) along the 'Backs' in Cambridge in 1975. Within weeks this magnificent tree was dead.
Galleries of elm bark beetle.

Plate 5 Powdery Mildews
Erysiphe pisi on pea (*Pisum sativum*) in Edinburgh; note the whitish, powdery covering of hyphae and conidia on the green leaves and stems.
Green islands induced by colonies of *Uncinula bicornis* on sycamore (*Acer pseudoplatanus*).
Sphaerotheca mors-uvae on gooseberry (*Ribes uva-crispa*); note the masses of brownish hyphae and conidia on the surface of the green fruit and leaves.
Magnified ascocarps (cleistothecia) and surface hyphae of rhododendron powdery mildew ('*Microsphaera* type').

Plate 6 Rusts

Honey-coloured, sweet-smelling pycnia of *Puccinia punctiformis* (creeping thistle rust) on elongated, paler green leaves of *Cirsium arvense.*
Orange aecia of *Puccinia* sp. (probably *caricina*) on the surface of nettle (*Urtica dioica*); note the distortion of the infected tissues.
Lesions with concentric rings of urediospores of *Puccinia graminis* f. sp. *tritici* (black stem rust) on wheat.
Teliospores of *Uromyces muscari*, together with green islands, on English bluebell or wild hyacinth (*Hyacinthoides non-scripta*).

Plate 7 Rusts continued and Smuts

Telial horns of *Gymnosporangium cornutum* on *Juniperus communis* near Walkerburn, Peeblesshire.
Pointed aecia of *G. cornutum* on a leaflet of rowan (*Sorbus aucuparia*), the alternate host.
Black ustilospores of *Ustilago violacea* on the anthers of red campion (*Silene dioica*).
White bud spores (= conidia) produced by budding of hyphae of *Urocystis primulicola* infecting the anthers of primrose (*Primula vulgaris*).

Plate 8 Wood-rots

Heterobasidion annosum causing death of young pine trees (*Pinus* sp.) being shown to students by the late John Rishbeth in Thetford Forest, Norfolk.
Basidiocarps of *H. annosum* pushing up through Sitka spruce leaf litter at Juniper Bank, Peeblesshire.
Meripilus giganteus basidiocarps growing from infected roots of beech (*Fagus sylvatica*) in Cambridge.
Ganoderma applanatum basidiocarps growing from a mature beech tree; note the masses of rusty brown basidiospores.

Editors' Preface

It is well said that the British are 'a nation of gardeners', and as such they must all be aware of the manifestations of plant diseases. Moreover, in recent years there have been very welcome signs that the *apartheid* separating notionally something called 'wildlife' from anything classed as a garden plant (or animal) has been largely broken down. It is therefore singularly appropriate that we include in the New Naturalist series this excellent volume devoted to the natural history of plant disease. It clearly satisfies both conditions included in our aims for the series: namely, it will 'interest the general reader in the wildlife of Britain by recapturing the inquiring spirit of the old naturalists', and it maintains 'a high standard of accuracy combined with clarity ... in presenting the results of modern scientific research'.

We could not have found better authors to undertake the task for us. As they tell us in their Foreword, their friendship and shared enthusiasm for plant pathology go back forty years, and the book represents a successful new collaboration. Throughout its pages we appreciate this joint enthusiasm, and we share the authors' hope that their many readers will gain, as they evidently did, 'the same immense pleasure and intellectual stimulus from the study of diseases'.

It is the convention of the Editors to entrust each book to a particular member of the Editorial Board, to whom falls the task of drafting the Editors' Preface to be published anonymously. For this occasion, I have asked that the rule be set aside, for the reason that will now be apparent. Within a day of my receiving the fully-corrected page proofs, David rang to tell me the news of Noel's death. Both authors were personal friends of mine, from their Cambridge days. I do not feel it appropriate to say more, but can wholeheartedly commend to you this collaborative work which can speak for itself.

S. Max Walters
July 1999

Authors' Foreword

Plant diseases are all about us, in the city and in the countryside. They often go unnoticed, except by the initiated, yet they are essential components of most natural plant communities. They are of immense social and economic importance, cause massive losses in agriculture, horticulture and forestry, and have sometimes changed not only the landscape but also the course of history.

The mechanisms by which plant pathogens – fungi, bacteria, viruses and their relatives and even some higher plants – interact with their hosts are endlessly fascinating and have led generations of scientists to try to unravel their mysteries. These efforts have contributed to many major scientific discoveries, including the structure of DNA, and have resulted in many novel products and technological innovations such as antibiotics, plant growth regulators ('hormones'), disease-resistant crops and techniques for producing genetically modified foods.

This book has been written in the hope that it will introduce to a wider audience our enthusiasm for, and enjoyment of, plant diseases. In the first three chapters we are concerned to understand the relationship between pathogen and host, a relationship that seems simple on the surface yet is often more complex than that between mammals and their pathogens (the animal system, based in large part on the circulatory system with circulating antibodies and leucocytes, is not available to the plant). The remaining chapters deal with the interaction between selected groups of pathogens and their hosts and the role, sometimes spectacular, of the resulting diseases in human affairs. Our greatest difficulty, as for most scientific authors, has been to find a suitable balance between technical accuracy, often involving unfamiliar terminology, and our desire to make the writing comprehensible to all. We hope that the finished product does indeed combine these two aims to the satisfaction of most readers.

Writing the book has been a true collaboration, and neither one of us can be regarded as the 'senior author'. It has revitalised a friendship and shared enthusiasm that began almost forty years ago, in October 1960, when we met in a lecture theatre, NFR as Professor and DSI as a first year student of plant pathology. Between us we span almost sixty years of work on plant disease, from simple cultural studies of parasitic fungi to the modern age of biochemistry and molecular genetics. In this period we have been closely acquainted with many great plant pathologists of the past and present; their views have so much influenced us that what we have written has inevitably derived, at least in part, from our contact with them.

The early studies of plant disease are easily accommodated in a natural history. More recent but equally important insights from biochemistry, genetics and molecular biology are perhaps less obviously part of such an approach. But if natural history is to go forward it must be ready to absorb, step by step, new scientific knowledge as it is developed, a way of thinking that the New Naturalist series seeks to promote. Moreover naturalists, who have played a significant role in the development of the science of plant pathology in the past, can continue to do so long into the future. Their contribution may embrace both detailed observation and the testing of hypotheses by experiment. It will be especially important, for example, in adding to knowledge of the role and distribution of plant diseases in natural plant communities, still poorly understood, and in unravelling details of the life cycles and host ranges of the innumerable pathogens that have not yet attracted the full attention of professional plant pathologists.

Finally, a word about plant diseases and conservation. Although the counterintuitive notion that plant pathogens should be conserved together with their hosts has hardly been mentioned in our book, it is a subject now dominating the scientific thinking of one of us, and may turn out to be just as important as the conservation of more obviously useful components of biodiversity. For those who wish to know more about this developing aspect of plant pathological thinking an appropriate reference (DSI, 1999) is included in the Bibliography.

We sincerely hope that all who read this book will gain the same immense pleasure and intellectual stimulus from the study of plant diseases as we have.

David Ingram & Noel Robertson
Aberlady and Walkerburn, Scotland
July 1999

Noel Robertson died peacefully on 2nd July 1999, just a few days after this foreword was written.

ACKNOWLEDGEMENTS

We have received help from many colleagues and former students during the writing of this book but acknowledge particularly help with difficult technical problems from: Clive Brasier, Don Clarke, James Duncan, John Friend, Chris Gilligan, Brian Harrison, Stefan Helfer, Paul Holliday, Anne Osbourn, Doug Parbery and Roy Watling. We also thank Monica Goldspink for checking the Latin and Greek derivations used in the Glossary. However, we take full responsibility for any interpretation or misinterpretation of the information and advice given so freely by all those we consulted. In a book of such scope the task of establishing a credible taxonomic framework has been greatly helped by reference to *A Dictionary of Plant Pathology* by Paul Holliday, *Microfungi on Land Plants* by Martin and Pamela Ellis and the many monographs and references mentioned in the Bibliography. The photographs are acknowledged individually in the captions and we are grateful to all who supplied them. However, we owe a particular debt of gratitude to Debbie White of the Royal Botanic Garden Edinburgh for all her hard work on our behalf. We also thank Mary Bates for the line drawings which are such an important element of the book. Finally we thank Alison Ingram, who typed and managed the manuscript for us, and without whom the book would not have been completed.

This book was written while DSI was Regius Keeper and then Honorary Fellow of the Royal Botanic Garden Edinburgh, and NFR was Honorary Fellow of the Royal Botanic Garden Edinburgh.

1

Blasting and Mildew
The Causes of Plant Disease

Introduction: plants, politics and pathogens

On 13 September 1845 Dr John Lindley, distinguished Professor of Botany in the University of London and editor of the *Gardener's Chronicle*, then a most influential scientific journal, held up printing to insert the following dramatic announcement.

> 'We stop the press with very great regret to announce that the potato murrain has unequivocally declared itself in Ireland. The crops about Dublin are suddenly perishing ... where will Ireland be in the event of a universal potato rot?'

This disease had come late in the season and destruction of the haulms of the potato – the stems and leaves – did not affect the yield significantly, for the tubers had already formed. Much to everyone's surprise, a good crop of apparently healthy tubers was lifted. But during harvest the weather was wet and the exposed tubers became infected by disease washed down from the decaying leaves, although no-one at the time realised what was happening. Within a few weeks most had rotted in store and the famine in Ireland had begun. The epidemic of 1845 was followed by a second, even worse, in 1846. This came early in the season and killed the haulms even before the tubers had started to swell, let alone reach maturity. Further equally devastating epidemics followed in subsequent years and all Lindley's worst fears were realised. Between 1845 and 1851 one and a half million Irish men, women and children died in misery from starvation and disease. A million or more people, driven to despair by the succession of appalling harvests, emigrated, mostly to America. A plant disease was changing the course of history, its economic effects causing not only social but also major political upheaval.

Sir Robert Peel, the Prime Minister in 1845 and leader of the Tory party, was seen, perhaps unfairly, as an aloof, unfeeling man, not noted for his Irish sympathies. Peel maintained that the early reports of starvation were exaggerated, and mishandled the problem from the start. When he was forced to acknowledge the gravity of the situation and do something about it he repealed the Corn Laws, which until then had protected the profits of English farmers by restricting imports of foreign grain. This move was ostensibly humanitarian, since it allowed grain to be imported into Ireland to feed the starving population, but was in reality highly political. Despite his passionate protestations to the contrary, Peel, it has been said, did not care a fig whether the Irish starved or not; he saw the famine as an opportunity to promote free trade, an element of Whig policy he admired and knew would operate to the advantage of the wealthy merchant and industrial class then rising to power, to which he belonged by birth.

He had delayed too long, however, on both counts: the measure came too late to do any good for the Irish, and the land-owning Tory voters were furious at the repeal of the Corn Laws. Peel and his party were thrown out of office and with the exception of some short breaks remained powerless for the next 20 years.

Dr Lindley, a journalist in touch with the practical world of gardening and an academic botanist, was well equipped to realise the potential damage an epidemic of disease could cause in an agricultural crop such as potato. Blight had been known for the three previous years on the eastern seaboard of North America, and had spread to ports where ships homebound from the Americas docked. It first manifested itself in Europe at Courtrai in Belgium, and from there spread to France and southeast England, but it was in Ireland, where the potato was the staple food of the peasant farmers, that the appearance of the disease would have disastrous consequences, and it was this knowledge that caused Lindley's alarm.

But what was the state of scientific knowledge at this time? How much was understood about plant diseases in the mid-nineteenth century, and why did the potato 'murrain' in Ireland have such a profound effect on naturalists and scientists alike, leading to the development of the modern science of plant pathology?

Diseases had been known on crops from the earliest times and various measures were constantly being tried in an effort to control them. During the seventh century BC, for example, the Romans sacrificed a red dog on 25 April each year in the vain hope of persuading the corn god Robigus to control the rust disease of wheat by chaining up the dog star, Sirius, which was thought to exert a malign influence on the crop. In medieval times farmers astutely observing that this same disease occurred where barberries were growing nearby, deduced that it must be barberry (*Berberis vulgaris*) that exerted a malign influence on wheat and consequently rooted it out from the hedgerows. Although they did not fully understand the connection we now know that the organism causing the rust disease of wheat has a complicated life cycle that involves a period of time growing on the leaves of the common barberry as well as on wheat itself (see p.178). Thus the medieval farmers were able to ameliorate the severity of the disease on their crops by breaking its life cycle.

Although in practical terms medieval observation was more successful than ancient stargazing, the scientific study of plant diseases only began in the seventeenth century. It advanced rapidly, however, and by 1751, when Carl von Linné (Linnaeus) wrote *Philosophia botanica*, he was able to list mildew, rusts, smuts and ergot. Studies by botanists in Britain, mainland Europe and America through the eighteenth and early nineteenth centuries did much to advance knowledge of these and other afflictions of plants. In particular the various forms of spores (the equivalent of seeds) occurring on the moulds emanating from the diseased areas of leaves and stems of afflicted plants were described in detail.

The true cause of disease remained in dispute throughout this time. Some, such as the great explorer of the Antipodes and gentleman botanist Sir Joseph Banks, believed that infections were caused by minute parasitic plants; others held that disease was caused by innate morbid factors, and that spores and

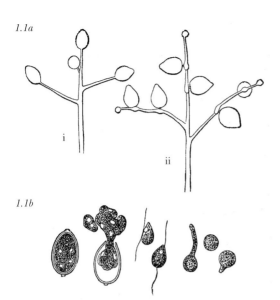

1.1a

1.1b

i

ii

Fig. 1.1 Drawings by Anton de Bary of *Phytophthora infestans*, cause of late blight of potatoes: (a) young (i) and mature (ii) sporangiophores with lemon-shaped sporangia (note that the first-formed sporangia are pushed to one side as the sporangiophores grow); (b) a sporangium and stages in the formation, release and germination of the biflagellate zoospores. The sporangia are approximately 20 μm in diameter.

other signs of the minute organisms they observed in diseased tissue were a consequence, not a cause of the infection, much as mould grows on stale bread or on old boots left too long in the damp. There was similar uncertainty about the cause of the potato murrain (now called 'potato blight' or in America 'late blight', to distinguish it from other diseases occurring earlier in the season). It was widely held at the time of the Irish famine that the cause of the disease was a kind of self destruction brought on by saturation with heavy rain (rarely in short supply in Ireland!), a plant equivalent of human dropsy.

Among the various experts brought in to advise on the causes of the epidemic was the Rev. Miles Joseph Berkeley, a country parson from King's Cliffe in Northamptonshire and 'a gentleman eminent above all other naturalists of the United Kingdom in his knowledge of the habits of fungi (moulds)'. In his seminal and classic paper entitled *Observations, botanical and physiological, on the potato murrain,* published in the *Journal of the Horticultural Society of London* in January 1846, he stated unequivocally that 'The decay is the consequence of the presence of the mould, and not the mould of decay.'

Others in America, Belgium and France had held this view previously, as Berkeley was careful to record. Indeed, an old surgeon of Napoleon's armies, Dr J.F.C. Montagne, then living in Paris, corresponded regularly with Berkeley on mould matters. It was Montagne's letters and the drawings they contained that led Berkeley to his ultimate conclusion, and it was Montagne who, on 30 August 1845, named the mould on potato as *Botrytis infestans,* now renamed *Phytophthora infestans* (Mont.) de Bary (Plate 1). The dispute continued, however, and even the enlightened Dr Lindley, that most eminent of botanical editors, continued to hold the view that the mould was the result of the decay and that the cause was a fatal combination of 'environmental and physiological factors'. It was not until many years later that the dispute ended with the publication of the definitive memoirs on plant disease by the great German plant pathologist Anton de Bary, in 1861 and 1863 (Fig. 1.1).

The pioneering studies of Berkeley and Montagne on the causes of potato blight became the model for studies of the diseases afflicting all other plants. With their work the modern science of plant pathology (the study of the causes, biology and control of plant disease) actually began. It has since been discovered that plant diseases may variously be caused not only by moulds (fungi) of many kinds, but also by bacteria, viruses and their relatives, and flowering plants. Organisms such as these which cause disease in plants are known as plant pathogens. This book will be primarily concerned with diseases caused by pathogenic fungi, for these are the most obvious to the naked eye and are most easily studied by the naturalist, but in Chapter 12 a wider view is taken of the other groups.

But why do epidemics of plant diseases, like the biblical 'blasting and mildew' (Amos 4.9) or blight on potatoes, develop from time to time with such devastating effect? To begin to answer this question we must first look at disease in natural plant communities.

In nature, wherever plants grow, from the top of the highest mountains to the meadows and woodlands of the valley floors, individuals of almost every species of plant present will be afflicted from time to time by a rust, a smut, a mildew or a rot. More often than not the casual observer does not register this, however, for it is a truism that most plants are resistant to attack by most potential pathogens.

Through evolutionary time, in nature, certain pathogens have evolved to attack and cause disease in particular host plant species. For example, in Mexico, where *Solanum demissum*, a relation of the domestic potato (*S. tuberosum*), evolved, *Phytophthora infestans* may be found on wild populations of *Solanum* spp. but does not normally cause catastrophic epidemics in them, although epidemics can and do occur from time to time. The success and survival of both fungus and host are ensured because the populations of each are genetically diverse, in the host for resistance to infection by the pathogen and in the pathogen for the ability to cause disease in the host. A balance between the two populations is achieved such that each may grow and reproduce to give rise to the next generation. Individuals or distinct populations may be lost from time to time but the species as a whole survives. Natural plant communities consist of many different species and separate populations of each species, including a large number of pathogens, all occupying different niches in space and time. In such situations disease is rarely obvious and a practised eye is needed to spot it. Yet it is always there – during a recent walk through an apparently disease-free meadow in East Lothian, for example, we spotted diseases on the stems, leaves, flowers or fruits of a multitude of grasses, as well as on species of *Cirsium*, *Malus*, *Tragopogon*, *Capsella*, *Silene*, *Tussilago*, *Rosa*, *Lamium*, *Symphytum* and many more, all in less than an hour!

Catastrophic epidemics happen when circumstances intervene to upset this balance. In agriculture and horticulture, for example, genetically uniform populations of crop plants are grown to facilitate good husbandry; if wheat or potatoes are to be cultivated and harvested efficiently all the plants in the field must grow and ripen at about the same time. If a particular pathogen mutates (changes genetically) to overcome the innate resistance to disease of such a uniform crop species the disease will spread unchecked, like wildfire, for all individuals will be equally susceptible. This happened in North America in the 1970s when mutants of the previously innocuous pathogen *Bipolaris maydis* began to cause disease on corn (maize – *Zea mays*) carrying a gene for male sterility used in the breeding of a large proportion of the corn varieties then grown in the United States. The resulting epidemic of southern corn leaf blight had devastating economic consequences that were felt in countries far beyond the USA. It also happens, less dramatically, on a regular basis whenever an indigenous pathogen mutates to overcome the resistance in any crop cultivar or to detoxify a fungicide applied to it (see below). Alternatively, instead of a mutation, a new pathogen or a new strain of a pathogen to which a crop has no resistance may be introduced from elsewhere, another country perhaps. This was the case with *Phytophthora infestans* in Ireland, and more recently in Britain with the new highly pathogenic form of *Ophiostoma* (*Ceratocystis*) *ulmi* (see p.134), cause of Dutch elm disease, that has changed the countryside so dramatically.

In such cases there is an initial build-up of disease around the point of introduction into the crop and then, once sufficient spores of the pathogen have accumulated, rapid development of the disease until the whole crop has been destroyed. When expressed quantitatively the development of the epidemic follows a typical s-shaped biological growth curve (Fig. 1.2 and Plate 1). The situation may be exacerbated if weather conditions favour the growth of the pathogen, in which case the slope of the curve will be steeper. Epidemics may be restricted to a single field or may spread over a much larger area. An epidemic occurring on a continental and global scale is called a pandemic.

When the balance is upset and epidemics occur, farmers may either attempt to prevent or control the disease by the application of chemical sprays or dusts, or may plant varieties of crops specially developed by the plant breeder to be resistant to disease, or may do both. In developing resistant varieties of a crop, genes for resistance are derived from wild populations of the species or from existing cultivated lines and introduced into the cultivated form (cultivar) of that species by a series of controlled crosses. For example, the cultivars of

Fig. 1.2 Disease progress curves: (a) an example of an s-shaped (sigmoidal) curve, typical of many diseases such as powdery mildew (*Erysiphe graminis*) or yellow rust (*Puccinia striiformis*) on grasses and cereals (Gramineae), in which the epidemic starts slowly, then accelerates as more spores are produced, before slowing down as the supply of healthy hosts becomes exhausted; (b) an example of a curve for a disease in which the epidemic increases and then subsides before initiating another cycle of infection (as in the occurrence of Dutch elm disease, caused by *Ophiostoma ulmi* on *Ulmus* spp.) or in which the host outgrows the disease (as in diseases of many agricultural crops). The scales for both curves are in arbitrary units. (Illustrations by C.A. Gilligan & B. Goddard, University of Cambridge).

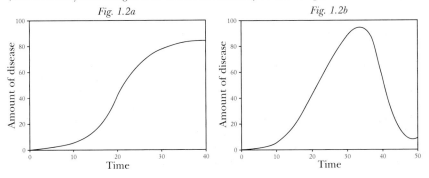

potato grown today carry resistance to potato blight derived from wild *Solanum* spp. and undomesticated potatoes discovered in South America; modern wheats are rendered resistant to attack by rusts because they have been developed by introducing resistance genes from older cultivars and sometimes from related wild grasses growing in the Middle East. Neither the use of fungicides nor the strategy of growing resistant cultivars of a crop offers a permanent solution, however, for most plant-pathogenic organisms are able to mutate very rapidly to become resistant to all but the most toxic of fungicides, or to become unrecognisable to plants carrying genes for resistance to a disease (see p.67). The fight against disease in agriculture is thus a continuous one in which new fungicides and new resistant cultivars must constantly be produced, adding very significantly to the cost of food production.

Because of their great economic and social significance, the diseases of crops have been widely studied since the nineteenth century, and we now know a great deal about the causal organisms and the ways in which they interact with their hosts. Diseases on wild plants, however, remain relatively unexplored. The eye of the naturalist abroad early in the day all too easily fixes on the showy flowers of goat's-beard or Jack-go-to-bed-at-noon (*Tragopogon pratensis*), completely missing the fact that individuals of this species may be heavily infected with disease. Our observations have shown that goat's-beard may carry no less than three fungal diseases – white blister, smut and rust – all at one time and still remain not only alive, but able to flower and set seed. Closer examination with a lens or microscope allows us to see that the pathogens involved are just as beautiful as the flowers of the host plant, whilst at the same time our minds are busy with the knowledge that any interaction between a plant and three parasites all at once must result from a most sophisticated set of ecological and physiological relationships. Close observation of the world of plant pathogens brings not only the intellectual challenge of exploration but also the excitement of discovering a group of organisms as fascinating as anything that may be encountered elsewhere in nature.

In the present book we hope to lead the general reader into this world. We shall draw our examples from the garden, for most of us first become aware of plant diseases on our antirrhinums, roses and cabbages; from agriculture, where diseases have their greatest economic impact and have therefore been most studied; and from natural plant communities, where much still remains to be discovered. Throughout we shall thus follow the advice of the American plant pathologist, J.C. Walker, who always told his students to 'keep one foot in the furrow'.

The plant destroyers: fungi, bacteria and viruses

Fungi

We have already seen that the early investigators of plant disease almost invariably found moulds associated with the lesions on the plants they studied. Although the dispute as to whether these organisms were a consequence rather than the cause of disease continued until late into the nineteenth century, studies of these minute, curious organisms were made, for their own sake, from the early seventeenth century onwards. Such studies were greatly aided by the invention of the compound microscope at the end of the sixteenth century, enabling the great microscopists of the seventeenth century, Hooke in

London, Leeuwenhoek in Leiden, Malpighi in Italy and others, to see for the first time that the moulds consisted of minute trailing threads that not only grew on the surface of the infected plant tissues but also penetrated them. These threads bore spores, singly or in chains or clusters, in a variety of shapes and sizes. We now know that the majority of plant diseases are caused by such moulds, or fungi as they are more correctly called, and the techniques of modern microscopy and biochemistry have revealed that they have a complex and sometimes beautiful structure (Fig. 1.3, overpage).

Many features distinguish fungi from both plants and animals, and they are therefore classified in a Kingdom of their own (but see p.20). Unlike plants, fungi lack the green pigment chlorophyll. They are therefore unable to carry out photosynthesis, the process which enables green plants to capture the physical energy of sunlight and transform carbon dioxide from the atmosphere into the chemical energy of sugars. Since sugars are essential for the growth of all living things, being 'burned' in respiration to release the chemical energy required for driving all basic metabolic processes as well as for building bodies, fungi must obtain their sugars, and indeed all other foods, from an external source. They have evolved two strategies to deal with this problem, saprophytism and symbiosis. Saprophytes are free-living, utilising nutrients that leak out onto the plant surface or the dead and decaying remains of plants and animals wherever these occur, especially in the soil. Evidence of the breakdown of dead organic remains by saprophytic fungi may be seen most dramatically in the compost heap, or when a pecked-at apple is left on the ground to decay, but goes on everywhere in a more dispersed way and is essential to the cycling of nutrients in all natural plant communities.

Symbiotic fungi, on the other hand, live in close association with and derive their nutrients from living organisms, usually plants. Such symbiotic associations may be mutualistic, or parasitic. In mutualistic associations both host and fungus benefit from the presence of the other. This is the case with mycorrhiza, where various fungi may associate mutualistically, in a variety of ways, with the roots of most plants. Examples are the vesicular-arbuscular mycorrhizas, virtually ubiquitous amongst flowering plants in the wild, in which the mineral scavenging properties of the fungal partner are utilised to aid the uptake into the plant of inorganic phosphates and other compounds from the soil. The fungus, meanwhile, benefits from a continuous supply of sugars from the photosynthetic activities of the host plant. Mycorrhizal associations are fundamental to the success of most natural plant communities, for essential mineral nutrients are always in short supply in the soil and some, notably phosphates, usually occur in complex forms not easily available to the unaided roots of plants (see pp.141–152).

The naturalist sees none of this, for it all happens below the soil, but the curious may find evidence of mycorrhizas in the strangely thickened and branched roots of pine (*Pinus* spp.) and beech (*Fagus sylvatica*) that may be found growing just below the surface layer of leaf mould in woodlands.

Symbiotic associations between plants and fungi are not always of this type, however, frequently being parasitic, with the plant deriving no benefit whatsoever from the fungus, and the fungus or parasite taking not only sugars but also other nutrients from its host. In such parasitic associations the fungus is said to be pathogenic if it causes symptoms of disease, and this is usually the case. In many instances, but not always, the host may be killed completely.

Fig. 1.3a

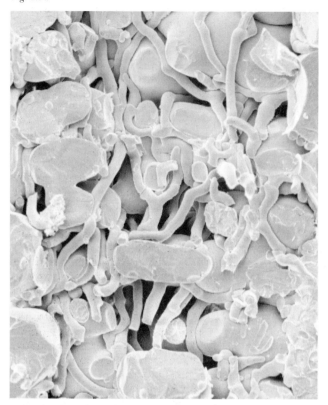

Fig. 1.3 Leaves of barley (*Hordeum vulgare*) infected with the yellow rust fungus (*Puccinia striiformis*): (a) the hyphae (approx 5 μm in diameter) of the fungus are clearly visible growing between the cells; (b) eventually the ornamented urediospores (approx 20 μm in diameter) form and burst through the cuticle of the leaf. (Scanning electron microscope photographs of freeze fractured tissues by N. Read, University of Edinburgh.)

Fig. 1.3b

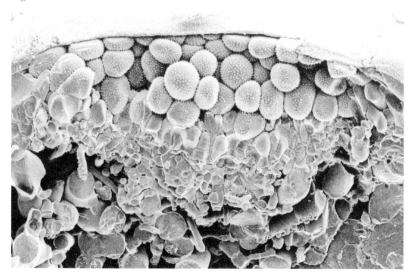

A mode of life in which the closely packed cells of living or dead plant tissues must be penetrated and tapped as a supply of food requires a very special structure, and so it is that most fungi are made up of microscopic, branching, thread-like tubes called hyphae, (sing. hypha) which form a mat or web termed the mycelium. These are ideal for burrowing between and through plant cells and enable the fungus to absorb nutrients over its entire growing surface. The large surface to volume ratio renders the fungus very susceptible to desiccation, which is why saprophytic fungi are rarely found in dry places and why rots and disease are always most prevalent in wet or humid conditions. Hyphae vary greatly in diameter, from 0.5 to1,000 µm according to species (1 µm = 10^{-3} mm), although most are about 1 to15 µm across and may be anything from a few centimetres to several metres long. The hyphae may or may not be divided up into individual cells by cross walls, the septa (sing. septum), and in some cases may be aggregated to form false tissues such as those which comprise mushrooms and toadstools and the root-like rhizomorphs of some wood decay fungi, which may be seen growing on dead or decaying trees in almost every woodland community. The term 'false tissue' is used to distinguish such structures from those of plants which are built up not by the aggregation of hyphae but by the division of cells in two or more planes. The hyphae contain the cytoplasm (the living cell contents) of the fungus. As in all living creatures, this is packed with minute structures, collectively referred to as organelles, such as mitochondria (the sites of respiration), ribosomes (for protein synthesis) and vesicles (for enzyme secretion) as well, of course, as the organising centres of the cells, the nuclei (which contain the deoxyribose nucleic acid, DNA, the genetic material of the fungus). The cytoplasm is surrounded by a membrane, the plasmalemma, which controls the uptake of nutrients and water and the extrusion from the hyphae of other substances such as plant cell-degrading enzymes and substances toxic to plant cells (phytotoxins).

The walls of the hyphae have a very complex layered structure that has been revealed by a combination of biochemical analysis and sophisticated electron microscopy. The inner surface is composed of fibres, the outer surface is smooth and amorphous and there are several layers in between. The fibrous material, which gives the wall great strength, is composed of polysaccharides (long chains of sugar related molecules), which may be cellulose (a major component of plant cell walls) or glucosamine-based chitin (a major component of the exoskeleton of insects) and other related substances. The amorphous components that bind the fibres together and provide a smooth outer surface to facilitate growth between cells are composed of other sugar polymers called glucans. Proteins may bind all these components together.

The bulk of the hyphal wall is rigid and capable of withstanding considerable internal pressure, rather like a green plastic garden hosepipe in which is embedded a network of strengthening material. At the tip of the hypha, however, in a zone less than 20 µm long, the wall layers are thin and plastic, with no strengthening fibres. This combination is the key to fungal growth, for high internal pressure appears to be essential for the extension of the hypha at the soft plastic tip. The high pressure is achieved by absorption of solutes through the semipermeable plasmalemma, leading to the uptake of water; the resulting internal pressure, with the plasmalemma pressing outwards against the hyphal wall, may be as high as 8 atmospheres (0.8 megapascals = 117.6 pounds per square inch!). New wall materials and cytoplasm are synthesised behind the advancing tip of the hypha and are carried forward to the tip by streaming of the cytoplasm. As the tip moves forward the newly made wall behind it thickens and becomes rigid. All the time enzymes are secreted from the tip out into the plant tissues through which the fungus is growing, to break down the cells, especially the cell walls, into simple sugar and other molecules suitable for absorption by the hyphae. A hypha may therefore be thought of as continually growing forward into a soup of nutrients in water, which are then absorbed in the region immediately behind the tip and utilised to fuel further growth.

The branching of a hypha usually occurs well behind the tip, after the wall has become rigid; a minute area of the wall where the branch is to form is softened by the internal secretion of enzymes, which loosen the chemical bonds between the wall components, allowing a branch initial to 'balloon' out. This then grows in exactly the same way as the hypha from which it arose. The continuously advancing and branching hyphae rapidly and efficiently permeate dead or living plant tissues of every type, extracting from them the nutrients they need. Some fungi grow so fast that colonies will reach a diameter of several centimetres overnight. Others, especially those that invade living plants, may take hours or days to advance a millimetre.

Some species of fungi, such as the yeasts, normally exist not in a hyphal form but as rounded cells, which grow by dividing or budding off new cells. Such forms, which do not usually infect plants, are well adapted to a free-living, benign saprophytic life on the surface of the plant, utilising nutrients that leak out from the cells. Those on grapes are at the heart of the fermentation process essential to wine-making because of their ability to convert sugar to alcohol. Some pathogenic fungi have no cell wall at all. Instead, membrane-bound bodies of fungal cytoplasm grow within the living plant cells, setting up a sophisticated relationship with the cytoplasm.

WHAT'S IN A NAME? – CLASSIFICATION OF FUNGI

We have already suggested that some fungi do not have walls at all, that those with walls form hyphae, which may or may not be divided into cells by septa, and that the chemicals that comprise the walls of hyphae may be very variable. There may be differing numbers of nuclei per cell, the number of sets of chromosomes (which contain the DNA) in the nuclei may differ, and reproduction may be effected by the formation of a variety of different kinds of spores. Also, although most fungi have two phases in their life cycles, sexual and non-sexual, the structures associated with these are very different in different groups or species. When all these factors are taken together clear patterns emerge, which allow the fungi to be divided into major categories or taxonomic groups as set out in Table 1.1.

The word 'fungi' has been used so far in this book to embrace all the groups listed in Table 1.1. Indeed, until recently these groups would all have been contained within a single Kingdom, the FUNGI. In modern classifications some groups previously regarded as FUNGI are placed in two additional Kingdoms, the PROTOZOA and the CHROMISTA. In the chapters that follow we will continue to use the word 'fungi' colloquially in its older, wider meaning. 'FUNGI' printed in capitals will be used in its modern taxonomic sense as in Table 1.1.

Table 1.1 The principal taxonomic groups of the 'fungi' which include parasites and pathogens (see Hawksworth *et al.*, 1995).

Kingdom PROTOZOA

Phylum	Class	Order
Myxomycota	Myxomycetes	6 orders of saprophytes
	Protosteliomycetes	Protosteliales
Plasmodiophoromycota	Plasmodiophoromycetes	Plasmodiophorales

Kingdom CHROMISTA

Phylum	Class	Order
Labrinthulomycota	Labrinthulomycetes	Labrinthulales
		Thraustochytriales
		(some members of both orders parasitic on marine plants)
Hyphochytriomycota	Hyphochytriomycetes	Hyphochytriales
		(parasites of fungi and algae)
Oomycota	Oomycetes	Olpidiopsidales
		Peronosporales
		Pythiales
		Sclerosporales
		+ 7 orders of saprophytes and animal parasites

Kingdom FUNGI

Phylum	Class	Order
Ascomycota	Classes not yet defined	Diaporthales
		Dothideales
		Erysiphales
		Hypocreales
		Leoteales
		Meliolales
		Perisporiales
		Rhytismatales
		Taphrinales
		+ 37 orders of saprophytes, lichen partners and animal parasites
Basidiomycota	Basidiomycetes	Phragmobasidiomycetidae: 5 orders of saprophytes
		Homobasidiomycetidae:
		Agaricales
		Boletales
		Ganodermatales
		Poriales
		Stereales
		+ 27 orders of saprophytes, mycorrhiza formers and doubtful parasites
	Teliomycetes	Uredinales

cont.

	Ustomycetes	Ustilaginales
		Exobasidiales
Chytridiomycota	Chytridiomycetes	Chytridiales
Zygomycota	Zygomycetes	Mucorales, mostly saprophytic but some parasitic on plants + 6 orders parasitic on arthropods
	Trichomycetes	4 orders parasitic or commensal on arthropods

All the major categories listed in Table 1.1 contain pathogenic as well as saprophytic species. Clearly pathogenic fungi are very diverse, which suggests that this mode of nutrition, as well as the hyphal habit, may have evolved independently in different groups at many different times.

The form and structure of the reproductive organs are the major characteristics used to define the groups, although the absence of a cell wall is the characteristic that distinguishes the Myxomycota and Plasmodiophoromycota, both now included in the Kingdom PROTOZOA, from the rest of the fungi. The members of these groups grow as a plasmodium, a mass of cytoplasm with many nuclei contained within a membrane, or as a similar structure, a pseudo-plasmodium, comprising many smaller masses of cytoplasm, each with a single nucleus. In both cases the structures have a slimy consistency, hence the common name slime mould. There are many thousands of species of slime moulds, all of which are saprophytic. The members of the closely related

Fig. 1.4 A stained and squashed cell of swede (*Brassica napus*) infected with *Plasmodiophora brassicae*, cause of clubroot disease. The nucleus of the host has just divided, prior to cell division, to produce two daughter nuclei. A multinucleate plasmodium of *P. brassicae* has divided with the host nucleus to produce two smaller plasmodia, each associated with one of the daughter host nuclei. The membranes of the plasmodia cannot be seen, but the nuclei, each about 3–5 μm diameter and a fraction of the size of the host nuclei, are clearly visible. (Light microscope photograph by I.C. Tommerup.)

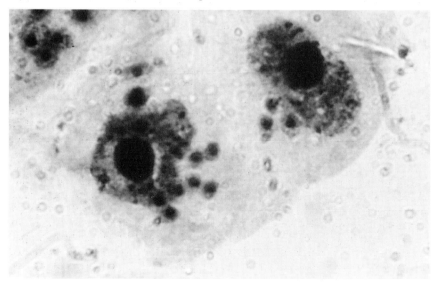

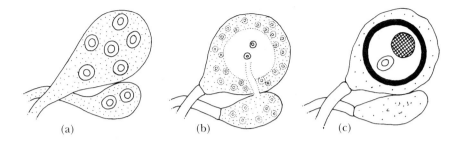

(a)   (b)   (c)

Fig. 1.5 Stages in the development of an oospore. (a) Compatible hyphae with diploid nuclei grow together and swell at their tips to form a large oogonium with a smaller antheridium attached to it. The hyphae may come from different mycelia containing nuclei of opposite mating type (heterothallism) or from a single mycelium containing nuclei with genes for both mating types (homothallism). (b) A reduction division (meiosis) occurs to produce haploid nuclei; a single haploid nucleus travels from the antheridium along a fertilisation tube and comes to rest alongside a single haploid nucleus within the oogonium. The remaining haploid nuclei in the oogonium move to the periphery of the cell and begin to degenerate. (c) The paired haploid nuclei fuse to form a diploid nucleus, a large lipid droplet forms in the oogonium and a thick wall develops to create the oospore (approximately 25–50 μm diameter). The antheridium and the remaining cell contents of the oogonium shrivel and die. (Diagram by Mary Bates.)

Plasmodiophoromycetes, however, cause diseases in the roots of plants and include *Plasmodiophora brassicae*, cause of the clubroot disease of wild and cultivated members of the cabbage family (Cruciferae). The multinucleate plasmodia, each surrounded by a membrane, grow within the living host cells, close to the nuclei (Fig. 1.4). These fungi thrive in wet conditions and reproduce by forming spores with a membrane but no cell wall and with two flagella (sing. flagellum), minute, whip-like threads that propel them through water. Swimming spores of this kind are called zoospores and are able to swim in a film of water on the surface of the plant roots and in the soil, from where they may infect another plant. The Plasmodiophoromycetes may also produce thick-walled resting spores (see p.77), able in the absence of a suitable host to survive adverse conditions for many years or even decades.

The fungi with walls fall into two Kingdoms, the CHROMISTA and the FUNGI. The CHROMISTA, although superficially resembling the members of the Kingdom FUNGI, have in fact more in common with the brown and yellow-brown algae and the diatoms, except that they are not photosynthetic. The plant pathogens within this Kingdom are all classified in the Phylum Oomycota (class Oomycetes), which embraces the orders Olpidiopsidales, Pythiales, Peronosporales and Sclerosporales and includes important genera such as *Olpidium, Pythium, Peronospora, Phytophthora* and *Sclerospora*, described in detail in Chapter 4. The walls of the hyphae contain cellulose rather than chitin as the major structural component. There are normally no septa, each hypha containing many nuclei, which are usually diploid (i.e. with two sets of chromosomes; diploidy is the normal state in the body cells of plants and animals). Asexual reproduction is by means of lemon-shaped or bean-shaped diploid zoospores with two flagella (Fig. 1.1b). The zoospores are contained before release in a spherical to lemon-shaped structure, the zoosporangium.

The possession of zoospores means that most fungi in this group require water to spread to a new food source, although some of the more advanced, plant-infecting members of the group have lost the ability to produce zoospores, the sporangium itself functioning as a single, wind-dispersed spore (a conidium – see below), a clear adaptation to a land-based life. Sexual reproduction takes place when specialised hyphae from different sexually compatible individuals of a species grow together. One swells at its tip to form a spherical structure, the oogonium. The other swells slightly, also at the tip, to form a club-shaped structure, the antheridium, which becomes attached to the side or base of the oogonium. At about the same time, the diploid nuclei undergo a form of division called meiosis, which halves the number of chromosomes. A fertilization tube forms and one of the haploid nuclei resulting from the mieotic division passes from the antheridium into the oogonium and fuses with a hapolid nucleus there, thus restoring the diploid state. In this way genetic material from two different individuals is brought together, generating variation within the species, just as in humans. A thick, resistant cell wall develops around the new diploid nucleus to produce a resting spore called an oospore (Fig. 1.5, previous page). This contains food stores, usually fatty substances (lipids), and is capable of surviving for many years in uncongenial conditions such as in the absence of water or the lack of a suitable host plant. When conditions improve and germination occurs to produce new individuals the chances of growing successfully are increased because of the genetic mixing that has occurred during the sexual process.

Within the Kingdom FUNGI, only the class Chytridiomycetes (often abbreviated to chytrids) produce swimming zoospores. Here the fungal body is not a hypha, but a microscopic rounded structure with chitin walls and root-like threads called rhizoids arising from its base. Asexual zoospores, spherical in shape and with one flagellum, are produced by fragmentation of the contents of the fungal body. Thick-walled resting spores containing reserves of food are sometimes formed, often associated with sexual reproduction, to enable the fungus to survive adverse conditions.

Fig. 1.6 Chytrids (Chytridiomycota) infecting the freshwater planktonic alga *Asterionella* sp. (a) Maturing chytrid bodies attached to the host's cell by threads (rhizoids) which penetrate the living cell contents. (b) The chytrid cytoplasm divides up to form uninucleate zoospores, each with a single flagellum. (c) Zoospores become attached to the surface of an uninfected cell. (d) New infections are established. (Illustration by Mary Bates.)

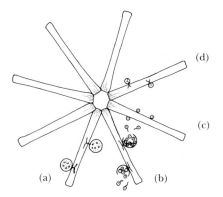

There are many plant-infecting chytrids, usually found parasitising the surface cells of roots or the cells of free living algae (phytoplankton) (Fig. 1.6). They eventually kill such plankton hosts but in higher plants are more likely to cause debilitation than death. One species, *Synchytrium endobioticum*, causes wart-like tumours on potatoes. Chytrids can also cause problems in agriculture and horticulture by spreading certain damaging plant-pathogenic virus diseases (see below).

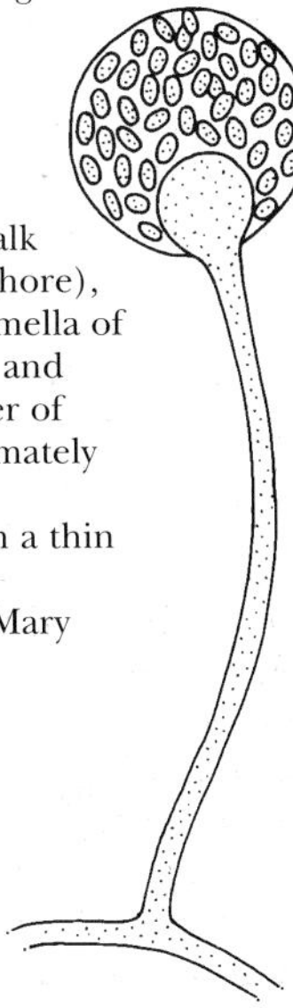

In older classifications another major group of the FUNGI, the Zygomycota, were placed together with the Oomycetes in a single group, the Phycomycetes, because they also normally lack septa in the hyphae. Contemporary information clearly indicates, however, that the Zygomycota are very different in almost every respect from the Oomycetes and merit a separate group. The nuclei within the hyphae are haploid and the asexual spores, formed in large numbers in sporangia, lack flagella and are therefore non-motile (Fig. 1.7) and are spread by the wind. The major structural polymer within the cell wall is chitin-chitosan. Sexual reproduction (Fig. 1.8) takes place when hyphae of different members of the same species fuse and the haploid nuclei within them also fuse to form a zygote nucleus (zygote is a term used to describe a nucleus in which the diploid state has been created by fusion of two haploid nuclei). A thick, often spiky wall forms around this and lipids are accumulated, thus forming a zygospore with considerable powers of survival in adverse conditions. Most

Fig. 1.7 A sporangium, in section, of a hypothetical member of the Zygomycetes (Zygomycota); note the long stalk (the sporangiophore), the central columella of the sporangium and the large number of spores (approximately 5 μm diameter) contained within a thin sporangial wall. (Illustration by Mary Bates.)

Fig. 1.8 Stages in the development of a zygospore. (a) Specialised branches form on compatible hyphae and grow towards one another. The hyphae may arise from different mycelia containing nuclei of opposite mating types (heterothallism) or from a single mycelium containing a mixture of nuclei of opposite mating type (homothallism). (b) and (c) The resulting gametangia fuse and the zygospore begins to form, supported by the suspensors, which may be the same or unequal in size and/or shape. (d) The haploid nuclei fuse to form a single diploid nucleus (zygote) and the wall surrounding the zygospore (approximately 50–100 μm diameter) thickens and darkens. (Illustration by Mary Bates.)

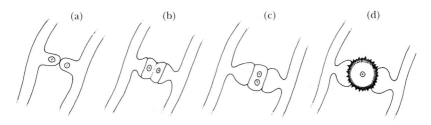

(a)   (b)   (c)   (d)

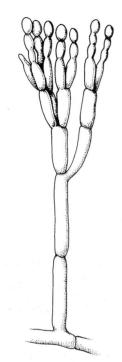

Fig. 1.9 Chains of conidia developing on a conidiophore of a *Penicillium* sp. Each mature conidium is approximately 3–3.5 µm diameter. (Illustration by Mary Bates, based on a drawing by Webster (1980).)

members of the Zygomycota are saprophytic moulds living only on dead tissues, the pin mould that grows on bread being a familiar example, but a few cause rots on damaged, mature or senescent plants.

A third major group of the FUNGI, the Ascomycota, are different again from the first two groups and include many plant pathogens. The hyphae contain chitin and glucans as the structural polymers. The nuclei are again haploid, although the hyphae are divided into sections by septa. These have pores in them, however, through which cytoplasm and nuclei may move quite freely. Each hypha, therefore, functions as a single cell with many nuclei, unlike the situation in most plants where each cell contains a single nucleus. The asexual spores are not motile and are not contained in a sporangium (Fig. 1.9 – see also footnote, p.100). Such spores, which look like dust to the naked eye, are termed conidia. They may be formed singly at the ends of hyphae, or in chains, or as complex heads of spores on special hyphae called conidiophores, according to species. Conidia may be various shapes. Often the shape reflects the method of dispersal: for example, those carried by wind are often spherical or thread-like, while those carried by water may be branched or hooked, enabling them to be caught inside bubbles of foam or in debris.

Sexual reproduction in the Ascomycota (Fig. 1.10) involves fusion of the hyphae of different individuals, followed by the fusion of haploid nuclei in pairs, one from each parent, followed by meiosis and division of the cytoplasm to produce a series of 4, 8 or more haploid spores called ascospores. The products of one fusion/division are contained in a sac called the ascus, which is frequently sausage-shaped and adapted to eject the ascospores forcibly from one end as the result of a build-up of internal pressure. The asci are usually protected by a complex fruit body, the ascocarp, which may be saucer-shaped, flask-shaped or spherical like a football. Ascocarps are usually only a millimetre or so in diameter, although in some groups they may be larger.

There are very many plant pathogenic Ascomycetes, including a large number that cause rots or tissue death (necrosis), and the ubiquitous powdery mildews (Erysiphales) that live on the surface of the host, drawing nutrients through special feeding branches called haustoria, which penetrate the cells of the leaf and stem surface. In Britain powdery mildews may be found on almost every wild and cultivated plant species as the summer wears on. They are visible as a fine white coating on the surface of the plant which, on examination with a hand lens, is revealed as a network of hyphae bearing chains of conidia. Later on in the season the ascocarps may be seen as small black or dark brown structures, just visible to the naked eye, all over the plant surface. The

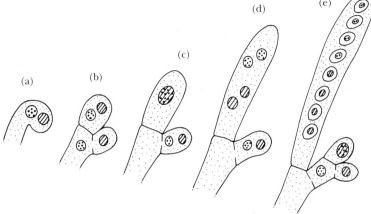

Fig. 1.10 Stages in the development of an ascus. (a) A hyphal tip containing haploid nuclei of opposite mating type bends over to form a crozier. (b) The two nuclei are walled off in the ascus initial. (c) The haploid nuclei fuse to form a diploid nucleus. (d) A reduction division (meiosis) results in four haploid nuclei and the ascus begins to elongate. (e) Each haploid nucleus divides vegetatively (mitosis) and eight uninucleate ascospores form and lie in the vacuole of the ascus. (Illustration by Mary Bates.)

infected plant, meanwhile, stays green and relatively healthy, providing the pathogen with a continuous supply of food. Powdery mildews are amongst the most devastating diseases of cereal crops in Europe, especially barley.

The hyphae of the last major group within the FUNGI, the Basidiomycota, also contain chitin and glucan as the major structural components of the hyphal walls. The septa may possess complex pores that are partially blocked by membranes, and although cytoplasm may move through the pores nuclei cannot, for they are too large. The hyphae are therefore effectively divided up into cells. The nuclei are haploid but in most species each cell normally contains a complementary pair of such haploid nuclei, making it functionally diploid. This state is said to be dikaryotic. Special bridges called clamp connections, which develop at cell division, ensure in some groups that only complementary nuclei (i.e. nuclei derived from spores of the opposite mating type or 'sex') are contained in each cell (Fig. 1.11, overpage). Asexual spores are not often produced, but when they are they take the form of dikaryotic conidia. Sexual reproduction occurs when complementary nuclei fuse within a dikaryotic cell and then undergo a meiotic reduction division to restore the haploid state. This occurs in specially adapted hyphal branches, the basidia, which produce at their tips groups of four uninucleate, haploid basidiospores, the product of one meiotic division (Fig. 1.12, overpage). Following dispersal, usually by wind, the haploid basidiospores germinate and the resulting hyphae, with uninucleate (monokaryotic) cells from different individuals of opposite mating type fuse so that the dikaryotic state can be re-formed.

Most of the familiar mushrooms and toadstools of fields and woodlands, as well as the bracket fungi that cause rots of trees, are formed by fungi of the class Basidiomycetes in the Basidiomycota. The hyphae mass together to form

Fig. 1.11 Stages in the
development of a clamp
connection. (a) A hyphal tip
containing a pair of
complementary nuclei (a
dikaryon). (b) A branch
forms and grows away from
the apex, with one nucleus
positioning itself within the
branch. (c) The branch
makes contact with the main
hypha and the nuclei divide
vegetatively (mitosis). (d)
The branch fuses with the
main hypha, the daughter
nuclei migrate to lie together
as compatible pairs and cross
walls (septa) form to seal off
the branch and create a new
dikaryotic apical cell.
(Illustration by Mary Bates.)

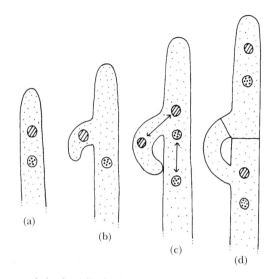

structures called basidiocarps, and the basidia form on the surfaces of the gills, tubes or teeth which may be observed on the underside of the cap. From this position they are easily dispersed by wind. Many of the toadstool-forming Basidiomycetes associate with plant roots as mycorrhiza, which aid the uptake of nutrients from the soil. Most people know the red-capped, white-spotted *Amanita muscaria*, the fly agaric, which forms mycorrhizal associations with birch (*Betula* spp.) and other tree species. Other Basidiomycetes, such as the honey fungus, *Armillaria mellea*, and *Heterobasidion annosum*, are destructive pathogens of trees.

The other large classes of the Basidiomycota, the Teliomycetes and the Ustomycetes, include important and widespread pathogens of many plants, the Uredinales (rusts) in the first and the Ustilaginales (smuts and bunts) in the second. The Teliomycetes and Ustomycetes do not form basidiocarps. Instead, the teliospores or ustilospores, which eventually erupt through the surface tissues of the infected host, germinate to form hyphae-like basidia which bear the basidiospores (Fig. 1.13). Many of the rusts have complex life cycles, with several spore stages that alternate between two different hosts. The smuts spend a good deal of their time growing systemically deep in the tissues of their host, invisible to all but the tutored eye. Both rusts and smuts have species that are amongst the most important pathogens of cereals worldwide, including black stem rust and bunt or stinking smut of cereals.

The final group within the FUNGI, the 'Deuteromycetes' or mitosporic fungi, has no official taxonomic status and is therefore not included in Table 1.1. It is mentioned here, however, because it is widely used by scientists as a convenient way of referring to fungi that have either lost the ability to reproduce sexually or do so only rarely, and cannot therefore be easily classified in the usual way. On the basis of their hyphal characteristics and their asexual spores (usually conidia) most are thought to be members of the Ascomycota, although it is not known what evolutionary forces have led to the loss of the sexual process. Many are pathogenic on plants, causing a variety of rots and necrotic lesions on leaves.

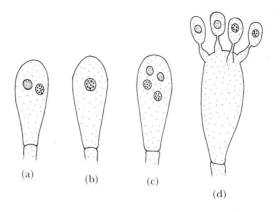

(a)   (b)   (c)

(d)

Fig. 1.12 Stages in the formation of basidiospores. (a) A young basidium containing two hap-loid nuclei of opposite mating type (a dikaryon). (b) The nuclei fuse to form a single diploid nucleus. (c) A reduction division (meiosis) occurs to produce four haploid nuclei. (d) Each nucleus migrates into one of the four basidiospores (approximately 10 μm diameter) which form at the tip of the basidium. Basidia are normally massed, together with sterile hyphae (paraphyses), in a fertile layer (the hymenium) which lines the surface of the gills, pores or spines of the fruit body (basidiocarp – usually a toadstool or bracket). (Illustration by Mary Bates.)

VARIATION IN FUNGI

Fungi are extremely variable, able to change genetically to adapt with great rapidity to changes in their environment. This has particular significance in diseases of crop plants since it means, for example, that pathogens can evolve rapidly to overcome disease resistance introduced into a crop by a breeder. There are two principal ways in which an organism can change its genetic make-up. Individual genes may change spontaneously (mutation), a comparatively rare occurrence, although certain types of mutation are increased by exposing cells to mutagenic chemicals or radiation. The other type of variation comes from recombination of the genes through meiosis and the sexual cycle. Each sexual cycle has an effect rather like shuffling cards between games, with the new combi-nations of genes interacting to give different qualities to the new organ-ism. Both sources of variation occur naturally in the fungi and are suffi-cient to account for the level of varia-tion found in sexually reproducing fungi; similar variation, differing in detail and precision, occurs in the bac-teria and the viruses.

A problem arises with many of the 'Deuteromycetes', however, in that although a sexual cycle rarely or never seems to occur, the group includes some of the most variable of the path-ogenic fungi. This variation was shown

Fig. 1.13 A basidium, with four basidiospores, arising from a two-celled teliospore of a rust fungus (Uredinales). The teliospore, approximately 50 μm long, is drawn in section to show the thickened wall. (Illustration by Mary Bates.)

by Guido Pontecorvo and others to arise when nuclei of differing genetic make-up are present in the same vegetative hyphae. It was already understood that most of the 'Deuteromycetes' carry genetically different nuclei in their hyphae. Such heterokaryons arise when two hyphae of different genotypes fuse, with subsequent mixing of their nuclei. Pontecorvo demonstrated that haploid nuclei in such heterokaryons may fuse to form a diploid nucleus in which recombination of genes between adjacent chromosomes can occur, similar to that which occurs in meiosis. These diploid nuclei are unstable and may break down to form haploid nuclei with a new genetic constitution, the variation produced being released when the new haploid nuclei enter the asexual spores of the fungus. Shortly after, Eric Buxton demonstrated that this process, termed the parasexual cycle, could be responsible for the creation of new variants of the plant pathogen *Fusarium oxysporum*, cause of wilt disease of many plants, and it has now been shown to be a widespread source of pathogenic variation in the 'Deuteromycetes' and the conidial (asexual) phase of the life cycles of members of the Ascomycota.

The importance of variation by pathogens will be seen later in the book (see p.175) when the problems of breeding crop plants with a stable resistance to disease, and the failure of fungicides in the face of pathogen variation, are discussed.

Bacteria and viruses

In addition to the fungi, two other major groups of microorganisms cause diseases in plants, the bacteria and the viruses.

Both plants and fungi, together with animals, belong to the Superkingdom EUCARYOTA. These are cellular organisms in which the DNA is contained within membrane-bound nuclei inside the cells. The bacteria, by contrast, belong to an entirely different, possibly more primitive, Superkingdom, the PROKARYOTA. Organisms in this Superkingdom are usually unicellular or have a very simple body consisting of little more than a chain or aggregate of cells, the chromosomes being dispersed within these, and not contained in a nucleus.

The PROKARYOTA probably evolved before the EUCARYOTA, more than 3,500 million years ago. They are very successful and have survived because of their capacity to evolve and change rapidly to meet the demands of a changing environment. Most, though not all, are incapable of synthesising their own carbon compounds and must therefore derive their energy either by oxidising inorganic chemicals or from other organisms. Many are saprophytes, whilst others live in association with living plants and animals as mutualistic parasites or pathogens.

Bacteria have been especially successful as pathogens of animals and humans, being the cause of such familiar diseases as syphilis (*Treponema pallidium*), anthrax (*Bacillus anthracis*), typhoid fever (*Salmonella typhi*) and leprosy (*Mycobacterium leprae*). Their capacity for rapid evolution means that they soon develop resistance to the antibiotics used to control them, necessitating a continual search by the pharmaceutical industry for new drugs to keep in check the diseases for which they are responsible. In spite of the many diseases they cause in animals, bacteria are far less important than fungi in causing diseases of plants in temperate regions, being poorly adapted to grow at low temperatures. In the tropics, however, they cause some devastating diseases.

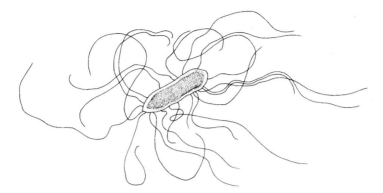

Fig. 1.14 A cell of *Erwinia amylovora*, the bacterium that causes fireblight disease of members of the rose family (Rosaceae). The rod-shaped cell, which is approximately 2 μm long, bears numerous flagella over its surface. (Illustration by Mary Bates; based on an electron micrograph by R.N. Goodman reproduced in Lucas (1998).)

Most plant pathogenic bacteria consist of rod-shaped, single cells (Fig. 1.14), and some have the capacity to develop into thick-walled resting spores, capable of surviving adverse conditions. They also often possess flagella, either over the whole surface or in groups at either end of the cell, which enable them to move independently in water. An exception is the genus *Streptomyces*, which is filamentous and lacks flagella. Each bacterial infection can arise from a single cell, with multiplication and subsequent invasion of the tissues occurring by simple division of the cells. Many bacteria have the capacity to divide every twenty minutes, which means that if the conditions are good (for the bacterium, not the host!) infection may develop very rapidly. Sexual mating and subsequent genetic mixing, if it occurs, results from the fusion of compatible cells, followed by meiosis and cell division in which the DNA from the two partners is distributed between the daughter cells.

Bacteria pathogenic on plants, like pathogenic fungi, secrete enzymes of various sorts that break down the plant body into soluble compounds, which may then be absorbed through the bacterial cell wall and the cell membrane to provide the energy and nutrients required for continued bacterial growth. Some, also like fungi, produce toxins which kill plant cells while others produce plant growth regulators (hormones) which may modify the host's growth and development. It has been found that the production of such substances is often controlled not by the DNA of the bacterium itself but by a short piece of parasitic DNA, not unlike a much reduced virus (see below), called a plasmid.

Most plant pathogenic bacteria grow or survive, in the absence of a living host, as saprophytes in the soil or on decaying tissues. When they infect a host they usually take advantage of wounds or natural openings such as stomata, and cause a rapidly developing rot that results in the death and complete decay of the host. An example might be *Erwinia caratovora*, which causes soft rots on a wide range of plant hosts or *E. amylovora*, which causes a destructive disease called fireblight on hawthorn (*Crataegus monogyna*) and other members of the rose family (Rosaceae). *Streptomyces scabies* causes corky lesions on potato tubers, called common scab. A small number of species have a more balanced relationship with their hosts, such as *Agrobacterium tumefaciens*, the cause of crown gall of many plants. The tumour-inducing factors produced by this bacterium are controlled by a plasmid that is transferred from the bacterial cell to the plant cell, where it associates closely with the plant DNA, thus 'transforming' the latter to tumorous growth. Study of this unique partnership has led to the development of the technology of genetic engineering in plants.

Another small group of bacteria has also developed symbiotic relationships with plants, but in this case the relationships are mutualistic rather than pathogenic, and are of great benefit to the infected plants. These are members of the genus *Rhizobium*, a group able to 'fix' nitrogen gas from the atmosphere and convert it into soluble nitrogenous salts. The process is controlled by a plasmid in the bacterium and is of great importance in maintaining the fertility of soil. Although many nitrogen-fixing bacteria grow freely as saprophytes in the soil, *Rhizobium* spp. enter the cells of the roots of plants belonging to the pea family (Leguminosae), where they cause small wart-like nodules to form by cell proliferation. Within the tissue of the nodule the bacterial cells take sugars and other nutrients from the plant and provide soluble nitrogenous compounds, fixed from the atmosphere, in return. Nitrogenous compounds are eventually returned to the soil and improve fertility. This association is used in traditional agricultural crop rotations involving legumes such as clover or beans, to reduce the need for the application of nitrogenous fertilisers.

Before leaving the bacteria, it is important to mention another group of the PROKARYOTA, the mycoplasma-like organisms called phytoplasmas and spiroplasmas. Mycoplasmas have been known for some time to cause respiratory and other diseases of animals, but the isolation of mycoplasma-like organisms from diseased plants is comparatively recent. They are smaller than bacteria and lack both cell walls and flagella. Most are spherical or oval in shape, but some are filamentous or helically coiled (spiroplasmas). The symptoms they cause (e.g. yellowing, necrosis, stunting or abnormal growth, especially of flowers) and the fact that they are normally spread from plant to plant by insects, usually leafhoppers, has sometimes led to diseases caused by phytoplasmas and spiroplasmas being wrongly attributed to viruses (see below).

The final major group of pathogens of plants are the viruses and viroids, which are quite unlike the fungi or the bacteria. Viruses are so small that their structure cannot be resolved even with a powerful light microscope. Some animal viruses, such as the smallpox virus, can be 'seen' as scattered light under a powerful light microscope, but no structure can be resolved. Viruses require the much higher magnification of the electron microscope for their resolution (the electron microscope uses a beam of electrons instead of light, and magnets instead of lenses, to reveal the structure of tissues, the image being displayed on a cathode-ray tube). Virus symptoms on infected plants, however, are very marked, including variously stunted growth, distortion of stems or leaves, tumorous growth, patches of dead cells (necrosis), yellowing or mottling of leaves (chlorosis) and colour breaking in flowers. Anyone who has marvelled at early Dutch paintings of colour break tulips will have seen the evidence of infection by mosaic viruses. Similarly the variegated *Abutilon*, widely grown as a bedding plant, derives its mottled appearance from infection with *Abutilon* mosaic virus. Some viruses remain close to the point of infection, causing limited symptoms such as spots or mottles, while others may be transported throughout the plant, usually by way of the phloem.

Viruses have no cellular structure but instead consist of a nucleic acid template, containing all the genetic information, encased in a protein coat (Fig. 1.15a and 1.15b). Viroids are even simpler, consisting only of nucleic acid without a protein coat. This very basic structure means that viruses can only exist as parasites of plants, animals or bacteria (those which infect bacteria are called phages), and have no capacity for an independent existence. The nucleic acid provides the basic codes that enable the virus to subvert the cellular machinery of the host to make further virus particles identical with the parent virus in every respect. In this way, the virus is replicated and its numbers increase. It is essential for the host cells to remain alive until the virus has replicated, and for this reason infected plants are rarely killed immediately, although their metabolism and biosynthetic machinery may be so disrupted

Fig. 1.15a Virus (x 160 000): rod-shaped particles of tobacco rattle virus – the long and short rods contain different parts of the genome. (Electron microscope photograph supplied by the Scottish Crop Research Institute.)

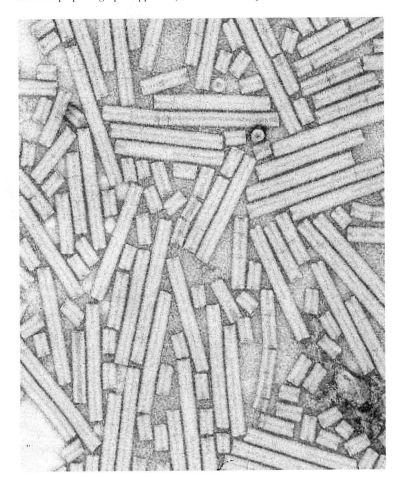

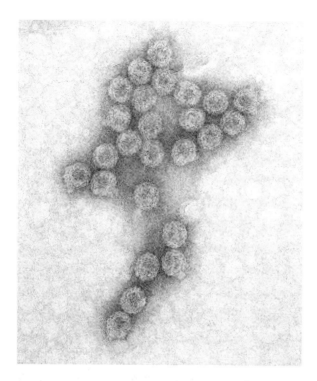

Fig. 1.15b Virus (x 160 000): polyhedral, near spherical particles of cauliflower mosaic virus. (Electron microscope photograph supplied by the Scottish Crop Research Institute.)

that growth is severely limited or distorted, leading to the kinds of symptoms described above.

Whether viruses evolved to this simple but sophisticated state by reduction of more complex microorganisms or whether they evolved from even simpler forms is not known, but the former explanation seems the more likely. Whatever their origins, they are very successful, existing in a wide variety of shapes and forms and causing a great diversity of diseases in animals, including the common cold, polio, measles and some forms of cancer, and also in plants, including cucumber mosaic, turnip yellows, beet mosaic and beet yellows.

Since viruses have no capacity for an independent existence they must be transported from host to host, to cause new infections and to survive, by an external agent of one sort or another. Some viruses, such as potato virus x, may be transmitted when leaves of adjacent plants rub against one another, but this is a rare occurrence. Usually plant viruses are transmitted by a carrier, called a vector, which feeds on or is a parasite of the host. Such relationships may be very specific, although they are not always so, and usually involve insect species such as aphids, leafhoppers or whiteflies. In some cases, such as lettuce mosaic virus, the relationship is very simple and involves particles from the host sticking to the surface of the stylet or mouthparts of the vector and being transmitted to a previously healthy plant during feeding. In other instances, such as beet yellows virus, the virus may enter the body of the vector, remain there for a few days and again be transmitted during feeding. A third category, including barley yellow dwarf virus, enters the insect and multiplies within its tissues before being regurgitated and transmitted to the plant host. In addition to insects, nematode worms and fungi may transmit certain viruses from one plant to another. For example, raspberry ringspot virus is transmitted by nematode worms attacking raspberry roots (*Rubus* spp.), while lettuce big vein virus and potato mop top virus are spread by the root-infecting fungi *Olpidium brassicae* and *Spongospora subterranea*.

Finally, plant viruses may be spread to new plants by being carried in seed or pollen, as a result of vegetative propagation or during grafting. Cucumber mosaic virus, for example, is carried in the embryos of seeds, while lettuce mosaic may be carried in pollen and passes into the embryo via the pollen tube during fertilisation. Almost any infected plant propagated by cuttings, bulbs, corms, tubers or grafting is likely to transmit its virus to its progeny, as in the case of the virus causing colour breaking in tulip, or many of the potato-infecting viruses.

Because viruses are usually investigated only with the aid of sophisticated instruments such as electron microscopes, or by using chemical techniques that recognise the protein coat, they are difficult for the naturalist to study. Moreover, many plant viruses remain latent in the plant, causing no symptoms at all, and even when symptoms do occur they may be difficult to detect except in a uniform and otherwise unblemished crop, in which case the infected plants stand out as being different from the rest. In natural plant communities, although virus infections are very widespread, the population of the host is so varied genetically and may be of so many different ages and conditions that it is almost impossible to tell with the naked eye whether an apparently abnormal plant is actually diseased or not.

Subsequent chapters will concentrate on diseases caused by fungi, although pathogenic bacteria, viruses and their relatives will be discussed again in Chapter 12. Flowering plants themselves may occasionally cause plant disease, a familiar example in Britain being the mistletoe (*Viscum album*), and these will also be discussed in Chapter 12.

Firstly, however, it is useful to examine in more detail how pathogens infect host plants and extract nutrients from their cells, and how potential hosts attempt to resist such attacks or restrict the growth of the pathogen. These processes, which will be described in Chapters 2 and 3, determine the nature of the symptoms of disease, knowledge that is of great importance to the naturalist wishing to identify infected plants and their pathogens in the field. Since some readers may not be familiar with diseases on wild plants, the examples used to illustrate the various concepts will be, by and large, diseases encountered on cultivated plants. The symptoms of diseases found in the garden are usually so pronounced and, because of the dense nature of the plantings, so prevalent as to be difficult to miss, even for the unpractised eye.

2

Invading the Plant

An early morning poke about in the garden during late summer or early autumn is an exciting experience for a plant pathologist, especially if the weather is wet and the owner has neglected to thin his vegetables or apply fungicide sprays to his fruit trees. Disease will be found everywhere: rots on the fallen apples, shot-holes in the cherry leaves, scab on the pears, black spot on the late-flowering roses, blight on the potatoes, downy mildew on the lettuces and cabbages, rust on the leeks and antirrhinums and a fine white covering of powdery mildew on almost everything, but especially noticeable on the vegetable marrows and roses. Even the underground parts of the plants may show symptoms if dug up: rots on the carrots, evil-smelling clubroot (finger and toe) on the cabbages and other brassicas, rot and scab on the potatoes and, if blight is present, a foxy red discoloration when cut open. The glasshouse will provide a rich harvest too: rots and wilt on the tomatoes, powdery mildew on the cucumbers and damping-off in the pan of seedlings sown too densely and then overwatered. And if the owner has neglected to weed his patch, even the weeds will be afflicted: rust on the groundsel and annual grasses, white blister on the shepherd's purse and bright red target spots on the dock leaves. The old sycamore in the far corner may have tar spot on its leaves, and the gap beyond may be the only reminder of the elms that once stood at the very end and shaded everything before falling victim to the devastating wilt, Dutch elm disease. The same trail of death and decay, albeit less obvious but no less varied, awaits the naturalist who explores the meadow or the wood, who studies the verges and hedgerows of a country lane, who climbs the high hills or descends to the sand dunes and salt marshes of the coastal plain. The remarkable, often unrecognised, fact is that plant diseases are everywhere, an integral and often even essential part of the natural scene, and the variety of symptoms resulting from these diseases is simply enormous.

This variety is not, however, without pattern. In every case it will be found that particular pathogens and groups of pathogens elicit the same type of symptoms, although the severity of the attack may vary according to the host species, the general state of vigour of the host and the prevailing weather conditions. For example, *Monilinia fructigena* (Fig. 2.1), a member of the Ascomycota (see p.100), will cause a brown rot if it infects an apple fruit injured, perhaps, by a pecking bird or a rubbing branch. It will, however, cause rot more quickly in 'Cox's Orange Pippin' apples than in 'Bramley's Seedlings', and it will always be more destructive if the weather is cool and wet. By contrast, the powdery mildews, *Erysiphe* spp. and other members of the Erysiphaceae (see p.156 and Plate 5), also in the Ascomycota, rarely seem to kill the host tissues, whether the host be an apple tree or a vegetable marrow, despite the presence of an enormous amount of fungal mycelium on the surface of the leaves. However, if the weather is dry the host will be much more stunted than usual, will probably produce a smaller crop, may lose its leaves

Fig. 2.1 An apple fruit (var. Golden Delicious) rotted by the necrotrophic fungus *Monilinia fructigena*. Note the wound where the pathogen gained entry to the fruit. (B. Goddard, University of Cambridge.)

early and will certainly be more heavily infected.

To understand what is going on it is necessary to know something about the ecology and physiology of the fungi and bacteria that cause plant disease. This information has been gathered over the last hundred years or more by plant pathologists studying crops of agricultural and economic importance such as wheat, potatoes, lettuce, grapes or apples; but the same principles apply and the same patterns are evident in diseases on wild, uncultivated plants.

Strategies of parasitism

Until about 20 years ago, plant pathogenic fungi and bacteria were classified, for the purpose of study, according to a scheme devised by the great nineteenth century German plant pathologist Anton de Bary. The basis of this classification was ecological, and fungi and bacteria were grouped according to whether they were pure saprophytes, facultative parasites or obligate parasites. Pure saprophytes grow and reproduce only on tissues that are already dead, and are completely incapable of infecting living plants. They are familiar, for example, as moulds on bread or cooked food. It is interesting to note that frequently they possess exactly the same armoury of cell-destroying enzymes and toxins as pathogens; clearly, therefore, the capacity to infect a host and cause disease requires more than the mere ability to destroy living tissues. Facultative parasites, such as *Monilinia fructigena* mentioned above, are able to infect living plants, kill them and then live on the dead tissues as saprophytes. Obligate parasites, however, such as the powdery mildews, are much more limited in their ecology and are unable to grow in nature in the absence of a living host (see Chapter 8 and Plate 5). P.W. Brian, in 1967, drew a distinction between two sorts of obligate parasites. Those that are restricted to a living host in nature, but can be grown on their own on an artificial culture medium in the laboratory, he called ecologically obligate parasites. *Phytophthora infestans*, cause of potato blight, fits this category (see Chapter 2 and Plate 1). In such cases obligate parasitism results from the inability of the pathogen to compete on dead host tissue with pure saprophytes, which invade the tissues after death. Others, those so specialised in their parasitism that they are unable to grow in the absence of a host, even in the laboratory, he called physiologically obligate parasites. The powdery mildews are of this type. By implication, the physiologically obligate parasites possess some biochemical or physiological deficiency that can only be supplied by a living host, and this prevents independent growth.

Fig. 2.2 Classification of plant parasitic fungi. After A. de Bary (1867) and P.W. Brian (1967).

FUNGI

OBLIGATE SAPROPHYTES (never parasitic)	PARASITES	
	FACULTATIVE PARASITES (parasites or saprophytes in nature)	OBLIGATE PARASITES (always parasitic in nature)
	ECOLOGICALLY OBLIGATE PARASITES (parasitic in nature, but can be cultured)	PHYSIOLOGICALLY OBLIGATE PARASITES (uncultured so far)

De Bary's classification was the basis of the revolution in the study of the ecology of pathogenic fungi led by Denis Garrett in the 1940s and 50s, and Brian's modification was a response to the growing interest in the 1950s and 60s in the physiology and biochemistry of the interaction between pathogens and their hosts, led in the UK by R.K.S. Wood of Imperial College, London and stimulated by Brian himself in the University of Glasgow. The de Bary classification as modified by Brian is still very helpful, and will be used later in the present volume, a summary being set out in Fig. 2.2 for ease of reference. However, for the purpose of considering the mechanisms by which fungi and bacteria interact with a host and cause the visible symptoms of disease, it is much more convenient to use another scheme, first devised by the Australian plant pathologist Neville White in 1957, defined in detail by David Lewis in 1973 and placed in an ecological context by Parbery in 1997. The basis of this classification is the mechanism by which fungi and bacteria obtain nutrients from their hosts, for there is a direct relationship between the strategy employed and the symptoms of disease that result.

In this scheme three broad categories of pathogen are recognised.

1. Necrotrophs

Having penetrated a host, these pathogens rapidly kill the cells and tissues and break them down with enzymes to release the sugars and other nutrients

required for growth and reproduction. They are therefore destructive parasites that cause symptoms such as damping-off, rots, wilts, necrotic spots on
leaves and so on, all of which result from the death and breakdown of cells and
tissues. *Monilinia fructigena* belongs to this category.

2. Biotrophs

Having penetrated a plant, these pathogens derive their energy from the living
cells and tissues by tapping into the plant's own sugar and nutrient supply. The
relatively balanced relationship required for this form of parasitism may last
for a long time, with the host plant not being killed for several weeks, months
or even years. The powdery mildews (Plate 5) are typical biotrophic fungi. The
host leaves may be covered with the surface-growing hyphae of the pathogen
bearing long chains of spores, yet still remain green and intact, albeit somewhat stunted and debilitated. Other biotrophs, such as the downy mildew, rust
and smut fungi, may permeate host tissues but the infected organs remain relatively green and healthy, only being damaged where spores erupt through the
surface layers or when the load of infection is too great. Symptoms include a
fine down of hyphae and spores on the green leaf in the case of downy
mildews, or pustules on leaves or other organs where the spores of rusts or
smuts break through the cuticle (the outermost layer), prior to release.
Yellowing and dead areas of tissue may indicate that the host cells in a limited
area have given up under the strain. Biotrophs also cause tumours and other
symptoms indicating disturbance of cell and organ growth.

3. Hemibiotrophs

These are intermediate in their parasitism and exhibit features of both
necrotrophy and biotrophy. An example is *Venturia inaequalis* (see p.105 and
Plate 4), the cause of apple scab, a disease that manifests itself as dark, scabby
spots on the leaves and fruit of apples. Spores infect the young leaves of the
apple tree in spring, forming a mycelium that grows biotrophically just below
the cuticle. The infected areas may be paler green than normal, but there is little more to see. Later, as the season advances, the hyphae may grow deeper
into the leaf, killing cells as they go, so that the relationship then becomes
necrotrophic and areas of dead tissue develop. Also, spores erupt through the
cuticle and a dark scab is formed. Eventually the tissues die completely and the
leaf falls to the ground, yet the pathogen continues to grow on the dead tissue
on the surface of the soil and, after surviving the winter, may produce spores
which will reinfect the new apple leaves in the spring.

It is probable that many of the plant pathogens thought by plant pathologists
to be necrotrophs are, in fact, hemibiotrophs with a very short biotrophic
phase (e.g. a few hours to a few days) at the start of their life history.
Unfortunately, the early stages of infection of too few diseases have been studied in sufficient detail to provide an unequivocal answer to this question. A better understanding of hemibiotrophy would greatly help in the interpretation
of the host specificity and symptom development of many plant diseases like
apple scab or light leaf spot of brassicas (*Pyrenopeziza brassicae*) that are still
poorly understood.

Necrotrophs and biotrophs are very different in their physiologies and food-
gathering strategies. The features that characterise these two groups merit further consideration, for this will help in understanding the nature of the

diseases they cause, and these are listed and discussed below, with some simplification for reasons of clarity.

The necrotrophs, cause of rots, wilts and necroses

1. Saprophytic growth

The pathogen is capable of saprophytic growth on the tissues it has killed, or on other substrates in the absence of a suitable host. The more extreme necrotrophs, like the *Pythium* spp. that cause damping-off of seedlings, may also be able to compete reasonably well with pure saprophytes, which grow on dead tissues and nothing else. They therefore equate with the facultative parasites of the de Bary-Brian classification. Increasing specialisation as a necrotrophic pathogen usually means a progressive loss of this 'competitive saprophytic ability' (a term coined by the fungal ecologist Denis Garrett) and some necrotrophs may therefore equate with the ecologically obligate parasites of de Bary-Brian. Competitive ability depends on the capacity to grow rapidly or produce antibiotic(s) and/or other substances such as the enzymes needed to kill or inhibit rival fungi or bacteria competing for a food source. This capacity is sacrificed in the evolution of increasing ability as a pathogen invading living host tissues, where competitors are fewer.

2. Weak parasitism

It is a paradox that the most destructive necrotrophs are capable of attacking only the young tissues of seedlings, or more mature tissues of plants that have been weakened by stress, injury or old age. Moreover, a large number of pathogen spores or a large amount of mycelium is normally required to gain entry to the host. This probably reflects the high energy levels required for killing host cells and tissues, overcoming in-built mechanisms by which pathogen attacks are resisted, and breaking down cell walls.

3. Wide host range

Most necrotrophic parasites are capable of attacking a wide range of host species. *Pythium ultimum* (see p.84), for example, not only causes damping-off of seedlings, but will also cause soft rots on such unrelated species as cabbages (*Brassica oleracea*), potatoes and cucumbers (*Cucumis sativus*). This may result from a lack of specialised nutritional requirements, but may also reflect the fact that the relationship with the host is unsophisticated and therefore has not necessitated the evolution of features geared specifically to growth in the tissues of a particular plant genus or species. Moreover, since host tissues are killed before the pathogen makes any appreciable growth in them, host resistance mechanisms triggered by infection do not have to be avoided using specific strategies (see Chapter 3).

　It is interesting to note, however, that the host ranges of necrotrophs are not infinite, with some plant species being resistant to each pathogen. Pre-formed fungitoxins present in the tissues of some but not all species may be responsible in some cases (e.g. in onion skins), and many toxic substances such as phenols accumulate more rapidly in some plant tissues than others as a result of wounding. Moreover, the parasite may lack the appropriate enzymes to release nutrients from the tissues of some potential hosts, a matter that will be referred to further below, or the environmental conditions may be unsuitable for infection to occur. Finally, as suggested above, many pathogens currently believed to be necrotrophs may in fact be hemibiotrophs, with a short biotrophic phase at the start of the disease cycle. Since a characteristic feature of biotrophy is considerable specificity in host range (see p.49), this could explain the limited host range of many apparent necrotrophs.

4. Production of plant cell-destroying enzymes

This attribute is especially important and is largely responsible for the cell death and cell and tissue breakdown that lead to the rots and necroses that are the typical symptoms of disease caused by necrotrophic pathogens (Fig. 2.1 and Plate 3). Investigations suggest that the production of such enzymes is normally triggered by the presence of the chemical to be broken down. They are then released from the tips of the hyphae or cells of the pathogen and may diffuse for a considerable distance into the host tissues. Cells are killed by the enzymes, thus negating any active resistance response, and nutrients are released to nourish the advancing pathogen by erosion of the components of the cell walls and the cell contents. The necrotroph thus marches through the host tissues into a nutrient-rich 'soup' that supplies all the energy required for its continued growth and reproduction.

A wide variety of enzymes may be produced by a necrotroph, each capable of attacking a different component of the plant cell. Indeed, the production of different enzymes may occur sequentially during growth of the pathogen as different cellular components are uncovered. Before considering the nature of these enzymes, however, it is necessary to examine the structure of the plant cell wall, for despite its apparent simplicity it is, in fact, very complex and research to elucidate its finer detail is still going on.

Plant tissues are made up of individual cells, each surrounded by a rigid wall, and joined to one another where they touch by a glue-like layer called the middle lamella. There may also be spaces between cells, lined with a film of fluid. The wall has two zones: an outer primary wall, produced as the cell grows, and an inner secondary wall, produced after cell expansion has ceased. Both contain fibres; those of the primary wall are loosely packed, while those of the secondary wall are densely packed. The fibres are embedded in amorphous (non-fibrous) material which is continuous with the middle lamella. The fibres consist of cellulose, a substance composed of bundles of parallel, unbranched chains (polymers) of glucan molecules. These may be crystalline and are extremely resistant to enzyme attack.

One major group of components of the amorphous material and the middle lamella are pectic substances. These too are polymers of sugar molecules and are found in decreasing amounts from the middle lamella, through the primary wall into the secondary wall. The basic chains are known as pectic acids (e.g. polygalacturonic acid) and are soluble in water. They are especially plentiful in the tissues of fruit and are responsible for the setting of fruit jam. Frequently these pectic acids combine with calcium ions (Ca^{++}) to give insoluble substances called pectates (polygalacturonates), or they may react with alcohol to give pectins (polymethyl galacturates).

A second major group of components of the amorphous parts of the wall are the hemicelluloses. These are complex forms of mainly linear, but sometimes branched, polymers of various sugars, mainly xylose. Proteins are also present throughout the wall, especially those containing a sugar component, called glycoproteins. There may also be enzyme proteins present with a role in regulating the growth of the wall. The hemicellulose molecules that coat the cellulose fibres and are bonded to them are in turn chemically bonded to the pectic molecules and the glycoproteins, binding the whole cell wall together.

Other substances may be deposited in the walls of cells modified during growth and maturation to perform special functions in the plant. Lignin, for example, a tough, three-dimensional amorphous polymer of complex molecules such as coniferyl, sinapyl and cinnamyl alcohols, is deposited in the cells that conduct water and gives strength to the xylem cells, which are the major component of woody tissues. Fatty, hydrophobic (water-repelling) polymers such as cutin are a major component of the waterproof cuticle that covers the outer surface of the plant. Another hydrophobic fatty polymer, suberin, forms corky layers in the bark and in wounded tissue, which it seals against infection by necrotrophs.

The major components of plants that are attacked by most necrotrophs are the pectic substances of the middle lamella and primary wall. Much of the research on the enzymes used by necrotrophs to break these down into simple sugars has been done with *Monilinia fructigena*, but the results are widely applicable. There are two major groups of enzymes involved, those that cut the chains (polymers) by a chemical process called hydrolysis and those that cut the chains by a different chemical process called transelimination. Both groups include enzymes that attack the chains at the ends (exo-enzymes) and enzymes that attack the body of each chain (endo-enzymes). Also, both groups contain enzymes that specifically attack the pectins or the pectates. Together, or in sequence, these pathogen-produced enzymes act upon the middle lamella and primary walls of the host to solubilise the amorphous materials, allowing unrestricted growth of the fungus through the tissues and releasing all the sugars required for continued growth and reproduction. No single necrotroph produces all these enzymes, however, and it is important to realise that the particular range and sequence of enzymes deployed, and the symptoms that result, vary according to the host and pathogen species involved and the complex interaction between the enzymes, the chemicals predominating in the wall and the environment.

The enzymes produced by the pathogen determine whether a rot will or will not occur, but the chemical and physical environment of the host 'shapes' the interaction. For example, tissues that are relatively acid are more favourable for the process of hydrolysis, whereas neutral to alkaline tissues favour transelimination. Transeliminase enzymes are far more destructive than hydrolases and more severe rots result from their activity. If the plant cell walls are rich in free calcium or magnesium ions, this leads to the inhibition of hydrolysis but the stimulation of transelimination. If the water level of the host tissues is high, rotting is more likely, for this facilitates diffusion through the tissues of both groups of enzymes, an observation that fits well with the fact that rots are usually more severe in wet weather conditions.

Finally, inhibitors of pectic enzymes may be present in host tissues, especially certain glycoproteins and the dark products of oxidised phenols formed when cells are killed. The presence of inhibitors may explain why some fungi cause firm rots and others soft rots, a difference which is very obvious in apples infected by either *Monilinia fructigena* or another necrotroph causing rots in apples, *Penicillium expansum* (Fig. 2.3). Both are capable of producing a wide range of pectic enzymes, yet *M. fructigena* causes a dark, firm rot in apple fruits while *P. expansum* causes a pale, soft rot. It is likely that the pectic enzymes of the former are inhibited by the large quantities of the dark oxidised phenols produced by wounded apple flesh, the brown discoloration we see when apples are cut, while the enzymes of the latter are unaffected because it has evolved to produce an inhibitor of phenol oxidation.

It has already been noted that the cells of tissues disrupted by pectic enzymes invariably die. This probably occurs because the plasmalemma, the membrane that surrounds all living cells, is disrupted by the shear forces at the point where the plasmodesmata, the thin threads of membrane and cytoplasm that link each cell with its neighbours, pass through the cell walls. It is likely that such shearing results from the movement of one cell against another as the whole tissue mass is destabilised following breakdown of the middle lamella. A further factor may be the accumulation of toxic substances released or formed during the breakdown of the cell walls.

Enzymes capable of degrading the other structural components of the cell wall, described earlier, may also have a role in the breakdown of tissue to release nutrients for pathogen growth. Although, as we have seen, hemicelluloses are widespread in the cell wall, pathogen enzymes capable of breaking them down (called hemicellulases) are probably only of secondary importance in most diseases, but assume greater significance in those cases where pectic enzymes are inhibited, as in the rot of apples caused by *M. fructigena*. Cellulose, the other major wall material, is a relatively intractable substance because of its fibrous and crystalline nature. For most pathogens causing rots, quite sufficient carbohydrate for growth and sporulation may be obtained from the pectic substances and hemicellulose. Nevertheless, cellulases are important in softening the cellulose during the penetration of the cell wall, where the fibres are so tightly packed that the pathogen cannot simply push through. In the case of some necrotrophic pathogens such as *Pythium* spp. the ability to produce cellulases in great quantities may allow the fungus to continue to thrive on the dead tissues of the host long after the pectic substances and hemicelluloses

Fig. 2.3 Apple fruit (var. Bramley's Seedling) infected by *Monilinia fructigena* (dark, firm rot) and *Penicillium expansum* (pale, soft rot). (B. Goddard, University of Cambridge.)

have all gone. The production of cellulases thus confers on such pathogens increased ability to compete with pure saprophytes for nutrients.

The most intractable of all the constituents of the wall is lignin, the major component of woody tissues. Some necrotrophs have, however, evolved the ability to parasitise trees and many of these produce enzymes capable of degrading lignin. Two of the most important are *Heterobasidion annosum*, (*Fomes annosus*) cause of butt rot of conifers, especially those grown for timber, and *Armillaria mellea*, the honey fungus, which is most troublesome in hardwoods, especially in gardens (see Chapter 11). Breaking down wood is a very complex process, requiring a large range of lignin-degrading enzymes, so these fungi are very slow-growing. Although poor as competitors because of this, they are not at too great a disadvantage since as pathogens they enter the dead wood ahead of competing saprophytes. They thus have an assured supply of foodstuffs that will last for a very long time indeed and on which they can grow and thrive long after the initial pathogenic phase is over. *H. annosum* and *A. mellea* are often known as white rotters because, after breaking down the brown lignin, they leave only a bleached skeleton of residual cell wall material. This is in contrast with colonisers of dead wood, called brown rotters, which degrade only the cellulose, leaving the lignin in place.

Factors other than the chemical complexity of lignin combine to render woody tissues particularly resistant to pathogen attack. The close-knit structure of wood creates a physical barrier to pathogen growth, although fungi and bacteria are quite good at growing along the water-conducting xylem vessels. The water content of living wood is often low, which means that wood-rotters need to be capable of withstanding considerable desiccation. The low nitrogen levels of wood require strict nitrogen conservation strategies, especially the reabsorption by the pathogen of any residual enzymes (rich in nitrogen) not used in cell wall breakdown. Finally, because woody tissue is protected by aromatic natural fungicides such as phenols, terpenoids, tropolenes, flavanoids and stilbenes, the potential pathogen must be capable of producing the enzymes that detoxify these substances. It is small wonder that there are so few wood-rotting pathogen species, and that they grow so slowly.

5. Production of pathotoxins

In addition to producing cell-destroying enzymes, some but not all necrotrophs produce toxins that kill host cells in advance of infection, thus negating active defence processes. These are called pathotoxins. This is analogous to the situation in certain diseases of humans: as long ago as the nineteenth century, Louis Pasteur showed that the symptoms of tetanus and diphtheria are caused largely by toxins produced by the bacterial pathogens responsible for these diseases. Toxins produced by necrotrophic plant pathogens may kill cells so rapidly that areas of dead tissue develop on leaves, stems or roots; or they may affect the plant's basic metabolism, causing yellow halos to form around the point of infection; or some may be transported in the plant, affecting the water relations of cells and tissues, thus causing wilting, often followed by death.

The study of pathotoxins is fraught with difficulty. When grown in culture in the laboratory, most pathogenic fungi and bacteria produce toxic chemicals, many of which will kill plant cells. These are not, however, true pathotoxins, but simply the by-products of metabolism in culture. Only after exhaustive

Fig. 2.4 Lesions of *Bipolaris maydis*, cause of southern corn leaf blight. The fungal toxin, T-toxin, causes cell death in plants carrying the Texas gene for male sterility (used in breeding hybrid maize). (P.H. Williams)

experimentation can it be concluded that a toxic chemical produced by a pathogen in the laboratory is a true pathotoxin and actually has a role in disease. The first pathotoxin to be identified with certainty was Victorin, a cyclic peptide with a terpenoid side chain produced by *Bipolaris victoriae*, cause of Victoria blight of oats (*Avena sativa*) in the United States. Many others have now been identified. Some pathotoxins, like Victorin and the toxin produced by *Ophiostoma* (*Ceratocystis*) *ulmi*, cause of Dutch elm disease, and T-toxin produced by *Bipolaris maydis*, cause of southern corn leaf blight of maize, are highly specific in their activity (Fig. 2.4). Others, such as the tripeptide derivative produced by the bacterium *Pseudomonas syringae* pv *phaseoli*, which causes yellow halos around the points of infection in the halo blight disease of French bean (*Phaseolus vulgaris*), are non-specific; the halo blight toxin (phaseolotoxin) will cause yellowing in a wide range of plant species to which it is applied artificially in the absence of the pathogen, even though the host range of the pathogen itself is limited to *Phaseolus* spp.. The effects of another non-specific toxin are often seen not in green plants but in the fruit bodies of a fungus, for

fungi as well as plants have their pathogens. The brown blotches that sometimes appear on older mushrooms (*Agaricus hortensis*, a saprophytic fungus)
are caused by a bacterium, *Pseudomonas tolasii*, which produces two toxic peptides. These chemicals disrupt the membranes of the mushroom hyphae and
kill them, but they are totally non-specific in their action and will kill the
hyphae of a wide range of other fungi as well as the cells of plants if applied
artificially.

In summary, necrotrophs are unsophisticated pathogens, intent on breaking
down cells and tissues to release carbohydrates and other nutrients, but in the
process indiscriminately killing young and weakened plants and causing rots,
necrotic spots and wilts. At the same time, however, they are highly sophisticated organisms that have evolved to exploit a particular ecological niche. In a
mixed micro-flora of pure saprophytes and facultative parasites, being the first
to colonise a virgin substrate and extract food from it confers a considerable
competitive advantage. By evolving the mechanisms to become pathogens, kill
their hosts and then break down the tissues to release foodstuffs, necrotrophs
have ensured that they occupy a lead position in the race for the energy to
reproduce.

The biotrophs

1. Normally unable to grow in the absence of a living host in nature

Most biotrophic pathogens lack completely the ability to compete successfully
with pure saprophytes whether in the soil, on the surface of a dead leaf or in
dead tissue they have killed. They appear to have sacrificed competitive saprophytic ability for a more sophisticated form of parasitism and therefore equate
with ecologically obligate parasites (Fig. 2.2), restricted to a living host by their
ecological specialisation. Many biotrophs, such as the *Erisyphe* spp. which cause
powdery mildews (Plate 5), cannot even be grown in culture, and therefore
equate with physiologically obligate parasites. This inability to grow outside living host tissues may in part be due to the loss of the ability to produce extra-
cellular, cell wall degrading enzymes. If the host cells are to remain alive to supply carbohydrates and other nutrients, the production of enzymes that break
down and kill cells would be a significant disadvantage, like killing the goose
that laid the golden egg.

The lack of the ability to grow as a saprophyte may also be due to increased
adaptation to the host environment in other ways. Hyphal membranes may be
very leaky, for instance, to facilitate nutrient exchange with the host, or the
pathogen may have lost the ability to make an essential chemical itself, thereby becoming dependent upon its host to supply it. Thus the biotrophic fungus
Puccinia graminis, cause of the black stem rust of grasses and cereals (see
Chapter 9), depends on its host for a supply of sulphur-containing amino
acids. *P. graminis* was regarded as a physiologically obligate parasite until a
group of Australian workers showed that it was capable of very slow growth on
a culture medium containing complex protein extracts and yeast extract (Fig.
2.5). Even on these rich food sources colonies took many months to reach a
size attained by other fungi in a few days. It was later shown that *P. graminis*
could not manufacture its own sulphur amino acids and that it was these that
were required from the rich medium. The slow growth was due to the fact that
the membranes were so leaky that nutrients were lost almost as rapidly as they

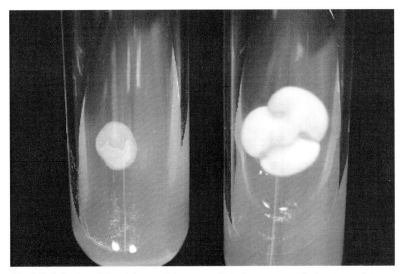

Fig. 2.5 Colonies of *Puccinia graminis*, cause of black stem rust of wheat, growing very slowly on nutrient-rich culture medium in test tubes. The largest colony has reached a diameter of approximately 1.5 cm in 48 days. (D.S. Ingram.)

were taken up. Growth in a liquid form of the culture medium was very much better than on a medium solidified with the jelly-like agar, perhaps because the former, which completely surrounded the hyphae, more closely resembled the intercellular fluid of the plant.

2. Minimal structural damage to the host

This results largely from the controlled production of cell wall degrading enzymes. Biotrophs produce such enzymes because they require them for penetration of the host and to facilitate growth between cells by dissolving limited areas of the middle lamella, but their production is triggered only by the presence of the chemical to be broken down and is then rapidly switched off again. Moreover, they are produced in only very small quantities. It may also be that in some cases such enzymes are actually bound to the walls of the pathogens' hyphae and are only active upon direct contact with host cell walls, having no capacity to diffuse further into the tissues. Those that are not bound in this way may be of higher than normal molecular weight (i.e. may be composed of very large molecules) so that they diffuse slowly and therefore move only short distances through cell walls from the point of production.

Tissues infected by biotrophs remain intact, the only evidence of disease visible to the plant pathologist often being the appearance of spores erupting through the surface of the leaf or other organ, accompanied by some chlorosis (yellowing).

3. Intercellular hyphae with haustoria

The hyphae of fungal biotrophs usually grow between cells and absorb nutrients from the intercellular fluid, but many also establish direct contact with the living contents of cells by forming specialised feeding branches called haustoria (Fig. 2.6). These penetrate the walls of the cells and induce the cell membrane lying just inside the wall to grow into loose folds. Then, as the haustorium grows into the cell, the membrane is unravelled to accommodate and enclose the expanding structure. A series of layers, the product of both the host cell and the pathogen, accumulate between the haustorium and the surrounding host cell membrane, and create an interface both for the uptake of nutrients by the pathogen and for the secretion into the host cell of substances that modify the cell's metabolism to benefit the pathogen. Haustoria increase very considerably the capacity of the pathogen to extract nutrients from living cells. This intimate contact may persist for a very long time in many biotrophic host-pathogen interactions such as the downy mildews (e.g. *Bremia lactucae* on lettuce (*Lactuca sativa*), (see p.91 and Plate 2) and the rusts (e.g. *Puccinia antirrhini* on antirrhinum, see p.176 and Plate 6). In the case of the powdery mildews (e.g. *Erysiphe pisi* on pea (*Pisum sativum*, Plate 5 and p.156)), because the mycelium grows only on the surface of the host's leaves and stems, the haustoria which penetrate the surface epidermal cells represent the only channel of communication between host and pathogen.

It should be added that some hyphal biotrophs, such as the smut fungi (see p.201),

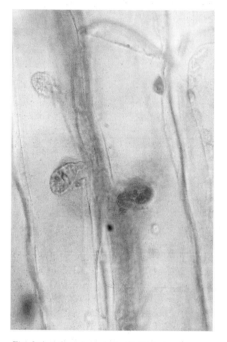

Fig. 2.6a (above) and *Fig. 2.6b* (below)

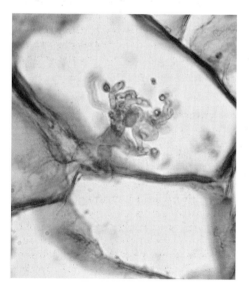

Fig. 2.6 Haustoria: (a) intercellular hypha and club-shaped haustoria of *Bremia lactucae*, cause of downy mildew of lettuce; (b) branched haustorium of *Peronospora farinosa*, cause of downy mildew of sugar beet. (Photograph (a), P.A. Mason, Institute of Terrestrial Ecology, Edinburgh; photograph (b), D.S. Ingram.)

do not produce haustoria but must absorb all their nutrients through the hyphae. Non-mycelial biotrophs establish direct contact with the cells of the host because the fungal body actually enters the cell and is surrounded by the host's membrane. This is the case with *Plasmodiophora brassicae*, the cause of clubroot on brassicas (see p.75). Again, a zone of interaction between the pathogen and the host's plasma membrane develops and creates an interface for the exchange of nutrients and other substances.

4. A very limited host range

Most biotrophic fungi are able to parasitise only a relatively narrow range of hosts, and in many cases may even have evolved races or strains capable of infecting only certain cultivars or genotypes of a host species. For example, *Plasmodiophora brassicae* only causes disease in members of the cabbage family (Cruciferae), to which brassicas belong. *Erysiphe graminis*, cause of powdery mildew in grasses and cereals, infects only members of the grass family (Gramineae). Moreover, *E. graminis* exists as a series of structurally identical strains, each called a *forma specialis*, and each capable of infecting only one genus or species within the Gramineae. Those attacking cereals are, respectively, f.sp. *tritici* (wheat, *Triticum aestivum*), f.sp. *hordei* (barley, *Hordeum vulgare*) and f.sp. *avenae* (oats). In each case there are races within each *forma specialis* that will affect only certain cultivars of wheat, barley or oats. Similarly, high levels of specialisation with respect to host range are found in many other biotrophs, including the rusts, smuts and downy mildews.

A high level of specificity is related to the degree of sophistication of the relationship of biotrophs with their hosts. This reflects not only nutritional specialisation and the fact that the requirements of the pathogen are closely integrated with the environment of a particular host, but also the sophisticated nature of the mechanisms that have evolved to enable the pathogen to bypass the resistance mechanisms of particular hosts without actually killing the cells. Over the millennia, each time a pathogen has evolved to overcome the resistance of a host, the host has responded by evolving yet more sophisticated mechanisms of resistance. This has been followed by further pathogen evolution and so on. The artificial production by plant breeders of cultivars of crop species carrying resistance to disease has speeded up this evolutionary tit for tat in some host-pathogen combinations, leading, for example, to the array of *formae speciales* and races seen in *E. graminis*.

5. Hormonal disturbance of the host

Most biotrophs appear to induce significant morphological disturbance of the host, which may take the form of tumours (e.g. crown gall on fruit trees – *Agrobacterium tumefaciens*), distorted stems or leaves (e.g. peach leaf curl – *Taphrina deformans*), swollen roots (e.g. clubroot of brassicas – *Plasmodiophora brassicae*) and so on (Fig. 4.1, p.76). In the case of crown gall it has been shown that a nucleic acid plasmid moves from the bacterium into the host cells and there induces the abnormal production of the growth hormone indoleacetic acid. In the case of the clubroot disease of brassicas, *P. brassicae* causes an increase in the levels of both indoleacetic acid and another growth hormone, cytokinin, in the brassica root cells, which increase in both number and size. In other cases, for example where host tissues do not respond to high levels of hormones by growing abnormally, the evidence of hormonal disturbance takes the form of green islands (Plate 6). These are areas of green tissue around the lesion that continue to look healthy long after the rest of the leaf has started to yellow and die. It is thought that production of plant growth regulators such as cytokinins by the pathogen, or by the host following stimulation by the pathogen, are responsible for this delay in the onset of senescence.

Almost all infections by biotrophs reveal some evidence of hormonal disturbance of the host tissue, which suggests that plant growth hormones produced by biotrophic pathogens or by hosts in response to biotrophic pathogens are very important in delaying the senescence of infected tissue and in stimulating host metabolism in such a way as to serve the needs of the pathogen. Plant growth hormones may also have a role in suppressing host resistance to disease.

6. Diversion of the sugar transport pathways of the host

It has already been seen that biotrophic pathogens do not obtain the sugars they need for growth and sporulation by breaking down host cells, as necrotrophs do. Because they require a supply of carbohydrates, however, they divert the normal sugar-transporting pathways of the plant for their own use. Instead of exporting the sugars made during photosynthesis to nourish the growing roots and stem tips, the infected leaf itself imports sugars produced in the plant's uninfected leaves. Within the infected leaf, the sugars travel towards the point of infection. In this way the fungus is assured of a continuous supply of food, but this is at the expense of the normal growth of the plant. This explains in part why one of the symptoms of infection of plants by biotrophs is stunted growth.

At the site of infection the sugars of the plant are taken up into the fungus and are there converted first to polyols (polyhydric alcohols) and then to the lipids, both storage forms of food that the plant itself is unable to utilise. This conversion is therefore a form of non-return valve, ensuring that once a fungus has extracted carbohydrate from the plant it cannot leak back again, even though it may not be required for immediate use. If the supply of sugars arriving at the infection site becomes too great for the fungus to convert to polyols and lipids it may induce the host plant itself to convert the sugar to starch, a storage form of carbohydrate in plants. Later, when the infected leaf is dying and the supply of sugar from other leaves is drying up as the plant ages, the fungus may induce breakdown of this starch to re-establish its supply.

The mechanisms by which biotrophs are able to change the pattern of sugar transport within an infected plant is not known, but it is thought that interference with growth hormone levels may be involved.

7. Increased metabolic activity of host cells around the site of infection

All plant pathogens, whether necrotrophs or biotrophs, cause an increase in the respiration rate of the host as it struggles to release sufficient energy to make both the fungitoxic chemicals needed to resist attack and the structural chemicals required to repair the severe damage caused by the pathogen. In the case of infection by biotrophs this increased respiratory activity may have another function in that it may supply, as by-products, chemicals required by the fungus for the conversion of plant sugars to polyols and lipids. The increased respiratory activity is accompanied by the enhanced formation by the host, in response to signals from the pathogen, of a whole range of other substances required by the pathogen for its nourishment. All this requires a massive increase in the formation of the enzymes that drive these processes and this, in turn, is reflected in an increase within the nucleus of the nucleic acids that control their formation. Examination of the cells of infected plants with a microscope reveals, almost invariably, that the nuclei are enlarged.

Ironically photosynthesis, which generates carbohydrates in green plants, is inhibited by biotrophs around the point of infection despite the frequent occurrence of green islands. The reasons for this are not known, but the problem is overcome by the diversion of the sugar transport pathways as described above.

In summary, biotrophy is a very sophisticated and very successful form of parasitism. The host cells remain alive for a long time and the pathogen is thus protected by the host's innate resistance mechanisms from the competition that would result if the tissues were invaded by pure saprophytes. In order to achieve this, however, biotrophic pathogens have sacrificed some of their ecological flexibility by reducing their ability to produce cell-degrading enzymes and toxins, thus limiting them to a parasitic mode of life. Once a host dies, since the biotroph is unable to compete as a saprophyte, it must survive in the form of resting spores or be dispersed to infect some other living host plant.

Hemibiotrophs, which appear to combine the features of both biotrophy and necrotrophy, may be the most sophisticated pathogens of all, for when the biotrophic phase of development comes to an end as the host succumbs to the ever-increasing demands of the pathogen, the pathogen remains in the tissue, moving into a necrotrophic phase of growth, and is thus first in the succession of fungi and bacteria that can break down dead tissues.

The evolution of biotrophy and necrotrophy

Theories abound as to how biotrophy and necrotrophy may have evolved. One school of thought suggests that necrotrophs are primitive in an evolutionary sense, having first evolved from saprophytes through the development of the appropriate enzymes and other factors required to overcome the resistance of plants to attack by fungi and bacteria. Subsequently, it is argued, biotrophs evolved from necrotrophs as they became increasingly dependent on a supply of sugars from the host's own photosynthetic processes, rather than breaking down the cell wall. The acquisition of the ability to produce plant growth hormones, which are said to play a central role in the diversion of the host's sugar-transporting pathways, is thought to have been a key evolutionary event. Such evolving biotrophs would be at a disadvantage if they produced enzymes that actually killed plant cells, and it is suggested that this ability was first suppressed by the accumulation of sugars at the infection site in the interests of maintaining the host tissues in a living and active state. The presence of the end products of a reaction are well known to inhibit the activity of the enzyme controlling it. At a later stage of evolution, selection of mutants with modified enzyme production more suited to the biotrophic existence would have occurred. According to this theory hemibiotrophs are fungi that have not yet reached the stage of becoming true biotrophs.

Another school of thought argues that biotrophs are evolutionarily primitive and that the earliest pathogens lived in close association with the living cells of evolving plants. It is further suggested that some biotrophs might then have acquired by evolutionary selection the enzymes for killing host cells, breaking them down and living upon them saprophytically. Such changes would confer a selective advantage in evolution for they would increase the range of ecological niches available to the pathogens concerned.

Since there are few fungal pathogens preserved within fossilised plant tissues it is difficult to accumulate sufficient evidence from the geological record to support either one of these hypotheses. Some have attempted to assemble evidence by the examination of living pathogens, but that available so far is not completely convincing and certainly does not allow an unequivocal distinction to be made between the two evolutionary strategies proposed. It is our belief that necrotrophy, biotrophy and hemibiotrophy have evolved many times since life on earth began, with evolution progressing from necrotrophy to hemibiotrophy or biotrophy and *vice versa* on different occasions in different pathogens interacting with different hosts. The genetic make up of the fungi and bacteria make them so plastic that very rapid evolution is possible. It is inconceivable to us that any one theory for the evolution of biotrophy or necrotrophy should be regarded as universally applicable. Moreover, hemibiotrophy is not necessarily a stage in the evolution of either biotrophy or necrotrophy; it may often be an evolutionary end point in its own right.

The infection process

We have not yet described how pathogens actually breach the outer surface of a potential host and establish themselves within the tissues. A consideration of the complex processes involved must begin with some observations on spores, for although mycelium growing from an infected plant often infects another potential host, spores are the most frequent initiators of the infection process.

Spores

The spore is the 'seed' of the pathogen. It may have a thick wall and be a resting spore, or it may be thin-walled, having recently been carried to the surface of the plant by wind or water. In either case it will contain food storage compounds such as the fatty lipids and will be packed full of cytoplasm.

The spore will usually be dormant, held in that state of suspended animation by a number of factors. For example, it may lack the nutrients it requires for germination. It may contain water-soluble inhibitors of germination, synthesised during its formation to prevent it germinating while still on the host plant on which it was produced, and will only be able to germinate when water is present to wash the inhibitors out. Alternatively, the fungal spore may be held in a state of dormancy called fungistasis, induced by the toxic products of other microorganisms (fungi and bacteria) growing as saprophytes on or near the surface of the plant. In the case of resting spores, more complex factors like the requirement for a cold shock or for a specific chemical signal from a potential host may be required to trigger germination. Such strategies have evolved to ensure that resting spores germinate only at a time when a susceptible host is likely to be present. Once the spore is provided with the appropriate nutrients, water and the necessary triggers of germination it may, if it is on the surface of a susceptible host, germinate and begin the process of infection (Fig. 2.7). The precise strategy utilised for infection varies from species to species, but a clear distinction can be drawn between the necrotrophs and the biotrophs.

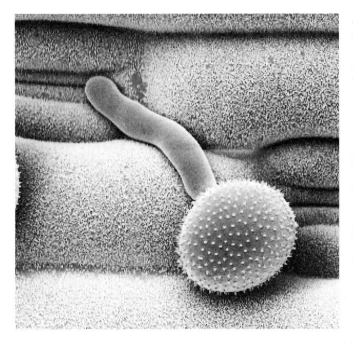

Fig. 2.7 A germinating urediospore (approx 20 μm diameter) of the yellow rust fungus (*Puccinia striiformis*) on a leaf of barley (*Hordeum vulgare*). The germ tube is growing over the epidermal cells towards a stomatal pore. (Scanning electron microscope photograph supplied by N. Read, University of Edinburgh.)

Cutinase

A key enzyme required for penetration is cutinase, which breaks down the cuticle of the plant. Recent research has shown that minute quantities of cutinase are produced by fungal infection hyphae. Once cuticle is encountered the cutinase begins to break it down into simpler substances. As soon as these are detected by the fungus it switches on the production of very much larger quantities of enzyme which then accelerate the process, enabling the cuticle to be breached. As soon as this process is complete, when no more breakdown products are detected, cutinase production ceases. The other cell wall degrading enzymes referred to above are then deployed to continue the penetration of the plant cell wall.

Some pathogens do not appear to produce cutinase. These either enter the host through natural openings such as stomata or wounds, or penetrate the cuticle by mechanical means – it is known that enormous pressures may be generated at the hyphal tip.

Infection by necrotrophs

As a rule, necrotrophic pathogens must build up a large mass of spores or mycelium at the surface of a host before infection can occur. This 'massing of the troops before attack' is necessary because considerable quantities of cell-destroying enzymes and sometimes toxins are required to kill and breach the host tissues, thus preventing the activation by the host of the machinery of resistance. Such a strategy is consistent with the unspecialised nature of the interaction between necrotrophs and their hosts. In the less specialised necrotrophs this build up of mycelium is not organised and may simply result

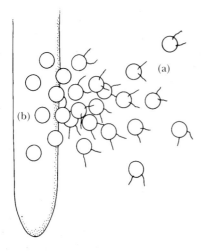

Fig. 2.8 The infection strategy of a necrotrophic *Pythium* sp., causing damping-off in seedlings. Zoospores (a) swim along a concentration gradient of leaked nutrients towards a wounded or weakened root. They cluster at the wound site or just behind the apex (b), germinate and send out hyphae which colonise and kill the root tissues beneath. (Illustration by Mary Bates.)

from the accumulation and branching of hyphae growing together, or of spores collecting on the plant surface. In more specialised necrotrophs, however, an organised infection structure may be involved.

Pythium spp., cause of the damping-off of seedlings, produce large numbers of zoospores with which to mount an attack on the host (Fig. 2.8). These are attracted towards the wounded or weakened roots of a potential host from which chemical nutrients such as sugars and amino acids have leaked into the wet soil that favours growth by this pathogen. The zoospores swim in the films of water around the soil particles, along the concentration gradient of the leaking nutrients, towards the surface of the root, a process known as chemotaxis. The zoospores cluster around the wound site or alternatively aggregate just behind the root tip, for this is the region from which nutrients leak most readily. They then germinate, using the leaked nutrients as a food source, and form a network of hyphae over the whole surface of the root, continuing to absorb leaked nutrients. This network provides a major energy source from which multiple penetrations of the root may be effected by individual hyphae. By contrast, *Rhizoctonia solani* (the asexual form of *Thanatephorus cucumeris*), also a cause of damping-off in seedlings, has a slightly more sophisticated strategy (Fig. 2.9). Here the fungus does not infect from a spore but from a sclerotium. The sclerotium is a consolidated mass of hyphae that has secreted a hard, water-resistant outer surface, thus enabling the

Fig. 2.9 The infection strategy of *Rhizoctonia solani*, a necrotroph that causes damping-off in seedlings. A hypha (a) growing from a sclerotium in the soil proliferates at a specific point on the root surface (b). This then serves as the energy source for an enzymatic attack on the root tissue beneath (c). (Illustration by Mary Bates.)

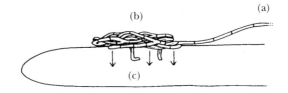

structure to serve as a kind of resting spore. The sclerotium is induced to germinate by chemical nutrients leaking from wounded or weakened roots, again
into wet soil, and the emerging hyphae then grow either towards the wound
site or to the point just behind the root tip, guided by chemotaxis. Once at the
roots, instead of forming a network of hyphae all over the surface, a mass of
mycelium aggregates at a single point and proliferates by branching. This
serves as the energy base for a massive attack on the root tissue below. Once
inside the root tissue, both *P. ultimum* and *R. solani* grow extensively by forming a network of mycelium that grows through and between cells, killing and
breaking them down indiscriminately.

Another type of necrotroph attack is that used by *Gaeumannomyces graminis*,
the cause of take-all disease of grasses and cereals (Fig. 2.10). Here the fungus
will be growing in soil on the dead tissues of its previous host. Using nutrients
from this foodbase, which may comprise old roots, stubble or straw, exploratory hyphae grow out through the soil until they encounter the roots of an uninfected plant. Nutrients transported along the hyphae from the food base are
then used to infect the new host and a bridgehead of infected tissue is established at that point. From that bridgehead, using the colonised tissue of the
new host as a food source, the pathogen sends out further exploratory hyphae,
thicker and darker than normal, over the surface of the root and these mount
a series of new attacks on the uninvaded areas of the root, establishing new
bridgeheads as they go. Eventually the whole root is colonised in this way.

An even more sophisticated infection strategy is that adopted by *Armillaria
mellea*, the honey fungus that causes rots in trees (Fig. 2.11, overpage). Again,
the fungus uses the previously infected host as a food base, but instead of sending out exploratory hyphae it forms a more complex structure constructed
from an aggregation of hyphae, called a rhizomorph. Rhizomorphs resemble
blackened roots, or even bootlaces, to the naked eye, and may be several millimetres in diameter. The hyphae within them are arranged in an ordered pattern, with outer protective layers and inner layers able to transport water and
nutrients. From the food base of a dead tree, *A. mellea* sends out rhizomorphs
that can grow great distances through the soil, sometimes many metres, until
they encounter the roots of a new host. There a rhizomorph attaches itself to

Fig. 2.10 The infection strategy of *Gaeumannomyces graminis*, a necrotroph that causes take-all
disease of cereals and grasses. Hyphae grow out from a food base such as a piece of buried
straw (a). When a hypha contacts the root of a potential host a bridgehead of infection is
established (b). Using nutrients from the newly colonised tissues, new runner hyphae grow
out and establish further infections until the whole root is colonised. (Illustration by Mary
Bates.)

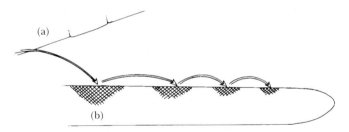

the surface of the new host tree and, using nutrients transported along its entire length, is able to muster the energy required to mount an enzymatic attack that will kill and break down the woody tissues. The fungus spreads through the wood and colonises it completely, eventually killing the whole tree. Further rhizomorphs are produced on the dying tree and grow out into the soil in search of new hosts. This specialised infection strategy means that once a tree in a garden, or a neighbour's garden, becomes infected, none of the surrounding trees is safe from attack.

Infection by biotrophs and hemibiotrophs

Biotrophs and hemibiotrophs, in contrast to necrotrophs, are normally able to infect from a single spore. In this case, a specialised infection structure is invariably produced and this goes through a series of stages in developing an interface with the living cells of the host. This kind of strategy is consistent with what we have already seen of the highly specialised nature of the relationship between biotrophs and their specific hosts.

Perhaps the most remarkable infection structures formed by biotrophs are those of *Plasmodiophora brassicae*, the cause of clubroot disease (Fig. 2.12). This non-hyphal fungus produces zoospores that are attracted by chemotaxis and follow a trail of leaked nutrients back to the roots of plants in the cabbage

Fig. 2.11 The infection strategy of the necrotroph *Armillaria mellea* (honey fungus), a necrotroph that causes rot in trees. A rhizomorph (a) grows out through the soil from the food base of a heavily colonised tree (b). Water and nutrients are translocated along the rhizomorph and provide the energy for an enzymatic attack on the roots of a new host (c). (Illustration by Mary Bates.)

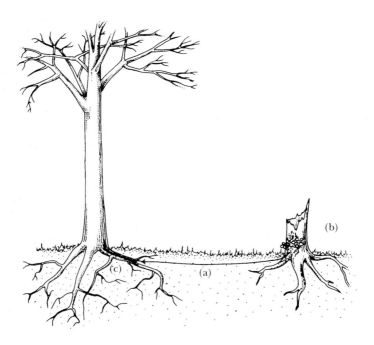

family (Cruciferae). There the individual zoospores attach themselves to the surface of the root hairs, become rounded and develop a thin but firm wall. From a point on the wall of this cyst a special hypha bulges out and glues itself to the surface of the root hair. It then swells slightly and ceases to grow. Attachment structures of this kind are called appressoria. At the same time, the contents of the zoospore begin a remarkable transformation. First a bullet-like structure, the stachel, forms within the cyst. Attached to the rear end of this is a large piece of membrane and this in turn is attached to that part of the cytoplasm containing the nucleus. Simultaneously, the large blobs of lipids that the zoospore contained at the beginning of the attack begin to break down very rapidly as their energy is released and consumed in the manufacture of a huge balloon-like body filled with liquid, called a vacuole. This increases in size very rapidly as water is pumped into it and pressure builds up within the cyst. Eventually this pressure becomes so great that the bullet-like stachel, together with the membrane, cytoplasm and nucleus from the cyst, is shot through the wall of the root hair and into the cell itself. As it goes the membrane of the root hair cell is pushed inwards and eventually closes around the pathogen cytoplasm and nucleus, thus establishing an intimate contact between host and pathogen. This whole process of infection takes only a few hours, with the actual penetration process itself, in which the pathogen is shot into the root hair, occurring over a period of minutes. The pathogen then takes up nutrients from the host root hair cell, its nucleus divides to produce many more nuclei and more cytoplasm is formed. The developing pathogen is called a plasmodium. As this process is repeated a very large plasmodium develops, occupying a significant part of the space within the root hair cell but still continuing to draw nutrients from it. Eventually the root hair plasmodium divides up into a mass of new zoospores. These are then released and reinfect the main body of the root by a process similar to that involved in infection of the root hairs. This second stage of infection is the one that leads to the development of the swollen roots so characteristic of clubroot disease.

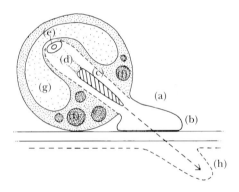

Fig. 2.12 The infection strategy of the biotroph *Plasmodiophora brassicae*, cause of clubroot disease of members of the cabbage family (Cruciferae). An encysted zoospore forms an appressorium (a) which becomes firmly attached to the root surface (b). Within the zoospore a bullet-shaped stachel develops (c). Attached to it is a membranous structure (d) which in turn is attached to a small part of the cytoplasm and the nucleus (e). Large lipid droplets (f) are metabolised to provide the energy for expansion of a vacuole (g) which drives the stachel, membrane, cytoplasm and nucleus in to the host root hair cell. The root hair cell remains alive and the membrane (h) is not pierced, but is pushed inwards and eventually encloses the pathogen. (Illustration by Mary Bates.)

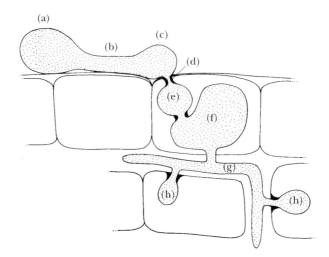

Fig. 2.13 The infection strategy of the biotroph *Bremia lactucae*, cause of downy mildew of lettuce (*Lactuca sativa*). A spore (a) germinates on the surface of a host leaf and forms a germ tube (b) which swells at the tip to form an appressorium (c) attached firmly to the cuticle. An infection peg (d) penetrates a living epidermal cell and forms first a primary vesicle (e) and then a secondary vesicle (f). An intercellular hypha (g) grows from the secondary vesicle and invades the leaf. The hyphae form haustoria (h) at intervals and these penetrate further living cells. (Illustration by Mary Bates.)

The hyphal biotrophs have evolved very different but equally sophisticated strategies for gaining entry to their hosts. For example, *Bremia lactucae*, the cause of downy mildew in lettuce, forms a series of infection structures called vesicles (Fig. 2.13). First of all a spore lands on the surface of a host leaf. If water is present a water soluble inhibitor of germination is washed out of the spore and sugars leak out from the surface of the leaf and provide the spore with sustenance. It immediately germinates to produce a special hypha called a germ tube, which grows over the surface of the leaf until it reaches a point where two cells of the outer layers of the leaf, the epidermis, actually meet. There it stops growing, glues itself to the surface of the leaf and swells slightly to produce an appressorium. From this a tiny infection peg emerges and produces powerful cell wall dissolving enzymes over a very small surface area, allowing the infection peg to penetrate the cuticle and the wall, after which enzyme production ceases. Once inside the cell the infection peg grows and expands to form a spherical structure called a primary vesicle. The contents of the spore then travel down through the germ tube and into the primary vesicle, and at this stage the point of penetration is plugged; in effect, the spore has been moved from the harsh environment of the surface of the leaf to the much more protected environment within the host cell.

Meanwhile the host cell does not die but is induced to produce far more plasmalemma than normal, and as the primary vesicle expands in the cell this membrane becomes unravelled to surround it so that the living cytoplasm of the host cell is in intimate contact with the fungus itself. At this point a new hypha begins to grow out from the primary vesicle, again inducing the host's

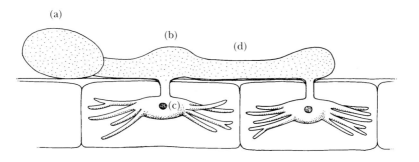

Fig. 2.14 The infection strategy of the biotroph *Erysiphe graminis*, cause of powdery mildew of cereals and grasses. A spore (a) germinates on the surface of a leaf and swells at the tip to form an appressorium (b) which produces an infection peg that penetrates the cuticle above an epidermal cell. A branched haustorium (c) forms in the living cell and extracts nutrients from it. A hypha (d) next arises from the appressorium and grows out over the leaf surface to establish a new infection, and so on. (Illustration by Mary Bates.)

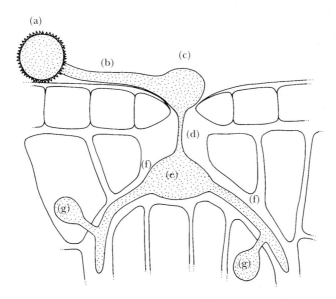

Fig. 2.15 The infection strategy of the biotroph *Puccinia graminis*, cause of black stem rust of cereals and grasses. A urediospore (a) germinates on the leaf and a germ tube (b) grows out over the surface until it encounters a stomatal pore. It then swells at the tip to form an appressorium (c) which in turn forms an infection peg (d) that penetrates the pore and expands to give a vesicle (e) within the sub-stomatal cavity. Intercellular hyphae (f) arise from the vesicle, producing haustoria (g) which penetrate the living cells of the host. (Illustration by Mary Bates.)

plasmalemma to grow abnormally. This hypha expands to become the secondary vesicle, which has an even closer contact within the host cell and takes up nutrients from it. The pathogen nuclei divide repeatedly and the cytoplasm increases to fill the expanding vesicle. Eventually the secondary vesicle itself forms a further hypha, which grows out of the initially penetrated cell and into the intercellular spaces of the lettuce leaf. This hypha branches and the resulting mycelium permeates the whole leaf, always growing between the cells. The hyphae take up nutrients from intercellular spaces but also form the haustoria that establish direct contact with individual living cells and increase the capacity for nutrient absorption.

Erysiphe spp., the cause of powdery mildews of a wide range of garden and wild plants, grow on the surface of the host (Fig. 2.14). In this case the spore germinates in the same way as *Bremia lactucae* spores, and the resulting germ tube grows and produces an appressorium that sticks to the surface of the host. An infection peg is again formed and the cuticle and cell wall of the host breached. In this case, however, no vesicles are formed. Instead the infection peg forms a haustorium directly, within the initially penetrated cell, and there begins to extract nutrients through it. The appressorium then produces a second hypha, which grows out and again produces an appressorium, which once more produces a haustorium in a further epidermal cell, and so on. Eventually the whole surface of the leaf is colonised and haustoria form a complex series of feeding branches that penetrate the surface cells.

A final example is *Puccinia graminis* (Fig. 2.15), the cause of black stem rust of grasses and wheat. Here the fungus takes advantage of the stomata, the pores in the leaf used for gas exchange, to gain entry to the host. The spore lands on the surface of a leaf and, if water and nutrients are present, germinates in the usual way. The germ tube then grows towards a stomatal pore, perhaps attracted by substances leaking from it, and begins to swell at its tip once it has encountered the pore. The resulting appressorium becomes firmly glued to the rim of the pore and then forms an infection peg that grows down through the pore itself into the substomatal cavity. There it swells to form a vesicle within the cavity and begins to take up nutrients from the intercellular fluid of the host leaf. The vesicle becomes a massive structure, full of cytoplasm and nuclei, and then produces hyphae which grow between the cells of the infected leaf, forming haustoria that establish direct contact for nutrient exchange with individual cells.

Many more examples of infection by necrotrophic, biotrophic and hemibiotrophic fungi might be cited. This small number of examples, however, gives the reader some idea of the huge range of strategies that may be utilised by pathogenic fungi to achieve the most difficult part of the process of pathogenesis, that of gaining entry to the host itself. Both nectrotrophic and biotrophic bacteria, lacking hyphae, usually enter their hosts through wounds or natural openings such as stomata. Viruses sometimes enter through wounds but are usually introduced into host cells by a vector such as an arthropod or a fungus.

In this chapter the development of disease in susceptible hosts has been considered. Plants have, however, evolved a variety of defence mechanisms to resist attack and only those pathogens that can overcome these are able to grow in a plant and cause the symptoms of disease. Defence against attack is the subject of the next chapter.

3

Resisting Attack

Although plant disease is everywhere and plants are constantly exposed in the air and soil to the spores and other reproductive structures of a wide variety of potential pathogens, most plants are resistant to attack by most pathogens. Just look around the garden or the countryside and you will find ample evidence that unless an epidemic is raging, disease is the exception rather than the rule. Take a close look at a rose border you have omitted to spray, for example: mildew may be present, but not on all the varieties; black spot may be there too, and if you are unlucky there may be rust on some of the older cultivars and species; but few of the plants will be infected by more than one pathogen, and even those that are infected will have limited the progress of the pathogen so that the lesions grow relatively slowly. Some of the species and/or cultivars will show more resistance than others to each of the pathogens, and some may even be immune.

Wherever you look, in the garden, in agriculture or in natural ecosystems, the pattern will be very similar. Interactions between pathogens and hosts are highly specific, and where disease does occur both susceptibility and resistance may be seen within host populations matched, as we shall see, by virulence (ability to cause disease in a specific host) and avirulence (inability to cause disease in a specific host) within pathogen populations. Finding an explanation for such specificity has occupied the minds of plant pathologists since the last century, and we are still a long way from having a complete picture of the mechanisms involved. In the pages that follow we will present our model of disease resistance in plants, based on the complex and often confusing body of evidence accumulated during the last hundred years or so. The model will not be correct in every detail, and we know that as new evidence comes to light the model will have to be changed to accommodate it. That is the nature of science: evidence is gathered, hypotheses are formulated and tested by experiment and new hypotheses follow.

The evidence and hypotheses are complex. If you would prefer to read about the diseases of plants, turn to Chapter 4, but if you would enjoy the challenge of getting to grips with one of the most exciting aspects of plant pathology, read on.

The first line of defence

Plants utilise a variety of mechanisms to exclude potential invaders. Firstly, the very structure of the plant body renders it relatively impregnable to attack. The above-ground portions are covered with a cuticle composed of the tough fatty polyester cutin. Water does not easily adhere to this hydrophobic surface, so that spores landing on the leaf are not easily wetted and spores arriving in water droplets may find no safe resting place. If wax crystals are also present on the surface, as is often the case, or a dense covering of hairs, water droplets are even less likely to stick. Even when spores do remain on the plant long enough to germinate, and sufficient moisture is present to sustain growth of the cells or hyphae, the potential pathogen must be equipped with the appropriate enzymes to dissolve cutin, or must be adapted to gain entry through wounds or natural openings like stomata. In older tissues and roots the process of penetration may also be frustrated by the presence of other complex hydrophobic polymers such as suberin, the principal component of the corky tissue of bark, or lignin, the main component of wood.

The poisoned chalice – pre-formed toxins

It is small wonder that, with such formidable barriers to circumvent, few potential pathogens ever gain entry to a plant. Those that do are then likely to have their further progress impeded by the inhospitable chemical environment of the cells and tissues. First of all, the pH may be too high or too low, or widespectrum toxins such as alkaloids, phenolic compounds or, especially in seeds, proteins, may be present to destroy all would-be pathogens except those with the appropriate antidote.

Take sitka spruce trees (*Picea sitchensis*), for example. Trees of this species have been shown to contain high levels of the stilbene glucosides astringin and rhaponticin. These chemicals are highly toxic to fungi and are thought to play a significant role in defence. However, some wood-rotting fungi such as *Heterobasidion annosum* (butt rot), *Armillaria mellea* (honey fungus) and *Stereum sanguinolentum* (red-stain rot) have developed the ability to produce enzymes that detoxify stilbenes and are therefore able to penetrate the wood and cause severe disease.

Similarly the cells of oat roots contain avenacin, a complex molecule called a triterpenoid saponin with a carbohydrate side chain. This binds to the cell membranes of fungi, including *Gaeumannomyces graminis*, the soil fungus responsible for the take-all disease of cereals and grasses (Gramineae). This means that oat, unlike wheat, is resistant to the common form of *G. graminis* (called f.sp. *tritici*). One form of *G. graminis* has evolved, however, to produce the enzyme avenacinase, which detoxifies avenacin by cutting off the side chain. This form, *G. graminis avenae*, is capable of infecting oat and causing a take-all disease that is just as destructive as the one seen on wheat.

Fighting back – general active defence

If a potential pathogen is able to breach the plant's structural defences and survive the toxic chemicals in the cells, a further line of defence may be brought into play. This involves the activation of new biochemical pathways that change the cellular environment, making it even more hostile to microbial growth. Firstly, however, the invader must be recognised as an alien. Whether specific chemicals secreted by microbes or structural molecules of their cell walls and/or membranes are recognised by the plant, or whether general stress is the signal, is not known. Fungi and bacteria secrete a wide range of chemicals during pathogenesis, such as enzyme proteins and polysaccharides, any of which might enable a plant to recognise them as aliens. Moreover, components of the cell wall such as proteins, glycoproteins and polysaccharides, as well as the fatty components of membranes, might also function as recognition factors. Different experiments with various host-parasite interactions have implicated all these factors, as well as metabolic poisons such as heavy metals and antibiotics, as 'elicitors of defence reactions'. It is probable that different groups of plants have evolved different mechanisms for recognising a wide spectrum of alien invaders.

Once recognition has occurred a chemical signal is activated in the plant cells, ultimately triggering the defence responses. The nature of this signal is again largely unknown, but various factors involved in signalling in animals, such as free radicals (especially superoxide), calcium, calmodulin, G-proteins, protein kinases and phosphatases have all been implicated in defence responses of different plant species. As with recognition, it is likely that plants have a relatively flexible signalling system to switch on general active defence responses in a wide variety of circumstances.

Following recognition, a 'hypersensitive response' almost invariably occurs (Figs. 3.1 and 3.2). This is a kind of cell suicide, activated by an invader, that has evolved as part of a suite of responses to limit or prevent further spread of that invader. It involves the rapid programmed collapse and death of the initially invaded cell and sometimes the cells adjacent to it. The dead cells turn brown almost immediately. The mechanism of cell death is poorly understood, but research with potato suggests that it probably involves release of the free radical superoxide within the invaded cell. This damages the membranes irreversibly and the cell collapses. Phenol-oxidising enzymes are then released and these convert cellular phenolic compounds to dark, toxic polymers such as melanin which are laid down in the cell wall. The hypersensitive response alone is probably sufficient to prevent the further growth of obligate biotrophs, which

Fig. 3.1 The hypersensitive response: cotyledons (first formed leaves) of a lettuce (*Lactuca sativa*) cultivar containing a gene for resistance, inoculated with droplets of water containing spores of an avirulent race of *Bremia lactucae*, cause of downy mildew. The invaded cells have died rapidly and turned brown. (I.C. Tommerup.)

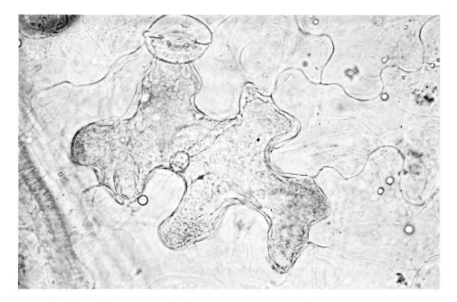

Fig. 3.2 A single cell of lettuce exhibiting the hypersensitive response to an avirulent race of *Bremia lactucae.* (D.S. Ingram.)

require living cells if they are to survive, but additional responses are required for the limitation of necrotrophic and hemibiotrophic pathogens.

The best studied of these is the rapid accumulation of new antimicrobial toxins of low molecular weight called phytoalexins. Phytoalexins were first discovered by K.O. Muller, working during the 1940s on the resistance of potato to the blight fungus *Phytophthora infestans.* Subsequently they were studied by I.A.M. Cruickshank, working on the resistance of pea to a variety of fungi, and then by many other plant pathologists. Much of the research centred on peas (*Pisum sativum*) and beans (*Phaseolus* and *Vicia* spp.) and their potential pathogens, because the young split pods, with their series of depressions where developing seeds had lain, provided a convenient experimental system for inoculating tissues with drops of water containing spores of a potential pathogen and then, after attempted invasion, for withdrawing them easily for chemical analysis.

It soon became evident that phytoalexins are normally synthesised by the living cells of the plant close to the point of challenge by an avirulent form of a pathogen. The response is a metabolically active one, requiring oxygen, and involving the synthesis of new enzymes rather than the passive conversion of the existing components of dead or dying cells. Furthermore, the production of phytoalexins is triggered by cellular messages synthesised following recognition of an invader as alien. A wide diversity of phytoalexins has been identified, each typical of a particular plant species or group. All are widely fungitoxic and completely non-specific in their activity against pathogens. Any specificity of resistance in which they are involved must therefore be embodied in the recognition process rather than in the mechanism of pathogen death.

The study of phytoalexins was important in the history of plant pathology, for it marked a change of emphasis away from simple observation of resistance reactions and towards the systematic investigation of the cellular and biochemical mechanisms involved.

In addition to the accumulation of phytoalexins, polymers such as callose, suberin and lignin, hydroxyproline-rich proteins and even silica may be synthesised rapidly and deposited in the walls of the living cells adjacent to the hypersensitive site, possibly in response to the general stress engendered within the tissues. These provide further structural barriers to pathogen growth and may also cross-link with the plant's cell wall components such as hemicelluloses and pectic substances, rendering them resistant to breakdown by fungal enzymes and therefore unavailable as a source of food. Finally, pathogenesis-related (PR) proteins may be synthesised and accumulate at the infection site. Some of these are enzymes such as glucanases and chitinases, capable of breaking down pathogen cell walls. Others bind to pathogen membranes and proteins, including enzymes, and exert an inhibitory effect in this way. They undoubtedly contribute to the toxic milieu in and around the hypersensitive cell.

PR proteins, together with lignin, are also synthesised at some distance from the infection site, even in different leaves on the same plant, and help to repel further invasion attempts by potential pathogens. This so-called systemic acquired resistance, first studied in members of the cucumber family (Cucurbitaceae), is triggered by salicylic acid, a phenolic acid produced or released during the hypersensitive response, but how the signal moves through the plant and how it activates the synthesis of PR proteins is not yet known. A related chemical, methyl salicylate, is volatile and may even be released as a gas by plants attacked by fungi to trigger resistance in adjacent plants. Methyl jasmonate, another gaseous messenger, may also be involved in triggering resistance responses in nearby plants. Such effects of volatiles could explain the efficacy of 'companion plants' in preventing disease. For example, the aromatic *Artemisia* spp. such as tarragon (*A. dracunculus*), southern wood (*A. abrotanum*) and wormwood (*A. absinthium*), which are said to prevent infection in adjacent plants, are known to emit significant quantities of methyl jasmonate.

There may be other products actively synthesised during the basic defence of plants against aliens, but they have not yet been discovered. Any one of the resistance responses is probably capable of preventing or restricting the growth of a potential pathogen, yet most appear to be activated together during defence. Plants thus seem to have adopted a 'belt and braces' approach in the evolution of mechanisms for protecting themselves against all possible invaders, which is not surprising given the great diversity of potentially pathogenic microorganisms. Furthermore, the accumulation of toxic factors such as phytoalexins and PR proteins within the hypersensitive cells prevents colonisation by secondary invaders and saprophytes, which might otherwise take advantage of the dead cells as a source of food.

As with the defence of a country, the defence of plants against potential invaders incurs a cost. For example, the synthesis of phytoalexins requires the consumption by the plant of considerable reserves of energy. The effect on the productivity of the plant (or yield in the case of a crop) will inevitably be significant.

The basic defence processes described so far are non-specific and are probably controlled by several genetic factors in both hosts and microorganisms. It is likely, however, that only relatively few genes are involved in triggering resistance and these activate a cascade of responses. This kind of resistance has been variously described as 'basic incompatibility', 'general resistance', 'non-host resistance' and 'species incompatibility'.

Breaching the barriers: necrotrophs

Microorganisms that have evolved to become pathogens of particular plant groups have somehow acquired mechanisms for overcoming the basic defence processes of those groups.

For many necrotrophs, like the *Pythium* spp. that cause damping-off of seedlings, the basic strategy is to kill the host cells with a combination of enzymes and/or toxins before the resistance responses are activated (see p.53). In addition, some necrotrophs have evolved the capacity to detoxify resistance factors such as phytoalexins and PR proteins, either by breaking them down enzymatically or by binding them chemically. For example *Fusarium oxysporum pisi*, cause of vascular wilt in peas, has been shown to produce an enzyme capable of breaking down the phytoalexin pisatin, known to be involved in the resistance of peas to attacks by fungi. It has been postulated that some necrotrophs may also produce suppressors of defence processes, although these have been little studied.

The host ranges of necrotrophs are generally very wide, but are nevertheless usually circumscribed in some way. This suggests that different necrotrophs have evolved mechanisms enabling them to attack some groups of plants but not others. Such broad specificity might be explained by the production of enzymes capable of degrading some but not all cellular components. For example, most necrotrophs are unable to attack the woody tissues of trees, this capacity being restricted to pathogens such as *Armillaria mellea* (the honey fungus), which have evolved the ability to produce lignin-degrading enzymes. Furthermore, the pathotoxins of necrotrophs may not be active against all species of plants. Victorin, the pathotoxin produced by *Bipolaris victoriae*, cause of Victoria blight of oats, is toxic only to the cells of oat cultivars bred using the rust-resistant cultivar 'Victoria' as a parent. The gene controlling rust resistance, it would seem, also causes oat cells to bind the toxin Victorin, causing them to die. Many other necrotrophs produce host-specific toxins. If suppressors of general active defence are produced, they may be active against some groups of plants but not others.

Once a necrotroph has established itself within a host its progress may be impeded, but not prevented, by the wound responses of the plant. These are the reactions that most plants activate, non-specifically, in response to mechanical damage caused by pathogens, herbivorous insects and animals, gardeners or the elements. They are similar to the non-specific defence responses described earlier, but may occur more slowly. They include stimulation of a biochemical pathway called the shikimic acid pathway, which leads to the synthesis of toxic phenolic substances, many of which may be oxidised to dark toxic products like melanins. This pathway also provides the building blocks for the laying down of lignin in the cell wall. In addition, callose may accumulate to wall off the damaged area. Finally corky layers may be formed, creating further structural barriers. Presumably, the interplay between a necrotroph's pathogenicity factors and the host's wound responses determines the speed with which the infection spreads and thus the ultimate size of the lesion.

In some cases evolutionary selection may have led to forms of plants in which various aspects of the wound response to particular necrotrophs is exaggerated, leading to a significant slowing down of the progression of the disease. The overproduction of resins in roots of some conifers infected with *Heterobasidion annosum*, cause of butt rot, is an example of this phenomenon, as is the laying down of corky tissue around infections of potato by *Streptomyces scabies*, cause of common scab. Many of the components of general active defence (see above) may have had their evolutionary origins in wound responses.

A more subtle approach: biotrophs and hemibiotrophs

Biotrophs and hemibiotrophs seem to have evolved different strategies from those of necrotrophs for dealing with the plant's basic defences, enabling them to grow within living cells and tissues and extract nutrients from them. Assuming the presence of a mechanism for penetrating the cuticle, one means by which this could be achieved would be to present no signals to enable the plant to recognise the invader as an alien. This would involve the evolutionary loss or structural modification by the pathogen of all such signal molecules. In this model, which is accepted by many plant pathologists, the plant's basic defence systems would simply not be activated.

However, given that basic defence processes in plants seem to be activated by a wide variety of factors, many of them non-specific or not even the product of a pathogen, and that biotrophs and hemibiotrophs make extensive growth in the host tissues and modify their metabolism in a multiplicity of ways, it is hard to imagine that basic resistance mechanisms would not be activated at some point. An alternative model, which takes account of this, might be one in which specific suppressors of basic resistance were involved in establishing a particular biotrophic or hemibiotrophic species as a pathogen of a particular plant group. There is, however, as yet only limited evidence to support this hypothesis.

Once a biotrophic or hemibiotrophic species in the past evolved the means of becoming a pathogen of a particular plant group, there would have been strong selection pressure upon the populations of the host for mutants possessing the ability to recognise the pathogen as an alien and to activate resistance. By virtue of their increased capacity to survive infection and therefore reproduce, the progeny of such mutants would increase in the population and the potential food source available to the pathogen would diminish. This in turn would place upon the pathogen population similarly strong selection pressure for new mutants that were not recognised and could thus colonise the mutant form of the host. The process would be repeated over countless generations of both host and pathogen, and this evolutionary tit for tat would result in an elaborate system of different variants of the host and races of the pathogen, creating a highly specific pattern of resistance and susceptibility in the host and of virulence and avirulence in the pathogen. This is exactly the kind of situation we find when we examine the genetic structure of populations of biotrophic and hemibiotrophic pathogens and their hosts. The evolutionary process is speeded up and the resulting host variety/pathogen race structure accentuated in crop plants such as wheat or potato that have been artificially bred for resistance to the pathogens.

The resistance responses involved in these highly specific interactions appear to be the same as those associated with general resistance: the hypersensitive response and the accumulation of other factors such as structural barriers, phytoalexins and PR proteins.

Historic experiments with flax rust

To understand the mechanisms underlying such an elaborate system of resistant varieties and races it is necessary to know something of the genetic basis of the phenomenon. Following the rediscovery, at the turn of the century, of Gregor Mendel's work on the inheritance of characters in plants, the Cambridge plant pathologist and wheat breeder Rowland Biffen showed that the inheritance of the resistance of wheat to the rust fungus *Puccinia graminis* conformed to Mendel's laws. It was not until the 1940s and 1950s, however, that H.H. Flor, working in the United States, provided a clear genetic explanation of what was going on. He worked with flax, *Linum usitatissimum*, and its rust, *Melampsora lini*, which is a short-cycle rust (see p.176) with no alternate host.

In his first major experiment Flor inoculated four different cultivars of flax – 'Abyssinian', 'Ottawa 770B', 'Kenya' and 'Bison' – with seven different races of *M. lini* and then recorded the results, as set out in Table 3.1. Bison was susceptible (S) to all the races of *M. lini*, but the other cultivars gave a variety of responses. Flor arbitrarily classified them as immune (I), highly resistant (R+), resistant (R), semi-resistant (SR) and susceptible (S).

Table 3.1 Interaction of flax (*Linum usitatissimum*) and races of flax rust (*Melampsora lini*). (After Ellingboe, 1984.)

Races of *M.lini*	Flax cultivars			
	Abyssinian	Ottowa 770B	Kenya	Bison
11	I	I	I	S
1	I	I	R+	S
17	I	I	R	S
3	R	I	SR	S
19	S	I	S	S
24	I	I	S	S
22	S	S	R	S

I = immune, R+ = highly resistant, R = resistant, SR = semi-resistant, S = susceptible.

Next he crossed each of the cultivars to 'Bison', the universally susceptible cultivar, to determine the number of gene differences between each cultivar and 'Bison'. He also intercrossed each of the cultivars to determine the number of gene differences between the cultivars, and he intercrossed the different races of *M. lini*. From all this a basic pattern emerged, illustrated by the simple combination of two cultivars and two races as in Table 3.2.

Table 3.2 The interaction between two cultivars of flax and two races of flax rust. (After Ellingboe, 1984.)

		Flax cultivars	
		Ottawa 770B	Bison
	24	I	S
Races of *M. lini*			
	22	S	S

From crosses with 'Bison' and 13 other cultivars, the progeny being challenged by 12 races of the fungus, it became clear that 'Ottawa 770B' possessed a single 'major' gene for resistance, which Flor designated an '*R* gene'. This was inherited in a simple Mendelian way and was dominant over susceptibility. Thus the two situations are designated *RR* and *rr* respectively. The letters are repeated because, being diploid, there may be two copies of the *R* genes in each host cell.

Crosses between Races 24 and 22 of *M. lini* gave the following result:

First generation (F1) – all avirulent on 'Ottawa 770B'
Second generation (F2) – 75% avirulent and 25% virulent on 'Ottawa 770B'

The inability to infect a resistant host, avirulence, was dominant over virulence, the ability to infect. Thus these two situations are designated *Av* and *av*, respectively. The letters *Av* are not repeated because the nuclei of *M. lini* are haploid (i.e. they possess only one copy of each chromosome). In summary then, these simple experiments showed that there was one gene difference between the two hosts and between the two races, the presence or absence of a resistance gene (*R* gene) and an avirulence gene (*Av* gene) respectively. Resistance was dominant over susceptibility and avirulence over virulence.

Further experiments by Flor and others revealed additional information about *R* genes and *Av* genes, as follows.

• Individual cultivars of flax may contain more than one *R* gene, each conferring resistance to a different race of *M. lini*.
• Individual *R* genes in flax can give more than one resistance reaction (phenotype), depending on the genes carried by the race of pathogen with which they interact. For example, the cultivar 'Kenya' (R4 gene) gave the reactions I, SR and S to races 11, 3 and 24 of *M. lini* respectively.
• Two or more host *R* genes may interact with one pathogen *Av* gene. For example, the cultivars 'Abyssinian' and 'Ottawa 770B' each possesses a different *R* gene conferring resistance to race 1 of *M. lini*.
• The *R* genes in the host may be arranged in 'allelic series'. That is, the different genes in the series are alternative forms of the others in the series, and occupy the same site on the chromosome. Flor found evidence for 19 *R* genes in flax, arranged in three allelic series. Subsequent analysis by others has revealed a total of 29 *R* genes arranged in 5 allelic series.
• Some of the genes may be linked together and therefore inherited together.
• 'Inhibitor genes' may be present in some flax cultivars and modify or completely inhibit the expression of particular *R* genes. Such inhibitor genes are normally dominant.
• The pathogen's *Av* genes are much more scattered about its chromosomes than *R* genes and are not arranged in such clearly defined allelic series, although many are linked together.

The gene-for-gene hypothesis

Two clear conclusions may be drawn from Flor's data. Firstly, single dominant genes in the host and pathogen control resistance (*R* genes) and avirulence (*Av* genes), respectively. Secondly, resistance reactions (phenotypes) result from the interaction of one *R* gene with a single, specific avirulence gene. Conversely, a susceptible reaction results if the *R* gene is absent or recessive in the host, or the *Av* gene absent or recessive in the pathogen. Thus the range of pathogenicity of a physiologic race of *M. lini* is determined by pathogenicity factors specific for each resistance factor possessed by the host.

This one-for-one relationship became known as the gene-for-gene relationship. It can be expressed in its simplest form as shown in Table 3.3.

Table 3.3 The gene-for-gene relationship.

| | | Host genotype | |
		R1 R1	*r1r1*
Pathogen	*Av1*	-	+
genotype*	*av1*	+	+

- = Resistant interaction
+ = Susceptible interaction

The presence of a resistance gene in the host does not of itself confer resistance to the pathogen. It only does so if the attacking pathogen strain possesses the complementary avirulence gene. Where a host possesses two or more *R* genes, it is only necessary for the attacking fungus strain to possess one of the complementary *Av* genes for resistance to occur. Thus an interaction involving three *R* genes and three *Av* genes might be expressed as in Table 3.4.

Table 3.4 The interaction beween cultivars of flax with three *R* genes and races of flax rust with three corresponding *av* genes.

| | | Host Genotype | | | | | |
		rr	*R1*	*R2*	*R3*	*R1,2*	*R1,2,3*
	av	+	-	-	-	-	-
	Av1	+	-	+	+	-	-
Pathogen	*Av2*	+	+	-	+	-	-
genotype	*Av3*	+	+	+	-	+	-
	Av1,2	+	-	-	+	-	-
	Av1,2,3	+	-	-	-	-	-

+ = susceptible interaction (growth of pathogen)
- = resistant interaction (no growth of pathogen)

The gene-for-gene hypothesis and all the other essential features of the interaction of *M. lini* with flax appear to apply, with minor variations, to most other biotrophic and hemibiotrophic plant-pathogen interactions in crops and in the wild. Among the interactions studied in particular detail are the following: *Erysiphe graminis* (powdery mildew) – cereals and other members of the grass family (Gramineae); *Puccinia graminis* (black stem rust) – cereals and other members of the grass family; *Bremia lactucae* (downy mildew) – lettuce and other members of the daisy family (Compositae); and *Phytophthora infestans* (blight) – potato and other members of the potato family (Solanaceae). Significantly, the gene-for-gene hypothesis also appears to hold true for the resistance of plants to biotrophic fungi that are not normally pathogenic on them – resistance of barley, for example, to *Puccinia graminis* f.sp. *tritici*, which is normally a pathogen of wheat but not barley!

Among the variations to the basic theme, in addition to those listed above, are incomplete expression of some genes, resulting in partial resistance and partial virulence; and the presence of modifier genes which modify the expression of an R gene or an Av gene. In some instances resistance genes may be recessive, as in the case of the *mlo* gene for resistance of some barley cultivars to *Erysiphe graminis*. This gene results from a mutation leading to a defect in the mechanism controlling calcium levels in barley cells. Since calcium ions are involved in regulating callose deposition, the mutant lines lose all control over this process and callose is produced very rapidly, walling off the fungus before it is able to extract nutrients from the host cells.

Implication of the gene-for-gene hypothesis

The implication of the gene-for-gene hypothesis for understanding the mechanism of host-pathogen interaction is that resistance, being controlled by a dominant gene, must normally involve a gene product in the host that recognises a product of a dominant gene in the avirulent pathogen. This in some way leads to the hypersensitive response and the synthesis of products such as phytoalexins, PR proteins and so on that we have already seen to be involved in basic resistance, and which also appear to be characteristic of specific resistance to biotrophs.

For most of us it is counter-intuitive that a dominant gene in a pathogen should control production of a factor that enables that pathogen to be recognised as an alien and therefore rejected by a potential host. The reason, however, is that the host has evolved to recognise a pathogen product that may have no role in pathogenicity *per se*, such as a wall or membrane component, or alternatively may be secreted as an aid to pathogenicity (a pathogenicity factor). It is only in the context of interaction with such a resistant host that we think of this as the product of an avirulence gene. In all other contexts the gene is simply a dominant gene specifying a product that is a structural component or pathogenicity factor of the potential pathogen.

A potential pathogen might evolve not to be recognised by a host simply by loss or modification of the gene specifying the product. Providing that such a change was not lethal the pathogen would regain its virulence. The evolutionary response of the host would be to recognise another pathogen product. This might be any similar product of the pathogen genome, which is consistent with the observation that avirulence genes tend to be scattered about the genome, rarely being in allelic series as R genes are.

The host product that recognises the pathogen must be linked in an elaborate way to the resistance response mechanism, which makes it much more likely that the product of a new mutant form of the host is related to the original product. This proposition is consistent with the observation that R genes tend to be arranged in allelic series in the host genome.

Recognition in gene-for-gene interactions

During recent years great effort has been expended by plant pathologists in attempts to identify the recognition factors involved in the highly specific gene-for-gene interactions. Various candidates were proposed, including DNA, proteins, glycoproteins and polysaccharides, but no definitive answers were obtained. A breakthrough came with the development of the procedures of molecular biology, which made it possible to isolate and clone resistance genes, especially those conferring resistance of tomato (*Lycopersicon esculentum*) to races of the hemibiotrophic pathogen *Fulvia* (*Cladosporium*) *fulva* which causes debilitation and eventually necrosis. The story now emerging is that the recognition factors in these hosts, and probably in many other plants exhibiting gene-for-gene resistance to pathogens, are proteins rich in repeated molecules of the amino acid leucine (the so-called LRR proteins). Such compounds could evolve as a series of closely related forms in the way that would be required in gene-for-gene interactions. It seems probable that such recognition factors are often located in the membranes of the resistant host cells, with the part of the molecule involved in recognition orientated towards the outside of the cells and the part involved in activation of defence reactions orientated towards the inside.

Although avirulence genes have also been isolated and cloned from a number of pathogens, including the *Avr9* gene of *Fulvia* (*Cladosporium*) *fulva* that interacts with the *Cf-9* gene for resistance in tomato, the mechanism by which the avirulent pathogen is recognised is still not known. The *Avr9* gene codes for a peptide protein of 28 amino acids, but precisely how this binds to the corresponding *Cf-9* LRR protein remains to be elucidated. Peptide avirulence factors have not been identified in all the host-pathogen interactions studied so far. Moreover, how binding is converted to a signal that eventually reaches the nucleus and activates the resistance responses is still not at all clear, although calcium ions (Ca^{++}) within the cell seem to be involved. Free calcium is known to be involved in the regulation of a number of cellular processes in plants. With the rapid advances in molecular analysis of plant pathogen interaction it is certain that significant new information relating to the operation of the gene-for-gene system will be revealed in the next few years.

Bacterial hrp genes

Recent research on the pathogenicity of certain plant pathogenic bacteria merits special mention, for it may throw further light on the process of recognition in plant-pathogen interaction.

In addition to possessing genes controlling the synthesis of cell-degrading enzymes, toxins and other general pathogenicity factors, bacteria have also been shown to possess clusters of additional genes that are essential for pathogenicity. Mutants in which the genes have been deleted are not only non-pathogenic, but also lose the ability to elicit a hypersensitive resistance response in host cultivars carrying *R* genes. The genes have therefore been designated '*hrp* genes' (hypersensitive response and pathogenicity genes). They have been studied most extensively in forms of *Pseudomonas syringae* that infect a variety of plant hosts, including tomato, *Xanthomonas campestris* pv *vesicatoria*, which causes bacterial spot on tomato and pepper (*Capsicum annuum* var. *annuum*), and *Erwinia amylovora*, cause of fireblight on members of the rose family (Rosaceae).

Some experiments have shown that *hrp* genes are not active when the bacteria are grown in culture on media rich in nutrients. They are activated immediately, however, when the bacteria are transferred to a starvation medium. This suggests that in nature it is the nutrient-deficient state of the region between the cells of the host, which a pathogenic bacterium encounters as soon as it has passed through a stomatal pore or other opening, that activates *hrp* genes. Specific compounds such as sulphur-rich amino acids normally found in the spaces between plant cells may also be important triggers. Only some of the *hrp* genes are involved in 'recognising' the host environment. Once activated, however, they switch on the rest of the *hrp* genes, some of which synthesise proteins important in building channels through the bacterial membrane to enable the bacterium to pump into the host cell the enzymes that degrade cell walls and proteins, and other substances important in pathogenesis.

Very recent research with *Pseudomonas syringae* pv. *tomato* has suggested that one of the pathogenicity proteins pumped into the host cell in this way is the protein product of the avirulence *Pto* gene. In tomatoes carrying the corresponding *R* gene, this is recognised inside the cells by the serine-threonine protein kinase, the primary product of the resistance gene, and the cascade of reactions associated with hypersensitive resistance is then switched on.

It seems possible that plant hosts have evolved to recognise specific bacterial proteins pumped into their cells during, and as an aid to, pathogenesis, and this leads to activation of their resistance responses and exclusion of the pathogen. It is possible that biotrophic fungi produce similar pathogenicity substances, perhaps secreted by haustoria, and that these also function in triggering resistance.

Suppression of resistance

It may be argued, especially from experiments performed with potato and *Phytophthora infestans,* and cereals and *Erysiphe graminis,* that virulence of a pathogen may in some circumstances result from the secretion by the pathogen of substances capable of suppressing host resistance responses. No chemical substance has been positively identified as a resistance suppressor, nor has a satisfactory explanation been produced to suggest how such a system, in which virulence rather than avirulence would be dominant, could be consistent with the genetic data supporting the gene-for-gene hypothesis. Nevertheless, it remains an intriguing possibility and it is hoped that future research may provide further evidence to support or refute it. One possibility is that suppressors may be involved in determining the basic compatibility between a host and pathogen, but not in mediating the race-specific gene-for-gene interactions.

In summary

We have seen how plants have evolved a variety of structural and chemical barriers and active processes to prevent or repel attacks by the vast majority of potential pathogens, and therefore remain healthy. We have also seen how some organisms have evolved mechanisms to overcome the plant's basic resistance processes and become pathogens, and how plants have in response evolved additional mechanisms to recognise these as aliens and repel them. And we have seen how pathogens may circumvent these additional lines of defence. Knowledge of such matters enables the plant pathologist to understand better the distribution and dynamics of disease in natural populations of plants (see Bibliography). It also helps plant breeders to develop ways of controlling disease in crops by selecting and breeding resistant cultivars. Finally, as more is learned of the molecular biological control of disease resistance and of pathogen variation, it makes possible the application of the techniques of genetic engineering to produce the disease-resistant crop plants of the future by isolating resistance genes from one species or cultivar and transferring them to another in wide crosses that could not be achieved by conventional plant breeding methods.

4

The Water Connection

This chapter deals with the fungi dispersed and propagated by means of swimming spores called zoospores. At their simplest they resemble primitive animal forms and, except in the absence of photosynthetic pigments, some algae, which are primitive plants. They are the cause of some very important economic diseases of the present day, including clubroot of brassicas and other members of the Cruciferae, potato blight and downy mildew of grape-vine (*Vitis vinifera*), and are of interest to the naturalist for their interaction with many wild flowering plants. *Phytophthora cinnamoni*, cause of jarrah dieback in the eucalyptus forests of Australia, is one of the most destructive pathogens known.

Zoospores move by means of one or more flagella, which push or pull the individual cells forward through the water by an undulating motion and a sort of rowing or swimming action. The flagella function in much the same way as those of the motile cells of Protozoa, algae and the sperm of animals. The flagella in all of these are apparently constructed in a closely similar fashion with a bundle of eleven fibres (or tubules), two situated centrally and nine equally spaced parallel to the outer surface. The fibres run the length of each flagellum, and are all contained within an undifferentiated background of flagellar substance. Each fibre is formed from a protein capable of contracting and expanding, the molecules of which are activated enzymatically. By sliding on one another they impart a rapid undulating movement to the flagellum. Although this mechanism is biochemically distinct from the mechanism for muscle function in animals it bears certain very close similarities to it. If zoosporic organisms were progenitors of the animals, as some believe, the seeds of muscular function were already present at the very early stages of evolution.

The first groups we describe are not always easy to see because they attack below ground. Roots become infected by zoospores swimming in the water held between the soil crumbs or in water held at the surface of the soil, and this results in a lack of vigour, discoloration or collapse of the above-ground portion of the plant.

Zoospore fungi without walls

If you live on acid soil and grow brassica crops in your garden you are almost sure to encounter the clubroot or finger and toe disease caused by the biotroph *Plasmodiophora brassicae* (Protozoa, Plasmodiophoromycota – Table 1.1, p.21). This was first recorded in Scotland in 1795, and was at that time thought to be caused by watering cabbage plants with the bath water of syphilitic patients because of the strange shapes assumed by the roots. In cool climates it is a common pathogen of the many vegetables derived from *Brassica* species, including cabbages, cauliflowers, broccoli, turnips, swedes, radishes, rape grown for sheep feed and oilseed rape. It also attacks many ornamental

Fig. 4.1 Distorted growth of a cabbage
root (*Brassica oleracea*) following
infection by *Plasmodiophora brassicae*,
cause of clubroot disease, which leads to
increased levels of growth regulators in
the diseased tissues. (D.S. Ingram.)

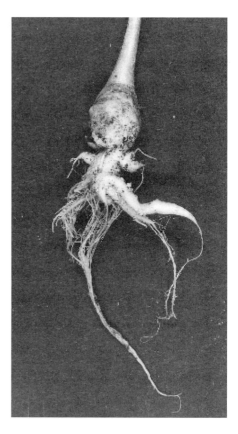

and wild members of the family Cruciferae. It can be recognised by a bronzing
of the leaves, stunting and wilting or collapse of the affected plants, which
when they are pulled up show swollen or clubbed and otherwise misshapen
roots (Fig. 4.1). Surprisingly for a soil-inhabiting fungus, *Plasmodiophora* is
genetically very variable and has produced a number of physiologic races,
attacking different *Brassica* species and their varieties. Some success has been
achieved in producing plants showing resistance to attack but none is completely effective.

As we have seen in Chapter 2 (pp.56–57), *Plasmodiophora* has evolved a
remarkable mechanism for infecting its host. First a single zoospore emerges
from a thick-walled, spiky resting spore (Fig. 4.2) in the soil. It has two
whiplash flagella which appear to be situated fore and aft of the zoospore body,
but in fact arise laterally, with one pointing forwards and the other backwards.
Having swum in the soil water towards a host root, being attracted by distinctive chemicals leaking from the cells, it becomes attached to the surface of a
root hair and differentiates within itself a bullet-shaped stachel contained within a membranous tubular 'gun-barrel'. The stachel, with a small fragment of
cytoplasm containing a nucleus attached to it by a membrane, is forcibly 'shot'
into the living root hair cell, where it becomes enclosed by the plasmalemma.
The stachel is powered by a vacuole within the zoospore that expands rapidly

as liquid is pumped into it, the energy being provided by stored lipids which are rapidly digested. *Polymixa betae*, a pathogen closely related to *Plasmodiophora* which attacks the roots of beet and its relatives (*Beta* spp.), has evolved an almost identical infection mechanism. *Polymixa* is important as a carrier (vector) of an important virus disease of sugar beet called Rhizomania.

Following infection of the root hair, *Plasmodiophora* moves towards the second stage of its life cycle. Its nuclei divide to form a multi-nucleate plasmodium. This eventually divides into 10 to 20 zoospores, each with a single, haploid nucleus. Some of the zoospores probably reinfect adjacent root hairs, but others appear to act as gametes ('sexual' spores) and fuse together in pairs to form larger cells each containing two nuclei, which then infect the living cells of the cortex, the body of the root. Drawing nutrients from the cells, the cytoplasm of the fungus enlarges and the nucleus divides, eventually producing another large multinucleate plasmodium in the enlarged tissues of the roots. It is during the development of these plasmodia that the nuclei from different parents come together in pairs and fuse to form large diploid nuclei. Eventually these undergo meiosis to produce haploid nuclei, each of which becomes surrounded by lipid-rich cytoplasm, a membrane and a thick wall, to produce a thick-walled resting spore covered with dead material that gives it its spiky appearance. When the root dies and rots the resting spores are released into the soil to begin the cycle again.

Fig. 4.2 A resting spore of *Plasmodiophora brassicae*, cause of clubroot disease of members of the cabbage family (Cruciferae). Note the spiky appearance of the thick outer wall (a), the large lipid droplets (b) and the single central nucleus (c). Each resting spore is approximately 3–5 µm diameter. (Illustration by Mary Bates, based on an electron microscope photograph of an ultra-thin section.)

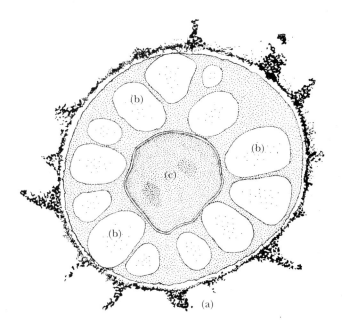

This brief excursion into the esoteric area of life cycles illustrates the complexity of studying the zoospore fungi that grow in the soil, a medium that makes direct examination very difficult. There have been well over a dozen different life cycles proposed for *Plasmodiophora* over the years. The studies that led to the one outlined here were facilitated by growing the host cells and the obligately parasitic pathogen together as infected tissue cultures in the laboratory. Although nobody has made a counter-proposal over the past 25 years, that does not mean it is correct in every detail – time will tell.

Infection by *Plasmodiophora* leads to a multiplication of the number of root cells and also an increase in size of the individual cells, caused by disturbance of the levels of the host's growth hormones, particularly the cytokinins and auxins. This is reflected externally in the galls and swellings which give the disease its common names, 'clubroot' and 'finger and toe'. Masses of starch also accumulate in the infected tissues, derived from the excess of sugars imported into the clubbed root to nourish the growing pathogen. An interesting feature of the life cycle is that the root hair phase can occur in plants of many different families, but the clubroot stage is restricted to members of the Cruciferae. The reason for this specificity is not known.

Control of clubroot is not easy. The most important element is an extended rotation so that the susceptible crop does not appear more than once in five or more years in the same field, thereby allowing resting spore numbers to decline, although some can survive for forty years or more. Particular brassica varieties show resistance, such as the swede cultivar 'Marian'. The acidity of the soil is also an important element in control, acid soils showing greater infection of susceptible crops than those that are more alkaline. However, from a farming point of view it is difficult to use this as a control measure, because in areas where potatoes are also being grown, raising the pH with lime encourages the development of another disease, potato scab (*Streptomyces scabies*). The extent to which clubroot occurs on wild members of the family Cruciferae is difficult to estimate and any records would be of interest.

Clubroot can easily be examined by the amateur with access to a compound microscope. Cut thin sections of clubbed root with a sharp blade or scalpel, mount in a drop or two of water, cover with a cover slip and examine with the microscope (see Appendix). The swollen cells, large plasmodia and masses of starch grains will easily be visible in younger tissues. In older tissues resting spores with their characteristic spiky walls can be seen at higher magnifications (x40 objective and above). Root hair infections can be produced by growing seedlings in silver sand and watering them with ground-up 'clubs'. If the sand is kept wet the spores will germinate and infect. Seven to fourteen days later the seedlings may be lifted, washed gently in water and examined with the microscope. The most easily seen stage is the cleavage of the plasmodium into groups of zoospores, which have the appearance of a row of pearls strung out along the root hair (Fig. 4.3).

Another crop disease that does not show above-ground symptoms is powdery scab of potato tubers caused by the biotroph *Spongospora subterranea*, a fungus closely related to *Plasmodiophora*. The clusters of resting spores, which form characteristic spore-balls, can survive in the soil for many years. They are induced to germinate by the renewed presence of the host, which releases stimulatory chemicals, causing the spores to produce biflagellate zoospores resembling those of *Plasmodiophora*. These infect the cells of the potato and

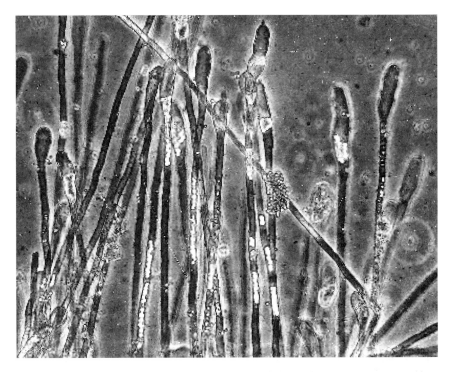

Fig. 4.3 Root hair plasmodia and zoosporangia of *Plasmodiophora brassicae*, cause of clubroot disease of the cabbage family (Cruciferae) in the root hairs of wallflower (*Cheiranthus cherii*). The individual spherical zoosporangia are approximately 6–10 μm diameter. (Phase contrast light microscope photograph, D.S. Ingram.)

form plasmodia, again like those of *Plasmodiophora*. The symptoms of the resulting powdery scab are cork-filled pitted lesions on the tubers from which spore-balls can be retrieved and readily identified with the aid of a microscope (Fig. 4.4, overpage). Look for *Spongospora* on freshly dug tubers. It appears as raised scabs which later become depressed shallow cavities as the surface breaks away to reveal the brownish balls of single-celled circular spores. Many other surface organisms, including common scab, caused by *Streptomyces scabies*, a member of the Actinomycetes, and black scurf, caused by *Rhizoctonia solani*, a member of the Basidiomycota, can be mistaken for powdery scab, so beware. There may be other symptoms, including a cankerous form in wet soils where whole tubers may be misshapen, making them unsaleable. The fungus can also attack tomatoes, producing small corky lesions on the roots.

Spongospora is of special interest because, like *Polymixa*, it is a vector of viruses, particularly potato mop top virus (see p.234). Another *Spongospora* attacks watercress (*Nasturtium officinale*). In the 1950s the English watercress industry expressed concern at the decline in productivity in certain beds. Investigation showed a characteristic bending of the roots and the condition was given the name 'crook-root disease' and shown to be caused by a hitherto unknown species of *Spongospora*, which was eventually considered to be a variety of the potato pathogen. The disease occurs on wild watercress, although it is hard to

Fig. 4.4 Spore balls (each approximately
40 μm diameter) of *Spongospora
subterranea*, cause of powdery scab of
potatoes (*Solanum tuberosum*), within a
single tuber cell. (Illustration by Mary
Bates, after Butler & Jones, 1961.)

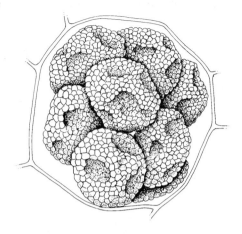

find until you get your eye in. The use of zinc 'frits', fragments of glass carrying zinc, which is toxic to *Spongospora*, controlled the spread of the disease,
which otherwise would probably have destroyed the watercress industry.

Plasmodiophora, Polymixa and *Spongospora*, having both zoospores and plasmodia without walls, are now classified along with primitive animal life in the
Kingdom PROTOZOA (see Table 1.1, p.21). The zoosporic fungi dealt with
next, from the moment that the zoospore encysts, possess walls of chitin and
glucans and are therefore classified in the Kingdom FUNGI as the
Chytridiomycota (*Synchytrium* spp. and *Olpidium* and its relatives – see Chapter
1 and Table 1.1, p.21). We shall also consider the Oomycota, which possess
hyphae with walls containing cellulose as the principal structural component.
They were once thought to be related to the Chytridiomycota but are now classified in an entirely separate Kingdom, the CHROMISTA. Included in the
Oomycota are such major groups of pathogens as the Peronosporales: *Pythium*
spp; *Phytophthora* spp; the downy mildews – *Peronospora* spp. and their relatives;
and the white blisters – *Albugo* spp.

Zoospore fungi with walls but without hyphae

Perhaps the most spectacular disease caused by a fungus in this group is wart
of potatoes, caused by yet another biotroph *Synchytrium endobioticum* (Fig. 4.5).
It was first identified in Hungary in 1895, but was probably present in Britain
from about 1860, for at that date a model of a potato tuber with cauliflower-
like growths, the characteristic manifestation of the disease, was presented to
the Royal Scottish Museum in Edinburgh. In these growths tiny thick-walled
spherical resting spores, resistant to decay, are eventually formed, and are
incorporated into the surrounding soil when the growths turn black and rot.
This ensures the contamination of the soil and the insidious contamination of
the surface of otherwise healthy 'seed' tubers before lifting; when planted elsewhere to produce a crop, these spread the disease.

Germination of the resting spores gives rise to spherical zoospores, each with
a single posterior whiplash flagellum. These swim to the surface of a developing 'eye' of a tuber and penetrate a surface cell, where they develop to produce

Fig. 4.5 A gall of *Synchytrium endobioticum*, cause of wart disease of potato (*Solanum tuberosum*), growing up from an artificially inoculated, pot-grown tuber of the cultivar 'Arran Chieftain' grown by Mary Noble. (D.S. Ingram.)

further infective zoospores and stimulate the division and enlargement of adjacent cells. A continuous cycle of vegetative reinfection gives rise to the very large outgrowths which are so characteristic and so destructive. At certain times some of the zoospores appear to act as gametes and fuse in pairs to give diploid zoospores which enter fresh cells and cause further cell division. Meiosis to restore the haploid condition is believed to occur immediately after germination.

By the early part of the twentieth century the disease was widespread in Britain and throughout Europe. In the beginning its spread was so rapid and mysterious that farmers who grew potatoes feared for their livelihood, and those charged with the oversight of the country's food supply feared for the future of the potato as a staple food. For example, 594 holdings in Scotland became infected between the first identification in 1907 and 1922. Gradually facts to aid the understanding of the disease were established. It became apparent that the disease was very severe in gardens and allotments (where rotation of crops was limited) and that fields adjacent to gardens were particularly prone to infestation. Moreover, across the British Isles the disease was most prevalent in the northwest and west midlands of England and in southern Scotland, areas of intensive potato cultivation. Once fields had been contaminated subsequent potato crops showed the disease and the severity of the infestation increased according to the frequency of the planting of potatoes. But none of this helped control the disease in any dramatic fashion and the situation looked hopeless until the report in 1920 of the observation by English growers that certain varieties of potato appeared to escape infection. Subsequent study showed that these varieties were in fact very resistant (popularly said to be 'immune') to attack, a resistance controlled by a single gene, with one or more subsidiary genes sometimes acting as modifiers (see p.69).

Once an understanding of the disease had been gained, the strategy for control was determined. Firstly a series of administrative orders was promulgated to control the planting of non-immune potatoes in designated areas and their movement about the country; secondly, planting susceptible varieties on land that had contained diseased plants was prohibited; and finally, resistant varieties grown on such land were not allowed to be used for 'seed'. Even the growing of susceptible varieties in gardens was prohibited (with special dispensations for some early varieties). As the genetics of inheritance of resistance and susceptibility was understood, it was possible to ensure that resistance was bred into all new potato varieties. Indeed, new varieties could not be released to commerce until they had undergone a statutory test for resistance and susceptibility to wart disease. These measures reduced the importance of the disease and allowed normal farming to proceed over much of the country. The effect of these measures can be seen in the period 1923–57, when only 223 fresh outbreaks were notified, and since then the disease has continued to decline.

In Canada the disease was first identified in Newfoundland in 1909, on the Canadian mainland in 1918 and at this time too in the USA, but in all these cases initial severe quarantine measures and the development of orders preventing the indiscriminate use of susceptible varieties prevented the spread of the disease.

The resting spores of this devastating disease lie a long time in the soil and some are still capable of infecting potatoes up to 50 years from the initial outbreak. The number of viable individuals gradually falls over this period but how many remain infective after periods in excess of 50 years is not known. Moves are afoot to de-schedule the fields earliest affected by wart disease, but the approach must be cautious. The disease still turns up on a regular basis in gardens when permitted non-immune susceptible varieties are grown, and many new owners of productive cottage gardens are horrified and mystified at the deformed tubers and cauliflower-like growths which are the products of their carefully planted early potato varieties, until they learn to grow only immune varieties.

Experimental work has shown that *Synchytrium endobioticum* will affect wild species of *Solanum* such as black nightshade (*S. nigrum*) and woody nightshade (*S. dulcamara*). It may have crossed over to the cultivated potato in Europe from such wild hosts, but evidence on this point is lacking. It is also known to exist in the Andes, but whether as a native or an introduction is not clear. The roots of tomato, a close relative of potato, can be infected to form small growths quite different in scale from the gigantic protuberances on the potato tuber, which is not a root but a stem. There are, however, no reports of the organism attacking other domesticated or wild members of the potato family (Solanaceae).

The genus *Synchytrium* contains a good number of other species that attack a variety of wild flowering plants in Britain; curiously, in most cases these attacks are at or near ground level and not below ground, and none of the hosts show infected growths of the extravagant size of the potato wart disease. Among the most striking of the infections are those of *S. taraxaci*, which forms yellow, orange or red bumps on the midrib of dandelion (*Taraxacum officinale*); *S. anemones*, which causes blackish galls on the above ground parts of wood anemone (*Anemone nemorosa*), and *S. mercurialis*, which attacks dog's mercury (*Mercurialis perennis*), producing glassy whitish-yellow swellings on the bases of stems and leaves, which can be discovered initially by running the fingers up and down the stems. If viewed with a microscope, thin sections of these galls can often be seen to contain large black resting spores. Lastly, look for *S. aureum*. This is a portmanteau name given to several *Synchytrium* species that have not yet been separated taxonomically. We have seen *S. aureum* on common chickweed (*Stellaria media*) and common loosestrife (*Lysimachia vulgaris*), where it forms myriads of small golden galls on the infected plants.

The biology and taxonomy of *Synchytrium* species on wild plants is still poorly understood and would be an excellent subject for study by the amateur naturalist with endless patience and access to a good compound microscope.

Another unforgettable disease resulting from infection by a unicellular organism related to *Synchytrium* is the crown wart disease of lucerne (*Medicago sativa*), caused by *Urophlyctis alfalfae*. This is seen in areas where lucerne is grown on soils that hold water (much lucerne is grown on very free-draining soils). The galls, which form at the crown of the plant, look like leafy lumps of cauliflower and contain elaborately decorated resting spores. The vigour of infected plants is much reduced.

A final genus, *Olpidium*, which, like *Synchytrium*, is a member of the Chytridiales, invades the surface cells of roots and leaves but causes minimal damage to the host. *O. brassicae*, for example, is widespread on the roots of brassicas and other plants (Fig. 4.6) and although it may cause problems for infected seedlings it has a limited effect on the adult plant. *O. radicale* is similar but is said to attack a range of host plants. The importance of these pathogens lies in their capacity to carry important viruses from plant to plant, including lettuce big vein, tobacco necrosis, cucumber necrosis and melon necrotic spot.

Most of the fungi described above can cause severe problems in crops; they contaminate the land and effectively limit its use for many years into the future; they can spread to new sites on the feet of animals or human beings,

Fig. 4.6 Resting spores (approximately 12 μm diameter) of *Olpidium* sp. in root cells of common bluebell, *Hyacinthoides non-scripta*. (Dr H.J. Hudson, University of Cambridge.)

on vehicles and implements, in blown soil or in water and on tubers and plants; but that being said, they have only comparatively limited powers of spatial movement.

Pythium, the secret destroyer

Pythium, a member of the Oomycota in the Kingdom CHROMISTA (see Chapter 1 and Table 1.1, p.21) has many of the characteristics of the non-mycelial fungi we have just been discussing. It is disseminated by zoospores but these, on reaching the surface of a host plant, germinate to form germ tubes which penetrate the tissues and produce a mycelium of ramifying hyphae without septa. These hyphae grow into and through the cells and, in their turn, break out into the surrounding water or air and bear zoosporangia, which release zoospores whenever free water is present. In some species the sporangia become detached and are distributed locally. Some are possibly wind-borne. They may also produce thick-walled oospores, resting spores which have very considerable powers of survival in the soil in the absence of a host. There are many surprising and interesting characteristics of *Pythium* species, but perhaps the most remarkable is their ability to destroy the tissues of the plants they invade.

In this they contrast with the zoosporic pathogens considered previously, which are all biotrophs. Having overcome the intrinsic resistance mechanism of the threatened host, they develop an intimate relationship with it. Minimum damage is inflicted and the living cells provide the pathogen with food, protection and the ability to multiply and produce spores over a long period of time. *Pythium* does not fit into this pattern of behaviour. It is a necrotroph, a plant destroyer which lies in wait for its host (either as an oospore or as mycelium growing saprophytically on dead tissue) and typically enters the tissue of a young seedling, causing disintegration of the middle lamella, death of the cells and in some cases disintegration of the cellulose cell walls. The consequent collapse of the host is called damping-off, and is most commonly seen in pots of seedlings contaminated by unsterile soil containing oospores of the pathogen (Plate 2). A beautiful crop of young seedlings will begin to die at the point of initial infection, and the collapse then spreads to all the seedlings in the pot until after some time almost nothing visible remains. This pattern may sometimes also be seen in cartons of cress seedlings purchased as salad and kept for a day or so on the kitchen window sill. The likelihood of damping-off occurring is increased if seedlings are stressed by being planted too close together, kept at too low a temperature or overwatered. The presence of excess water reduces the availability of oxygen to the roots and also facilitates the dispersal of the zoospores. *Pythium* species have a very wide host range, a characteristic of necrotrophs, although there are limits, as will become apparent.

The *Pythium* species associated with damping-off include the commonest, *P. debaryanum,* a species on which many taxonomists cast doubt but which is often referred to; *P. ultimum,* which also causes a tip blight of garden peas and watery wound rot of potato tubers, and *P. aphanidermatum,* which additionally occurs as a storage rot of vegetables and causes cottony leak of cucumbers. Each can cause damping-off, death and destruction of seedlings of almost any plant species while at the same time being implicated in more specific attacks on adult tissues of particular hosts. Other *Pythium* species cause root rots of plants as diverse as wheat, sugar cane (*Saccharum officinarum*) and garden violas, or rots of storage organs as varied as ginger (*Zingiber officinale*) rhizomes and sugar beet (*Beta vulgaris*) roots. It is possible that in seedlings resistance mechanisms are poorly developed and the susceptible tissues are not protected by a thick cuticle and corky skin. In older plants, where resistance and protective layers are well developed, particular strategies must be evolved to circumvent these in each host. Hence a measure of host specificity is achieved. A new dimension is given to the genus by the discovery that some species are even capable of attacking shrimps and young fish, and can become important impediments to good home aquarium management and efficient fish hatchery production.

Some *Pythium* species also appear to have the ability to cause disease without infecting the host plant. Toxins are produced at the root surface and these cause reduced growth of roots and collapse of leaves. This characteristic may be responsible for the occurrence of a number of diseases in which *Pythium* spp. interact synergistically with other pathogens such as *Fusarium oxysporum* (see Chapter 6) to produce diseases more severe than those caused by either organism alone.

Infection by *Pythium* species is very difficult to identify in wild plants. Damping-off can easily be produced at home, however, by sowing seeds of cress (*Lepidium sativum*) very thickly in a pot of old unsterile soil, placing in a cool corner and overwatering (or even better, standing the pot in a dish of water). Soon after the crowded seedlings have started to grow disease will probably set in. Infected seedlings should be mounted on a microscope slide in a drop of water and squashed under a cover slip. The wide hyphae without septa can then be observed with a microscope. Sporangia and oospores may sometimes also be seen.

If some pieces of infected tissue are placed in a dish of sterile tap water with a few healthy seedlings the hyphae, which look like strands of cotton wool, will grow out and attempt to invade the new hosts. As zoosporangia form, zoospores may be released into the water.

Into the air

In many *Pythium* spp. the zoosporangia are produced in water as slightly modified hyphae from which an undifferentiated protoplasmic mass surrounded by a membrane is extruded. The zoospores are differentiated inside this before being discharged to infect neighbouring plants. Other *Pythium* spp. have well differentiated zoosporangia which are often globular. Sometimes, as in *P. ultimum*, these are borne on the ends of hyphae which may project a short distance into the air. Being detachable, they are taken up in air currents and dispersed. Water is required, however, before they can germinate, release zoospores and infect a new host.

The closely related genus *Phytophthora* in the Oomycota (see Chapter 1 and Table 1.1, p.21) is characterised by the zoospores being fully formed in the sporangium before being discharged. Its sporangia are distributed in a variety of ways. Those of some species, like *P. cinnamoni* on *Eucalyptus* spp., are released in the soil and their zoospores distributed through the soil water to attack neighbouring plants. Others, such as those of *P. palmivora* on cacao pods (*Theobroma cacao*), are formed on the upper parts of the plant, detached by rain and distributed by splash and drip. At least two species, *P. infestans*, the cause of potato blight, and *P. phaseoli* on beans (*Phaseolus* spp.), have sporangia that are readily detached without the intervention of water, being removed from the plant into moving air currents. *Phytophthora* species can be found in a variety of situations in farms and gardens worldwide. Some are necrotrophs, but others show characteristics of hemibiotrophy or biotrophy (Plates 1 and 2).

A typical necrotrophic *Phytophthora* carried in soil water first appeared in strawberry fields in Scotland in 1921. During the economic depression that followed the First World War, encouragement was given to the establishment of horticultural holdings on which the growing of soft fruit, particularly strawberries (*Fragaria* x *ananassa*) and raspberries (*Rubus idaeus*), proved a useful enterprise. Then, quite suddenly, areas in strawberry fields became yellow with dieback of the plants and eventually the diseased areas spread to wipe out whole crops. Examination of the roots showed a dark red discoloration of the xylem, and the disease was named 'red-core' ('red stele disease' in the USA). A characteristic *Phytophthora*, *P. fragariae*, was isolated from the roots. Although chemical means were found to limit the disease and resistant varieties such as 'Auchincruive Climax' were eventually bred, the effect of the disease was to cause a relocation of strawberry growing and sometimes of strawberry growers from Lanarkshire in the west of Scotland to Perthshire and Angus in the east. The disease was contained in these new areas by strict statutory attention to the production of disease-free planting stock.

In the USA red core disease was found on strawberries in the 1930s and a similar disease appeared in the Pacific northwest in the 1950s. This, or something very like it, attacked raspberries in the 1980s in Scotland, other European countries and Australia in quick succession, but how it travelled such vast distances is not known. A great many names have been given to the necrotrophic *Phytophthoras* causing diseases of soft fruit, such as *P. fragariae*, *P. megasperma*, *P. erythroseptica* and *P. cryptogea*, but no definitive name has yet been agreed. There is evidence that the disease organisms of strawberries and raspberries are distinct but closely related and that they probably both arose on the Pacific coast of the Americas. These diseases are now among the most important of soft fruit worldwide, and control is being sought by strict monitoring of planting stock. At the same time, a search has begun for resistance in new varieties.

Another *Phytophthora* that infects from the soil, *P. cinnamoni*, is extremely important. In gardens and nurseries in temperate climates worldwide it causes a dieback of dwarf conifers and various species of *Rhododendron, Erica* and *Calluna*, with characteristic attacks on the growing roots, yellowing and dieback of the above ground portions and sometimes very characteristic brown segments in conifers. *P. cinnamoni* is also responsible for other diseases, and has been recorded on over 1,000 host plants. It is a serious problem on avocados (*Persea americana*) in California and has become particularly important, not to say devastating, as the introduced cause of dieback of the jarrah forests of Western Australia, where it is threatening the very survival of important natural ecosystems over vast tracts of country. Spread of the disease in the jarrah forests is especially rapid along roads and tracks, the spores being carried by vehicles and in water torrents during rainy periods.

Other necrotrophic and hemibiotrophic species of *Phytophthora* do not attack from the soil but produce their sporangia on the surface of twigs or fruits, and are important as the cause of severe rots in tropical crops. They include *P. heveae* and *P. meadii* on rubber (*Hevea brasiliensis*) and *P. palmivora* (a name covering a wide taxonomic group) attacking palm buds (Palmae), pepper vines (*Piper nigrum*) and other tropical species. In cacao, *P. palmivora* causes 'black pod disease'. The melon-shaped fruits, full of beans, are normally harvested as the pods turn orange or red. The beans are then removed and fermented to get rid of the surrounding pulp and improve the flavour. When *P. palmivora* is present, infection of the green pods leads to the development of lesions which rapidly spread to involve whole fruits. Cacao pods are rich in phenolic substances; as these are released from the dying cells they are rapidly oxidised to form dark compounds which turn the infected pods completely black. Sporangia are borne on the surface of such pods, washed off in drips or splashes and thereby carried to adjacent pods. There is a relationship between rainfall, yield and the rate of infection: in periods of heavy rainfall, the more crowded the fruits are the more vulnerable they are to infection. Traditionally control has been by means of chemical sprays, but frequent picking not only reduces the possibility of transmission but also removes infected pods before the infection involves the beans they contain. As we write *P. palmivora* is again threatening world chocolate supplies.

The *Phytophthora* species discussed so far produce their sporangia on simple, often very short stalks. In contrast, the hemibiotrophic (some would say biotrophic) potato blight fungus (*P. infestans*) and *P. phaseoli* (important on Lima bean, *Phaseolus lunatus*) both have elongating sporangiophores which swell at the end to produce a lemon-shaped sporangium which is pushed to the side to allow a new apex to grow forward to form the next sporangium. This process is repeated to produce a long, jointed sporangiophore with sporangia attached laterally (Fig. 1.1, p.13). The attachment is very delicate, and the sporangia are released when the sporangiophores, with their differentially thickened walls, twist rapidly with changes in atmospheric humidity. It is easy to see this movement if an infected leaf is observed with a low power microscope or lens in a drying atmosphere.

On arrival at the leaf surface of a leaf of potato or tomato, the sporangia germinate. If water is present and the temperature low, swimming zoospores are released and these rapidly encyst and produce tiny germ tubes (Fig. 1.1, p.13). At higher temperatures zoospores do not form and direct germination occurs by means of a much larger germ tube. The germ tube, however produced, first swells at the apex to form an appressorium which becomes cemented to the leaf surface. An infection peg then penetrates the leaf either by breaching the young epidermis or by passing through a stomatal pore. After entering the leaf the hyphae grow out from a central point without at first producing any effect visible to the naked eye. The microscope reveals that they grow largely between the living cells and sometimes, but not always, produce small, peg-like haustoria which penetrate the cell walls without killing the cytoplasm. The cells often respond by encasing the haustoria in callose, presumably in an attempt to limit the growth of the pathogen. After about three days the centre of the infected area takes on a water-soaked appearance and branching hyphae bearing sporangia appear from the stomatal apertures. Since stomata are commonest on the lower leaf surface, this is where the fungus is most in evidence, particularly at the edge of the lesion. Later, the centre of the lesion collapses and becomes necrotic, while the sporulating edge advances until, depending on the genotype of the host, the whole or part of the leaf has been colonised. Clearly the fungus grows and sporulates in the living tissues of the potato or tomato and may therefore be regarded as a biotroph. Whether it then goes on to kill the cells, growing and sporulating on them as a hemibiotroph would, is not known. It is possible that the cells of the host simply die from the stress imposed by the demands of the fungus, which then dies with them. It is capable of growth on artificial media in the laboratory, but whether it grows saprophytically in nature is a different matter.

When sporangia are washed down into the soil from the leaves, *Phytophthora infestans* may also infect potato tubers through the 'eyes'. Again, the hyphae grow between the cells, once more producing characteristic peg-like haustoria. The tuber responds to the presence of the pathogen by encasing haustoria in callose and laying down lignin and other phenolic substances in cell walls adjacent to the infection, giving cut tissues a characteristic bright fox-red colour. The fungus may overwinter in infected tubers, growing into the new shoots as they emerge in spring. Sporangia produced on these shoots infect adjacent healthy plants. *P. infestans* is also capable of producing male and female sex organs in the tissues of the plant, especially the tuber. Here the haploid male nuclei, in a poorly differentiated male organ, the antheridium, pass into the larger, spherical female organ, the oogonium. This becomes a diploid resting structure, the thick walled oospore (Fig. 4.7), which overwinters in or on the soil, ultimately producing sporangia containing zoospores with one or more nuclei. Until recently this phase had rarely, if ever, been seen in Britain but the situation may now be changing (see below).

The mystery of meiosis

A biotrophic organism that can be artificially cultured is of great interest to experimentalists attempting to puzzle out the nature of parasitism. If, for example, mutants of the fungus could be produced which either varied in pathogenicity or in some attribute thought to be important in pathogenicity, they would shed a light on the nature of pathogenicity itself. On the assumption that, like many simple organisms, *Phytophthora* was haploid for most of its life cycle with a short diploid phase in the resting spore formed after sexual union, many attempts were made by the authors and others during the 1960s

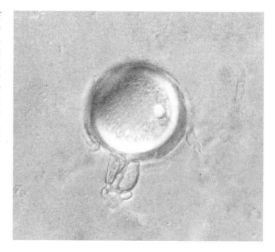

Fig. 4.7 An oospore of *Phytophthora infestans*, cause of late blight of potatoes (*Solanum tuberosum*). Note that the antheridium, the 'male' sex organ, encircles the base of the oospore (approximately 40–50 μm diameter). (Photograph supplied by the Scottish Crop Reseach Institute, Dundee.)

to create mutants by irradiating the zoospores with ultraviolet light or by treating them with chemical mutagens. These were all singularly unsuccessful. Meanwhile Eva Sansome, from cytological work in itself difficult because of the minute size of the nuclei involved, claimed that the *Phytophthora* species and their allies (*Pythium* speices and downy mildews – see below) were diploid organisms which only produced haploid nuclei by meiosis immediately before the final sexual union. Her work was ignored for many years but, thanks to the persistence of other scientists, has now been proved correct. This explains the difficulty in detecting mutants, for most mutational changes are deletions of genetic information from one but not both copies of a gene within a chromosome and are therefore masked by the unmutated copy of the gene on the sister chromatid. Once the position of meiosis in the life cycle was understood, more detailed genetic analyses were made. These suggest that the *Phytophthora* species and their allies in the Oomycota are quite distinct from all other fungi (see below).

The travellers

We have seen how soil-borne fungi with swimming spores can mount attacks in localised areas and can spread by the movement of resting spores with the soil and by the dissemination of zoospores in surface water. Among the *Phytophthora* species, many behave as soil-borne fungi. Others, which attack above-ground portions of the plant, spread through the canopy by splash and drip but can only move further afield in infected planting material. In *P. infestans* and *P. phaseoli*, the possession of aerially disseminated zoosporangia introduces a new dimension to the capacity to spread. *P. infestans* can spread locally in all the ways demonstrated by the species we have just described. In addition it can move around the globe in tubers used as planting material. Moreover, its capacity for aerial spread allows it to travel swiftly and widely from an original focus of infection such as a plant growing from an infected tuber (Plate 1).

  We know that the cultivated potato probably had its origin in the high Andes and was imported into Europe in the sixteenth and seventeenth centuries, becoming an agricultural crop in the eighteenth century. There is no evidence that *P. infestans* was carried with it and potato crops in Europe and in America were apparently free of the disease until the early nineteenth century. The disease probably appeared for the first time in 1843 on the eastern seaboard of the USA and Canada, either in planting material or more likely by aerial spread. There is good documentary evidence that it then appeared in Europe in 1845, almost certainly carried in tubers for planting. It was first noted at Courtrai in Belgium in late June to early July. From that point it spread eastwards into France and Switzerland, westwards into southern England (16 August) and southern Ireland (6 September) and at about the same time into Scotland south of the Highlands, all by aerial spread. We cannot observe aerial spread on this scale now because there is annual local reinfection from infected tubers, but there is clear evidence for extensive aerial spread in Holland, where potatoes on islands at a distance of 11 km from the mainland have developed blight in the absence of any infected planting material. In Britain there is good circumstantial evidence in most years for the simultaneous appearance of discrete outbreaks of blight over an area of several square kilometres from a single original focus.

Where did potato blight originate? A centre of origin for *P. infestans* has been identified in Mexico, where a number of wild species of *Solanum* have been found infected by the disease. But an historical connection between this area and the centre of origin of the cultivated potato in the Andes has not been made and recent outbreaks of potato blight on modern varieties in South America are considered to have been imported with those varieties. Historical evidence is accumulating for an association between blight and the original Andean home of the potato, but the matter is by no means concluded.

The travels of potato blight did not finish with its establishment as a world-wide pathogen. After its appearance in Europe and the USA and its step by step colonisation of other potato-growing regions, references to its sexual reproduction are scarce, contrary to expectation. There are one or two notes of oospores having being observed, but so rarely as to suggest that the fungus was functioning around the world without the intervention of a sexual cycle, a situation which made its ability to produce new races in response to the genes for resistance introduced into cultivated potatoes by plant breeders a matter of wonder. When the fungus on wild potatoes in Mexico was studied in detail it was found to produce abundant oospores following the conjugation of oogonia and antheridia, but this only happened when fungi of two mating types, A1 and A2, were brought together. *P. infestans* was therefore heterothallic. (This is the condition in which mycelia of opposite mating type are required to conjugate to produce the zygote. In fungi like *Phytophthora* spp. each mycelium is capable of producing both male antheridia and female oogonia, but these will only conjugate with an appropriate organ from mycelium of opposite mating type). Experimental crosses showed that the *P. infestans* in the rest of the world was all of the A1 mating type. Then in the early 1980s the A2 mating type was observed in Holland, the UK, Egypt and subsequently worldwide. It was often present at fairly low levels, suggesting that it could in the past have been found anywhere if sufficient search had been made, but in some areas such as southern Brazil and the former Soviet Union it was very common. Was this a migration of the A2 mating type, perhaps aided by the export of potatoes from Mexico, was it a mutation from A1 to A2, or had the A2 mating type been present all the time and overlooked? The balance of opinion seems to favour the migrational theory. What is strange, however, is that there does not seem to have been any increase in the capacity of potato blight to vary after the 'arrival' of the A2 mating type. One might have expected the capacity for sexual variation to have led to the appearance of highly pathogenic races of potato blight which threatened the present level of resistance in commercial potatoes. This has not happened so far, although there are reports of the identification in the USA of a new strain of potato blight that is able to resist control by all current fungicides. It is probable that *P. infestans* is capable of parasexual recombination (see p.29) and that this is responsible for variation, but the critical evidence to prove this point is still being collected.

Since local epidemic spread of potato blight each summer is dependent on the weather, forecasting methods have been developed which not only predict the progress of the epidemic but also indicate the appropriate times for effective spraying to control the disease. Probably the earliest attempt to formulate a set of predictive rules was that of two Dutchmen, Lohnis and van Everdingen, in the early 1920s. They found that at least four hours of dew during the night, a minimum night temperature of 10°C, with mean cloudiness the following day of 0.8 and at least 0.1 mm of rain on the same day led to a cycle of infection. Later, attempts were made to simplify the criteria and adapt them for particular geographic areas. As understanding increased and meteorological data became ever more sophisticated, elaborate computerised systems came into being, and most recently the forecasting system has been turned round to predict not when blight will occur but when it will not occur. The use of any of these systems allows the grower to apply fungicide only when it is likely to be needed, lessening costs and environmental contamination.

The downy mildews

The downy mildews, although closely related to *Phytophthora infestans*, all differ from it in having sporangiophores that, although branched, are of limited growth, sporangia being formed at the apex of each branch (Figs. 4.8 and 4.9). They are also obligate biotrophs, incapable (so far at least) of growth even on complex culture media. All are grouped together in the order Peronosporales. Some members of this group, such as *Pseudoperonospora* spp. and *Plasmopara* spp., produce zoospores, although the 'sporangia' of others, such as *Bremia* spp., appear to germinate exclusively by means of germ tubes. Even so, the traces of a zoosporangial exit pore are usually visible in the wall of the 'sporangia' of these species. All the downy mildews form well-developed haustoria in the living cells of their hosts.

Many of the downy mildews are important pathogens of crops worldwide. The downy mildew of vines, *Plasmopara viticola*, is another traveller from America; introduced into Europe in about 1878, it caused severe epidemics in

Fig. 4.8 Sporangia and branched sporangiophores of *Peronospora parasitica*, cause of downy mildew of the cabbage family (Cruciferae). Note that the sporangia (which rarely produce zoospores and therefore function as conidia) are formed at the tips of the branches of the sporangiophore. The sporangia are approximately 20 μm diameter. (Light microscope photograph, D.S. Ingram.)

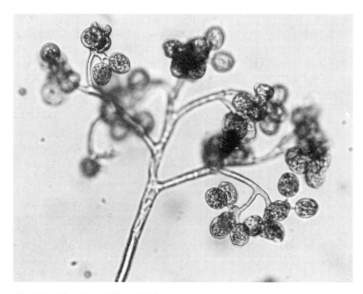

Fig. 4.9a (above) and *Fig. 4.9b* (below)

Fig. 4.9 Sporangia and sporangiophores of *Bremia lactucae*, cause of downy mildew of lettuce (*Lactuca sativa*) and its relatives in the daisy family (Compositae). In (a) the sporangia (which do not form zoospores and therefore function as conidia) are still attached. Each is approximately 20 μm diameter. In (b) the sporangia have become detached and the ends of the branches of the sporangiophore can be seen to be expanded into discs, each with 4–6 pegs around its margin. The sporangia were once attached singly to each peg. (Light microscope photographs by Dr P.A. Mason, Institute of Terrestrial Ecology, Edinburgh.)

European vineyards, leading to loss of production and worthless grapes. The development of control measures followed from the accidental discovery by Millardet in France in 1882 that vines sprinkled with verdigris (copper acetate) – it is said to make them taste bitter to those who might steal them – did not catch the disease. His invention, Bordeaux mixture (copper sulphate and lime), could be produced to stick to the vines and minimise any copper toxicity. It went on to become one of the main fungicides used to control not only downy mildews but also potato blight in the first half of the twentieth century.

Another world traveller in this group is *Peronospora tabacina*, the blue mould of tobacco (*Nicotiana* spp.). It has been present in the USA and in Australia for a long time. Because of the greater variability of the fungus and the greater spread of resistance factors in the Australian *Nicotiana* spp., it is postulated that the fungus may have had its origin there. In the USA it is a disease of the seed bed. Young tobacco seedlings being grown for planting out are particularly prone to attack, and if the weather is warm and moist whole seed beds may succumb. In other countries the disease may be more important in field crops of tobacco. The exact form of the attack is determined by the interaction of environmental and cultural conditions. Infection may be minimised by strict attention to hygiene, by the use of chemical fumigants for seedlings and by spraying with fungicides, but care must be taken to avoid damage to the quality of the tobacco leaf by these procedures.

In 1958 the disease was identified in southern England on *Nicotiana* spp. and some 45 months later it had spread through the tobacco-growing areas of southern Europe, North Africa and west Asia. It reached eastern Iran in 1961–62 but does not appear to have invaded India, tropical Africa, South America or China. There can be little doubt that this spectacular piece of travelling was largely by aerial transport of the sporangia, although whether the original introduction into Britain was by this means must remain in doubt. Some believe that it entered the country on tobacco leaves or seeds imported by an amateur tobacco grower.

Two genera of downy mildew frequently encountered by the naturalist on wild and cultivated species are *Peronospora* and *Bremia* (Figs. 4.8 and 4.9 and Plate2). *P. parasitica* may be observed on leaves of cultivated brassicas such as cabbage, Brussels sprout (*Brassica oleracea* var. *gemmifera*) and cauliflower (*Brassica oleracea* var. *botrytis*), as well as on wild brassicas and other members of the Cruciferae such as shepherd's purse (*Capsella bursa-pastoris*) and wallflower (*Cheiranthus cheirii*). Diseased plants often have swollen and distorted stems, usually covered with a grey-white 'down' of sporangiophores and sporangia. On leaves, infection may lead to yellowing around the lesion, although the tissues remain alive. Sometimes necrosis occurs in the older, central region where the tissues have apparently succumbed to the pathogen's demand for nutrients. A 'down' of sporangiophores appears on the living tissues of the underside of the lesion. If cross sections of infected tissue are cut with a stiff-backed razor blade or scalpel (see Appendix), mounted in water and examined with a microscope, the hyphae of the fungus are seen to be coarse, without septa (typical of the Oomycota) and growing between the cells. The haustoria, which penetrate the living cells of the host, are club-shaped or lobed (Fig. 2.6, p.48).

Peronospora farinosa, another downy mildew (Fig. 4.10), is frequently encountered on the leaves of members of the beet family (Chenopodiaceae), in the

Fig. 4.10 Peronospora farinosa, cause of downy mildew of sugar beet (*Beta vulgaris*) supplied by G. Scott. The plant is systemically infected. The leaves are distorted, paler green than normal and covered with a fine grey 'down' of sporangiophores and sporangia. (D.S. Ingram.)

garden or farm on beet and in the wild on orache (*Atriplex* spp.), goosefoot (*Chenopodium* spp.) and sea beet (*Beta maritima*). Infected leaves become thickened and distorted as a result of disruption of the hormonal control of cell division and cell enlargement in the host (Fig. 4.10). They also appear paler green than normal, perhaps because cell enlargement does not lead to any increase in the number of chloroplasts (organelles containing chlorophyll) per cell. The sporangiophores and sporangia are a dull grey colour. *Peronospora farinosa* may also become systemic in infected plants, growing through the living tissue to infect all parts.

Other *Peronospora* spp. that may be encountered by the naturalist include the following.

P. arborescens is commonly seen on Welsh poppy (*Meconopsis cambrica*), on which it appears to be occasionally perennial in the root stock, and on other cultivated poppies. The sporangia are purple in colour, giving the lesion a purplish tinge. *P. destructor* causes long yellow lesions on onion leaves (*Allium cepa*) and can seriously reduce the crop. The sporangia are violet-coloured. *P. grisea* occurs on *Veronica* species and is especially obvious on brooklime (*V. beccabunga*). The leaves have a steely, curled look (presumably resulting from hormonal disturbance) and bear dingy violet sporangiophores on lower surfaces.

Bremia lactucae is the cause of downy mildew of lettuce and its relatives (Plate 2). The disease is often observed as a white, crumbly-looking down of sporangiophores and sporangia on the lower surface of unsprayed lettuce plants. The lesions are angular, being confined by the veins of the leaf, and the tissues remain quite green for several days following infection, eventually turning yellow or even necrotic when the fungal growth is too much for the host. The disease may easily be seen on wild lettuce and on ragworts (*Senecio* spp.).

It is very easy to see *B. lactucae* inside the host tissues if you have a microscope. Find a young, infected lettuce leaf, preferably on a seedling, and cut a 1 cm square with a pair of scissors. Place this in a drop of water on a microscope slide and crush by rolling gently with a glass rod or round pencil. Add a little more water and a cover slip. At the lowest (or highest, depending on which way up you have placed the leaf piece) plane of focus the branched sporangiophores may be seen emerging from the stomata. The tips of these open out as a saucer-shaped disc with 4 or 5 pegs around the rim, each bearing a sporangium (the sporangia may become detached during preparation) (Fig. 4.9). At a different plane of focus the relatively broad, non-septate hyphae may be seen growing between the living cells of the leaf, producing at intervals short, club-shaped haustoria which penetrate the cell walls, invaginating the plasma membrane to establish an interface for nutrient and chemical exchange (Fig. 2.6). Occasionally the sexually produced thick-walled oospores may be seen.

There are numerous other downy mildew genera occurring on wild and cultivated hosts. Amongst the best known are the following.

Pseudoperonospora. This genus resembles *Peronospora* except that the sporangia germinate to produce zoospores. *P. humuli* causes perhaps the most serious disease of cultivated hops (*Humulus lupulus*) and also occurs on wild hops. Greyish sporangia may be seen on the lower surfaces of the leaves and cones, matched by yellow patches on upper surfaces. Infection eventually becomes systemic in the root stocks. *P. urticae* causes greyish-lilac downy patches of sporangia on the undersides of the yellowed leaves of nettle (*Urtica dioica* and *U. urens*).

Plasmopara. This genus again resembles *Peronospora*, but the sporangia germinate to produce zoospores. *P. viticola* occurs on vines and other Vitaceae. It is only occasionally seen in Britain but is endemic in the USA and occurs widely in European vineyards. *P. nivea* occurs on bishop's weed (*Aegopodium podagraria*) and chervil (*Chaerophyllum temulum*), both members of the carrot family (Umbelliferae). *P. pygmaea* occurs on wood anemone.

The tropical downy mildews

In the tropics or subtropics an additional group of downy mildews, the Sclerosporales, occurs on grasses and cereals. Within this group the genus *Sclerophthora* forms its sporangia in a similar fashion to *Phytophthora* and discharges its zoospores in a manner somewhat between that of *Phytophthora* and *Pythium. Sclerospora,* on the other hand, produces thick, tree trunk-like sporangiophores with the sporangia attached to a crown of small branches; the whole structure has an elephantine quality about it which is quite unmistakeable. Further taxonomic groupings are made on the ability of the sporangia to function as conidia without the production of free-swimming zoospores.

These fungi are widespread on grasses in the tropics, and *Sclerospora graminicola* is a serious problem on bulrush millet (*Pennisetum americanum*) and foxtail millet (*Setaria italica*), where the flower spike may become leafy and other parts of the plant stunted. The oospores, which have highly ornamental walls, frequently develop in the seeds of the host and it is by this means that the disease is spread.

The white blisters

Finally, mention must be made of another genus of the Oomycota related to the downy mildews but distinct from them. *Albugo* has aseptate hyphae and produces biflagellate zoospores from its zoosporangia. Like the closely related downy mildews, it is an obligate biotroph. The commonest British species, *A. candida*, causes white blisters (sometimes called white rust) on infected members of the family Cruciferae, such as cabbage, cauliflower and horseradish (*Armoracia rusticana*). In the wild it also causes spectacular, blister-like infections on shepherd's purse (Plate 2). Infection usually leads to swelling and distortion of the stems, flowers, seed pods and leaves due to hormonal disturbance of the host tissues. Early on in the disease the tissues remain quite green, but eventually raised, shiny white blisters appear as the chains of sporangia are produced on short club-shaped sporangiophores immediately below the cuticle (Fig. 4.11). Eventually the cuticle bursts and masses of white sporangia are produced, to be borne on the wind or in rain splash to new hosts. Zoospores are produced on germination of the sporangia. Within the host tissues, if examined in section, the hyphae can be seen growing between the cells, producing almost spherical haustoria which are attached to the main hyphae by narrow necks. The oospores, if present, have a brown, warty wall that appears to be folded like the skin of a rhinoceros.

Albugo candida often occurs on shepherd's purse in the same lesions as the downy mildew *Peronospora parasitica*, with spores of both fungi being produced on the infected tissues. The coexistence of the two pathogens appears to be

Fig. 4.11 Chains of sporangia (a), emerging zoospores (b) and a single zoospore (c) of *Albugo candida*, cause of white blister of members of the cabbage family (Cruciferae). Each sporangium is approximately 15–20 μm diameter and the released zoospores are approximately 8–10 μm diameter. (Illustration by Mary Bates, based on drawings by Webster (1980).)

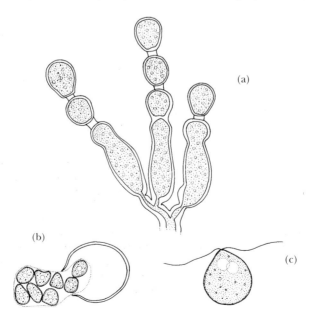

temperature-dependent. As the temperature rises in summer the balance may be tipped in favour of *Albugo* and the *Peronospora* dies out.

Another, less common, white blister is *A. tragopogonis*, which infects members of the family Compositae such as salsify and goat's-beard or Jack-go-to-bed-at-noon (*Tragopogon* spp.) and ragworts. *A. bliti* infects members of the purslane family (Portulaceae). Another *Albugo*, *A. ipomoeae-panduratae*, occurs in warmer climates, infecting sweet potato (*Ipomoea batatas*), and the similar *A. ipomoeae-aquaticae* attacks water spinach (*Ipomoea aquatica*).

The contribution to science

Not only are the zoospore-forming fungi important pathogens in their own right, but attempts to understand and control them have led to significant advances in the science of plant pathology.

In Chapter 1, reference was made to the first attempt to understand that fungi were the cause rather than the result of disease and death in plants, and to the role played by *Phytophthora infestans* in the development of the early hypotheses. It was Anton De Bary's definitive study of potato blight in the 1870s that finally clinched the argument in favour of the fungal pathogen.

One of the first demonstrations of a clear-cut inherited resistance (or immunity) to disease was shown in the wart disease of potato (*Synchytrium endobioticum*), an immunity which has been long-lived and of great practical utility. The hope that a similar immunity or resistance could be found against *Phytophthora infestans* led to crosses of *Solanum tuberosum* with wild potatoes such as *S. demissum*, using some elegant breeding techniques to overcome the differences in chromosome number between the parents. The observed breakdown of resistance to blight in these hybrid potatoes and, as new genes for resistance were introduced, their failure, in turn, contributed to the concept of *R* genes, an understanding of allelic series of resistance genes and studies of the genetics of virulence and avirulence. Perhaps, too, it provided a background against which the gene-for-gene hypothesis could be clarified in flax rust by Flor (see Chapter 3). Later, potato breeding with semi-domesticated potato species led to a further understanding of the broad-based resistance that is now widely used to control disease. Within the last ten years studies of *Bremia lactucae* on lettuce and *Peronospora parasitica* on the Cruciferae have added significantly to our understanding of the genetics of resistance and virulence (see Chapter 3).

The interaction between *Phytophthora infestans* and resistant cultivars of potato was used for early studies of the biochemical basis of resistance. This work led to the identification of the first phytoalexins in the 1940s and 50s, later shown to be the terpenoid compound rishitin and its relatives. Now, work with *Peronospora parasitica* on thale cress (*Arabidopsis thaliana*) is contributing to molecular biological studies of recognition of aliens by potential hosts.

The development of forecasting of epidemics from meteorological data began with *P. infestans*, as did the study of the nature of epidemics and the use of chemical sprays for their control. There is also a number of less immediately practical matters, the study of which began with the zoospore-producing fungi. Taxis in zoospores, for example, which is the way zoospores are attracted or directed to roots, surfaces and so on; the complex series of interactions mediated by sex hormones which controls the mating behaviour in these organisms, and the importance of sterols in controlling the production of sexual spores. Perhaps, however, the most interesting of all these esoteric matters studied in the zoospore-producing fungi is the question of their evolutionary origins and taxonomic position. There has always been considerable uncertainty surrounding the relationship of the Oomycota to other groups of fungi. For example, it has long been recognised that they have much in common with certain groups of algae: diploid life cycles; walls containing cellulose (and glucans); biflagellate zoospores; distinctive sexual structures and mechanisms of fertilization; and the induction of sporulation by mineral solutions. On the basis of such evidence many have argued that the Oomycota should not be regarded as true fungi at all. There has also been much debate about the relationship to one another of the genera within the Oomycota – *Pythium*, *Phytophthora* and the downy mildews.

Now, molecular biological techniques are making possible definitive studies of these matters. The new techniques allow comparison of the DNA of different organisms, especially those parts of the genome that are highly conserved (i.e. have remained relatively unchanged through evolution). Most important of these conserved pieces of DNA is the ribosomal RNA-gene repeat (rDNA for short). Ribosomes, subcellular bodies involved in the synthesis of proteins, probably evolved from bacteria very early in the history of life on earth.

Ribosomal DNA has now been sequenced (analysed in detail) for most major groups of living organisms. The rate of mutation within the DNA and the resulting evolutionary divergence seem to provide an excellent 'clock' to measure the progress of evolution of all living things. Computer programs have been developed to compare the rDNA sequences of different organisms and to generate phylogenies (evolutionary trees). By these means it has been found that the Oomycota are not in any way related to the rest of the members of the Kingdom FUNGI (Ascomycota, Basidiomycota, Chytridiomycota and Zygomycota) but instead belong to a new Kingdom, the CHROMISTA, which also includes the brown algae (seaweeds) and diatoms (unicellular planktonic algae with silica 'shells'). Thus *Phytophthora* and its allies are no longer regarded as fungi and it has become fashionable to call them 'pseudo-fungi'.

The 'spacer regions' of DNA that separate the main rDNA genes mutate more rapidly than the rDNA genes themselves. These mutations persist in populations of individuals carrying them, since they confer no disadvantage (because the spacer DNA appears to have no other function than to separate the main genes). Studies of such mutations provide a means of determining the 'relatedness' of one organism to another below the level of the Kingdom, especially at the level of the genus and species. Recent research of this kind has shown that *Phytophthora* is a clearly defined genus with sporangial form determining the main groupings within it. In contrast, *Pythium* is a very diverse assemblage, encompassing at least four well-separated groups, each about the size of the genus *Phytophthora*. The genus *Peronospora* (a downy mildew) has been shown to be little more than an obligately parasitic offshoot of *Phytophthora*.

All this rDNA work has had a valuable practical spin-off in disease diagnosis. By isolating and sequencing rDNA it is possible to detect within a day very low levels of infection of hosts by different *Phytophthora* spp., making it possible to apply control measures in good time. In Scotland, for example, early diagnosis is leading to better control of *Phytophthora* on the commercially valuable raspberry crop.

An underwater world

Finally, it is important to mention the great diversity of zoosporic fungi that parasitise algae. Studies of such organisms in fresh water has been pioneered by Dr Hilda Canter-hund at the Freshwater Biological Association in the English Lake District. For an enthusiastic amateur with a good microscope, however, a whole world remains to be discovered in almost any ditch, pond or lake. Here will be found underwater biotrophs and necrotrophs, resistance reactions, including the hypersensitive response, and epidemics of disease. References that provide a window on this world are included in the reading list to Chapter 4.

5

Curls, Scabs, Spots and Rots

The pathogenic Ascomycota (see Chapter 1 and Table 1.1 (p.21)) are not as dependent on water as the zoosporic fungi. Their sexual structures, the sac-like asci that contain the haploid ascospores, are usually protected from the extremes of the environment by a fruit body, the ascocarp, composed of fungal tissue. The details of ascus and ascospore form, the way in which the asci are arranged, the mechanism of ascospore discharge and the structure and shape of the ascocarp determine the place of these fungi in modern classifications. The asexual spores*, the conidia, lack flagella and are usually distributed by wind or insect carriers. As with all pathogenic fungi, both the sexual and asexual spores remain vulnerable until they have germinated and sent new hyphae into the moist internal environment of their plant hosts.

After germination of the spore on the surface of the plant, the germ tube usually forms an appressorium and infection peg, which grows through the cuticle, where it expands to form a subcuticular hypha. This either spreads further under the cuticle or grows down between the epidermal cells and colonises the tissues of the leaf or stem. There the subsequent relationship with the host may be biotrophic, hemibiotrophic or necrotrophic according to species.

When the ascocarps are formed they have a comparatively long life, and the capacity to survive the long periods of desiccation that may occur before high humidity once more makes conditions favourable for them to shoot their spores through the boundary layer of still air that covers all surfaces into the turbulent air in which they are dispersed. This gives them a certain independence from their moist substrate and also the capacity to spread spores over considerable distances to new hosts.

We deal here with the Ascomycota that cause leaf curls, scabs, spots and rots. Those that cause wilts and powdery mildews will be considered in Chapters 6 and 8 respectively.

Stripped for action: the Taphrinales

The first group of parasitic Ascomycota we describe are not in fact recognisable from the introductory description because they do not form an ascocarp. Instead, following infection, the fungal mycelium spreads between the leaf cells, eventually forming a well marked layer beneath the cuticle. This

---

*The asexual spores were once referred to as the 'imperfect' stage of the life cycle. The more modern terminology, however, is anamorph. Similarly, the sexual spores were once referred to as the 'perfect' stage of the life cycle, but the modern term is teleomorph. The terms anamorph and teleomorph will be used throughout the remainder of the book to refer to asexual and sexual stages of fungal life cycles, respectively. Anamorphs and teleomorphs are often given separate Latin names, especially where the teleomorph was not discovered until long after the anamorph had been described and named. For example, *Aspergillus* is the anamorph of the teleomorph *Eurotium*.

produces large numbers of asci, which then poke through the cuticle onto the leaf surface. The relationship with the host is biotrophic. Although haustoria are never produced, hyphae may penetrate the living cells of the leaf and become surrounded by the cell membrane. Invariably, infection leads to abnormal growth of the tissues of the host.

One of the most spectacular members of this group is *Taphrina deformans*, the cause of peach leaf curl disease (Plates 3). Affected peaches (*Prunus persica*) and their relatives, such as nectarines (*P. persica* var. *nectarina*) and occasionally almonds (*P. amygdalus*), have greatly distorted, blistered and incurled leaves in the spring. As the disease develops the leaves become patchily red or purple, fading to yellow and finally brown; in their red or purple phase they look at first sight like strips of meat hanging on the tree. Before the colours fade the upper surfaces of infected leaves become covered with a greyish bloom. On examination this is seen to be formed by the serried ranks of asci breaking through the cuticle to appear naked on the surface. Infected leaves fall prematurely, and it is the resulting massive loss of functional leaves that weakens the tree and interferes with fruiting.

The elongated asci on the leaf surface each produce eight ascospores, which sometimes bud within the ascus or after being forcibly discharged, thus increasing the inoculum. Infection by ascospores requires humid conditions and peaches only escape infection in dry, semi-arid regions such as the southwest USA, and in dry irrigated areas such as occur in Washington State in the northwest USA. In glasshouses in Britain careful watering and humidity control are required to ensure freedom from the disease without spraying.

Confusion over the method of infection delayed the development of effective control measures for peach leaf curl for many years. It was at first believed that the fungus infected the young shoots and was present inside the buds, so that spraying had to be directed at the emerging leaves, with critical timing. When it was discovered that leaf infection was initiated from spores overwintering on the surface of the shoots the efficacy of winter spraying with fungicides was easily demonstrated. Here we see the importance of a thorough understanding of the infection process for the development of good control measures.

It is not easy to identify clearly defined lesions in the pathogenic relationship between *Taphrina deformans* and its hosts. Indeed, when the other common *Taphrina* spp. on wild plants are examined, and there are over a hundred of them worldwide, some delimitation of infection may be seen, but no marked, visible tissue reaction to contain the infecting mycelium. With *T. populina*, which commonly occurs on the leaves of poplars (*Populus italica* and *P. nigra* x *serotina*) in Britain and mainland Europe, the infected portion of the leaf expands more than the uninfected portions to form an inverted cup, its mouth facing downwards (Fig. 5.1, overpage). Masses of golden yellow asci are produced within this cup on the underside of the leaf. There may be two or three such infected areas on each leaf, with no regular size, although they never occupy the whole of the leaf area. Other species attack the individual catkin scales of poplar, alder (*Alnus* spp.) and birch, causing swelling and abnormal growth. Still others attack the fruits of cultivated and wild plums (*Prunus* spp.) to form 'pocket plums' with swollen, misshapen hollow fruits, the surfaces of which are covered with a bloom of asci (Fig. 5.2, overpage). These occur variously in Britain, in mainland Europe and also, with slight taxonomic

Fig. 5.1 (above)
Taphrina populina
causing distortion of
leaves of poplar (*Populus*
sp.) in Cambridge.
(D.S. Ingram)

Fig. 5.2 (right) 'Pocket
plums' resulting from
infection by *Taphrina
padi* on bird cherry
(*Prunus padus*) at Juniper
Bank, Peeblesshire,
Scotland. (Debbie White,
Royal Botanic Garden,
Edinburgh.)

differences, in the USA. Here the fungal infection is limited by the size of the organ it attacks and from which it does not spread.

Other *Taphrina* spp. cause witches' broom disease on a number of trees such as birch (*T. betulina*) (Fig. 5.3) and *T. wiesneri* on cherry *(Prunus avium* and *P. cerasus*) in Britain and on other genera in mainland Europe and the USA. In these cases infection of the bud or developing shoot leads to a proliferation of the apex and ultimately a spectacular much branched structure, the eponymous witches' broom.

The enclosed orders

Turning now to those members of the Ascomycota where the asci are developed and protected within an ascocarp (Fig. 5.4, overpage), we find first a group of fungi that were once called quite simply the 'Plectascales' or 'Plectomycetes', and which have completely closed ascocarps (cleistothecia) containing rounded asci (Fig. 5.4a). The mature ascospores are released into the environment by the decay of the enclosing structures. On the whole the members of this group are not pathogenic to plants but are members of the saprophytic soil flora, obtaining nutrients by degrading various dead natural substrates. A few, such as the Eurotiales (known as *Penicillium* and *Aspergillus* in the anamorph stages) cause soft rots of fruits and storage organs such as bulbs and corms, especially if these are over-mature or wounded. The pale brown rots caused by *Penicillium expansum*, decorated with the blue-green conidia of the pathogen, are frequently seen on ripe apples damaged by birds or wasps (Fig. 2.3). Similarly the blue-green conidia of *P. italicum* may be seen on damaged oranges in the fruit bowl and may spread to neighbouring fruit, while other *Penicillium* spp. are found as basal rots of narcissus (*Narcissus* spp.), tulip (*Tulipa* spp.) and hyacinth (*Hyacinthus orientalis*) bulbs after winter storage.

Fig. 5.3 'Witches' broom' of birch (*Betulina* sp.) resulting from infection by *Taphrina betulina*, near Stobo, Peeblesshire, Scotland. (D.S. Ingram.)

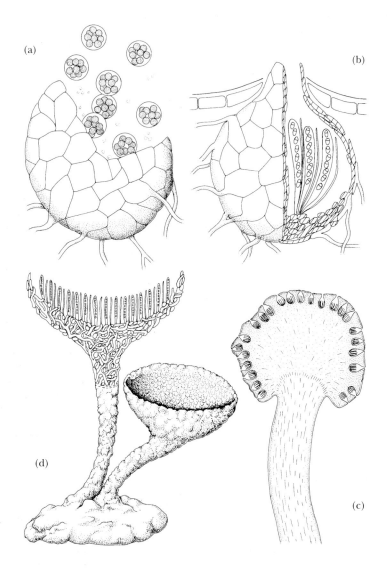

Fig. 5.4 The diversity of ascocarps produced by members of the Ascomycota: (a) a cleistothecium, as in *Erotium*; (b) a perithecium, here embedded in host tissue, as in *Venturia*; (c) perithecia embedded in a perithecial stroma, as in *Claviceps*; (d) an apothecium, here arising from a sclerotium, as in *Sclerotinia*. (Illustrations (not to scale) by Mary Bates.)

The Eurotioles are distinct from the remaining fungi with protected asci, which were once called the 'Pyrenomycetes' or flask fungi. A clearly defined group of these have dark-walled ascocarps (perithecia) with longish necks, giving them a flask-like appearance (Fig. 5.4b). The ascospores are shot out from the elongate asci through a hole at the top of the neck, caught up by the wind and dispersed to new hosts. A parallel group, the Hypocreales, sometimes forms its (mostly) coloured ascocarps on or in a thickened tissue called a stroma (Fig. 5.4c) and is associated with a range of well-known anamorph genera such as *Fusarium, Verticillium, Gliocladium* and *Trichoderma*. The fungi in this order also produce a range of metabolites that are poisonous or react with the metabolism of other fungi or green plants with sometimes spectacular effects.

A third group, in which the elongate asci are formed on the surface of an open disc or saucer (the apothecium), were once called the 'Discomycetes' (Fig. 5.4d). Most are saprophytic, but the Sclerotiniaceae are pathogens.

Apple scab – classic 'Pyrenomycete' disease

Venturia inaequalis (Dothideales) contrasts sharply with *Taphrina* in all the details of its pathological life history. It is the cause of apple scab, which has probably been with us since the Garden of Eden (Plate 4). The cosmetically attractive fruits that we find in the supermarket, polished and shining yellow, green or rosy red, are a credit to the plant pathologists who devised the technology to control the disease. We can still find infected fruit and leaves on crab apples (*Malus sylvestris*) and on eating apples (*M. pumila*) in gardens, in tumbledown orchards and on some country market stalls displaying produce from these sources. In the orchards of our ancestors unblemished fruits were the exception.

Infected fruits and leaves show circular or irregular, brownish, velvety (later corky) scabs, 5–10 mm across. These do not appear to affect the flesh of the apple or the deeper tissues of the leaf. They may, if the infection occurs early enough in the development of fruit, cause cracking of the surface, sometimes allowing the entry of rotting organisms. Sometimes infections may not develop immediately into lesions, but cause scabs to appear later in the season or on apples in store. Scrapings from the scabs, when examined with the microscope, are found to contain characteristic one or two-celled conidia. These asexual spores are known by the anamorph name, *Spilocaea pomi,* and are responsible for much of the summer spread of the pathogen.

The infection is initiated afresh each season. Much of it, and almost all in certain situations, comes from ascocarps which are formed in the spring by the pathogen growing saprophytically on overwintered leaves infected the previous summer. They release their ascospores to be carried by the wind to the surface of the newly emerging leaves, fruit buds and eventually young fruits. The ascospore (or later the conidia) germinate on the leaf surface and form an appressorium from which a germ tube grows through the young cuticle to produce hyphae. These grow between the cuticle and the upper wall of the epidermal cells, eventually forming a layer, several hyphae thick, from which are developed the conidia (*Spilocaea*). The fungal growth eventually ruptures the cuticle, exposing the spores to the air. Further infection takes place on both leaf surfaces, that on the upper surface being strictly limited in area while that on the lower surface spreads along the veins and midrib to give a less clearly defined infection. With some varieties and in some situations infection of young twigs leads to cracking and cankering of the twigs, and overwintering of conidia within these cracked areas.

In living leaves, even when heavily infected, the fungal mycelium is largely restricted to the outside of the leaf between cuticle and epidermis. Contact with living tissue may occur later as intercellular hyphae begin to penetrate further into the leaf. The situation changes again when the leaves die and fall to the ground. The subcuticular mycelium now begins to colonise the whole dead leaf and then, as winter passes, the mycelium aggregates in certain areas and ascocarps are formed. Eventually these produce asci and ascospores, which are released through the beak-like neck of the perithecium which emerges on the surface of the leaf. The ascospores consist of two unequal cells, either or both of which may germinate to initiate a new infection and begin the cycle again.

Venturia thus has a restricted saprophytic phase, which follows from the foothold obtained in the leaves in its biotrophic phase; from this foothold it can expand into dead tissue ahead of possible competition from other saprophytes. However, it never grows as a free-living saprophyte, competing successfully for nutrients with other saprophytes on non-host tissues.

Control of apple scab has been attempted in four ways. Firstly, attempts were made to prevent infection by spraying with fungicides, beginning each year at bud-burst. The first fungicide used was Bordeaux mixture, but it had the disadvantage of russeting the fruit and was replaced by lime sulphur, which could also be toxic in some situations. Elaborate spraying schedules were developed relating to the time of opening of the fruit buds, designed to maintain a complete cover of the fungicide on fruit and leaf surfaces throughout the growing period. Insecticides were also incorporated, and the floor of apple orchards could become like a desert with the weight of toxic material applied to the trees and incidentally to the soil below. Gradually new organic fungicides and insecticides were produced and methods developed for applying them at low volume in less water, thus reducing the cost of each application. Methods of cultivating and pruning the trees changed too, restricting their height so that there was less need for powerful spraying machinery projecting large quantities of liquid high in the air.

Secondly, attempts were made, both in the USA and Britain, to develop a predictive system to forecast the annual development of the disease epidemic. By trapping the fungal ascospores produced in an apple orchard in the spring it was possible to correlate the numbers of spores with rainfall, relative humidity and temperature and from this develop a method for predicting the time of appearance of the first spore showers and the time when the first spray should be applied. Later infection was routinely controlled by spraying at seven-day intervals. The critical factors controlling the initial spore discharge from asci were rainfall and temperature. This approach was also valuable in making growers aware of the natural history of the disease, an awareness which was translated into good management practices in a number of areas.

Thirdly, an attempt was made to control the discharge of the ascospores from the leaves on the orchard floor. This was pursued more energetically in the USA than in Britain, where the perceived importance of stem lesions in varieties such as Cox's Orange Pippin directed attention from the leaves as the main source of overwintering inoculum. The demonstration in the USA that substances such as dinitro-orthocresol could eradicate ascocarps on the orchard floor introduced a new element into the control of the disease.

A final approach was to attempt to develop varieties resistant to the disease. Observations showed differences in the intensity of attack on different apple varieties, some not being infected at all. Moreover, certain wild species, such as *Malus sikkimensis*, were apparently resistant to attack by the pathogen. It was hoped to use this resistance in the breeding of commercially acceptable apple varieties.

Scientific interest in breeding for resistance was fuelled by the possibility of an even greater prize. The similarity between the fruit bodies and asci of *Venturia* and those of *Neurospora crassa*, the bread mould, with which a great deal of classical genetic analysis had been carried out, suggested that detailed genetic analysis of *Venturia* might perhaps lay bare the metabolic steps in the interaction between pathogen and host.

Not surprisingly, no successful resistant commercial varieties of apple emerged from this breeding programme. The market makes great demands on potentially successful varieties. The fruit must look good, travel well, store well, taste good, yield well, outclass the already established varieties, have a good orchard habit, not be subject to biennial bearing, stand up to spraying and picking and so on. To meet all the commercial requirements and to incorporate resistance to disease in a crop with a life cycle as long as that of the apple is a tall order indeed. Moreover, the genetics of resistance proved to be hard to handle; by 1988 genes controlling virulence had been identified at no less than 19 loci!

As far as the genetic analysis of pathogenicity went, however, some interesting facts emerged, although the great breakthrough still eluded the investigators. It was found that matching genes for virulence and avirulence at particular loci gave either a 'lesion' or a 'fleck'. In the first the fungus became established in the leaf, spread for some distance under the cuticle and sporulated freely. In the second it germinated and produced appressoria but thereafter the progress of the mycelium was sharply restricted by a hypersensitive reaction (see Chapter 3). Although the further detailed analysis of pathogenicity proved impossible, enough had been done to suggest that modern molecular genetic methods using this system as a base could expand our knowledge of the interactions between pathogen and host.

Other *Venturia* spp. that infect members of the rose family (Rosaceae) are: *V. pirina*, which infects pears (*Pyrus communis*) in a parallel fashion to *V. inaequalis* on apples; *V. crataegi*, which affects hawthorn; and *V. cerasi*, which occasionally attacks cherries. There are also three *Venturia* spp. that occur on willows (*Salix* spp.) in Europe, of which *V. saliciperda* is a serious problem on *Salix alba*, *S. amygdalina* and *S. babylonica*, especially where they are planted as specimen trees in areas with moist climates.

A relative of *Venturia*, *Hormotheca robertiani*, is a pathogen of wild herb Robert (*Geranium robertianum*) throughout Europe. It is found on the upper sides of the leaves, especially the older ones, and consists of a layer of mycelium one cell thick, which spreads under the cuticle without destroying the underlying cells. There is no obvious external sign of the parasite at this stage and no lesion. Eventually the fungal layer begins to thicken in discrete areas and builds the flattened ascocarps in which the asci are formed. This subcuticular fungus seems almost to have adopted the strategy of the mutualistic fungi (see Chapter 7) in reducing the activities which might signal its presence and elicit a resistance reaction.

Another interesting variation of the pathogenic pattern shown by *Venturia* is seen in the life cycle of *Apiognomonia erythrostoma*, the cause of cherry leaf scorch, which is widespread on *Prunus* spp. in mainland Europe. In Britain it is particularly prevalent on the wild cherry *P. avium*, especially more recently

Fig. 5.5a

Fig. 5.5b

Fig. 5.5 'Take-all' of wheat caused by *Gaeumannomyces graminis* at Cambridge, England: (a) an infected root (right) beside a healthy one (left); (b) whiteheads on artificially infected plants (right hand group) beside healthy plants (left hand group); (c) dark runner hyphae on the surface of infected roots. (P.R. Scott, CAB International.)

Fig. 5.5c

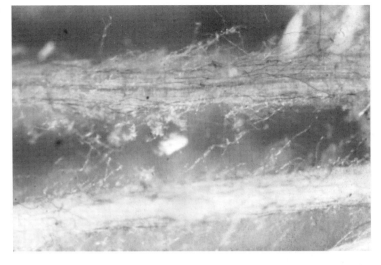

in the south of England. Its interest for us lies in the leaf infections that occur as the leaves are first opening but do not show until they are mature in the middle of summer, when small lesions appear. Infected leaves then fail to form an abscission zone, and instead of falling off the tree in the autumn can be found hanging as brown and slightly shrivelled flags throughout the winter. Ascospores are formed on these leaves in the spring. Here we have a delay in symptom expression, for the fungus is apparently latent in the leaves until midsummer. The trigger for symptom expression is not known. The other interest lies in the failure of the leaves to absciss, which suggests that the fungus is interfering with the normal hormone control of leaf abscission in the plant.

Take-all, the mysterious assassin

Take-all is one of the most important foot-rot diseases of cereals worldwide (Fig. 5.5a). In conditions favourable to the disease it causes the death of large patches of maturing cereals, the individual plants showing 'whiteheads' of ears that have failed to fill (Fig. 5.5b). A spectacular cause of crop failure in the USA, Canada and Australia, it occurs widely in Britain and mainland Europe, although with less devastating results. Even so, some say that the yield of wheat may often be reduced by 10% as a result of take-all infection. *Gaeumannomyces graminis*, the causal organism (classified in the Magnaporthaceae; its order has yet to be determined) attacks the roots of wheat and barley and also of certain wild grasses. Oats are attacked by a *forma specialis*. Characteristic dark hyphae (Fig. 5.5c) spread over the surface of the roots of infected plants and send short non-pigmented branches into their tissues, destroying them. Once the pathogen reaches the endodermis, the layer of cells surrounding the central vascular region of the root, individual cells appear to slow down the progress of infection by encasing the invading hyphae in wall material, forming structures called lignitubers. A build up of dark hyphae around the stem and lower leaf bases is also characteristic of the disease. The ascocarps are not common, but when they appear on the lower leaf bases they are flask-shaped and immersed in the tissues. The asci project their elongated ascospores into the air through the short, open, beak-like necks positioned to one side of the ascocarp.

Our knowledge of the natural history of pathogenic fungi derives from intensive work on a limited number of commercially important diseases of crop plants and the extrapolation of this knowledge more widely. Each study yields an insight into a different facet of disease and there is no doubt that the study of *Gaeumannomyces graminis* has allowed the construction of a theoretical framework for understanding the ecology of a great range of soil-borne plant diseases and has elucidated the nature of pathogenesis in a wide range of disease situations.

Much of this success can be attributed to the percipient work of Denis Garrett, who devoted his life to research on this organism and developed important generalisations about soil-borne diseases from his studies. His work was characterised by extreme simplicity and great attention to detail. Others had shown the causal relationship between *G. graminis* and take-all, but when he started work in 1929 the disease was still full of mystery. It appeared in some fields and not in others; its presence seemed to be related to soil type, but healthy crops could be grown on the suspect soils; application of nitrogen fertilisers sometimes favoured the disease and sometimes the plant; it was a parasite but it also seemed to exist saprophytically in the soil. How could these anomalies be reconciled?

What Garrett did was to recognise that the variability of conditions in the field made it virtually impossible to draw meaningful conclusions from field observations alone. He therefore determined to create conditions where as many environmental factors as possible were strictly controlled. Instead of spending time building complex equipment he chose to use cheap glass tumblers or jam jars, in which he placed field soils that had been dried and remoistened to a standard. Nutrients were added to these soils in standardised form, whilst moisture loss and temperature fluctuations were controlled by frequent weighing and topping up of the containers, placed in standard microbiological incubators. In these controlled conditions inoculum of the fungus was allowed to colonise standard lengths of straw buried in the soil, and these were later graded for colonisation and subjected to biological tests for infectivity. These were methods which any one of us could reproduce in our own homes, their one disadvantage being the sheer hard work required to achieve maximum uniformity. Many of the research students who worked with Garrett looked up wistfully from the straw they were painstakingly cutting into exact lengths to their neighbours whose electronic machines seemed to produce results with effortless ease. But Garrett's methods, carefully used, and accompanied by much thought and hypothesising, yielded valuable results and information.

The fungus survives the intercrop period on the roots or stubble debris of previously infected crops and spreads to neighbouring plants, mostly in early summer. It is reduced by adequate rotation but, where cereal follows cereal, as is now common in many agricultural systems, as little as 1% of infected plants can give rise in the following year to a serious loss of crop. The disease is most serious on light, loose alkaline soils poor in nutrients but reasonably moist. The critical point in the life cycle seems to be the saprophytic phase when the fungus is surviving in the dead crop residues. At this time the fungus is very sensitive to competition from other soil saprophytes, and its survival is also very dependent on the amount of nitrogen available to support fungal growth. Hence nitrogen applied early, even the year before, will encourage the survival of the fungus. Conversely, however, if infection has begun, nitrogen-manuring of the growing crop in early spring will enable the plant to produce new roots to replace the infected ones and to recover from the disease.

Based on this framework, farming methods that favoured high yields of wheat and barley and at the same time discouraged infection with take-all were worked out and put into practice. Today the disease is rarely a problem except in situations where agricultural systems are changing from a standard rotation to the serial cultivation of wheat or barley. It has been found, quite independently of Garrett, that take-all increases, in spite of the appropriate agronomic measures, for the first three or four years of serial cultivation, at which point the amount of take-all begins to decline to levels where it no longer threatens crop yields. The cause of this take-all decline is probably a build-up in the soil of fluorescent bacteria and other microorganisms which are competitors of *Gaeumannomyces* and other fungi.

Here we see a pathogen able to infect a range of cereal and grass species but sharply recognised and excluded by others. Where plants are attacked, the amount of inoculum and its vigour, which Garrett called 'inoculum potential', appear to be important in the success of the attack, and this vigour is determined by the success of the saprophytic survival of the organism in cereal and grass residues. But, and here is one of the unexpected bonuses of Garrett's studies, the saprophytic phase is sensitive to the presence of other soil-inhabiting organisms, and either grows more slowly or stops growing completely. When a range of root-inhabiting organisms is examined, the parasitic forms have nearly all lost their ability to compete in a soil environment. Garrett went on to theorise further from these observations and to suggest that the qualities that make for good saprophytic existence (i.e. rapid growth, the release of large quantities of exogenous enzymes and the production of antibiotic substances) are in fact the features which make for easy 'recognition' of the fungus by the host and its exclusion by the mechanisms already alluded to in Chapter 3. These and others of his ideas, too numerous to iterate here, have become the coinage of modern plant pathology and demonstrate how, with simple tools and careful thought, great advances can be made. Perhaps we should note, in conclusion, Garrett's remarkably shrewd approach to his choice of pathogen for a lifetime of study. Not for him the pure 'academic' approach but, as he wrote in the preface to his book *Pathogenic Root-Infecting Fungi* (1970): 'I need stress no further this economic aspect of their [root-infecting fungi] importance, except to remark that more information, more co-operation and more financial assistance are available for the study of an organism of economic importance than for one that is not; all university biologists must now be aware of this – though some may refuse to act upon it.'

The colourful Medicis, a poisonous bunch of fungal pathogens

We turn now to some members of the Ascomycota with coloured ascocarps. Those discussed here are grouped in the families Hypocreaceae and Clavicipitaceae.

A very familiar member of the Hypocreaceae is the coral spot fungus (*Nectria cinnabarina*), which appears as small cinnabar-red cushions (0.1–0.5 mm in diameter) on the surface of dead branches of a variety of trees and shrubs (Fig.

5.6a). It is especially common on lime (*Tilia* spp.), horse chestnut (*Aesculus hippocastanum*) and sycamore (*Acer pseudoplatanus*). Examination of the cushions, the *Tubercularia* anamorph, reveals small oblong conidia borne all over the cushion surface. The ascocarps (about 0.5 mm in diameter) are much harder to find but invariably appear where, for example, pea-sticks of sycamore have been left in the ground over winter. Then the cinnabar-red cushions are replaced by the dark red clusters of ascocarps at soil level (Fig. 5.6b).

With this behaviour the fungus looks like a good saprophyte, but it can also behave as a parasitic necrotroph. It frequently causes a dieback of soft fruit, especially redcurrant (*Ribes rubrum*) and gooseberry (*R. uva-crispa*) bushes, and it can also attack young trees, causing spectacular dieback of individual branches in the summer. The fungus apparently cannot make a direct entry into the plant through leaves or stems, and seems to enter through pruning

Fig. 5.6a

Fig. 5.6b

Fig. 5.6 Nectria cinnabarina, cause of coral spot of dying trees and shrubs: (a) the cinnabar-red conidial cushions (0.1–0.5 mm diameter); (b) dark red ascocarps (approximately 0.5 mm diameter); both at Juniper Bank, Peeblesshire, Scotland. (Debbie White, Royal Botanic Garden Edinburgh.)

wounds and broken branches. From these infection sites it enters the woody tissues of the xylem, spreading upwards and downwards as well as laterally. When, in the redcurrant, it spreads laterally sufficiently to girdle the stem, the whole bush collapses. The invasion of the plant through the vessels of the xylem and their apparent blockage, either by interaction with the poisonous products of the fungus or by the hyphae themselves, foreshadows the wilt fungi, many of which are related and which are described in detail in Chapter 6.

Nectria galligena, a close relative of the coral spot fungus, is the cause of canker of apple trees (Fig. 5.7). The conidial pustules (anamorph name *Cylindrocarpon mali*) are here buff-coloured and produce either unicellular spores or sickle-shaped spores with one to five septa. Infection of the apple branch takes place through small cracks in the bark caused by natural leaf fall, by woolly aphids (*Eriosoma lanigerum*) or by the apple scab fungus *Venturia inaequalis*. The pathogen spreads into the bark tissues and is temporarily impeded by the production of corky layers. There is no regrowth of host tissue under the original infection and the limited annual reinfection of the stem growth on the margin of the infected area leads to the development of a crater-shaped canker. Some cankers girdle the stem, with the subsequent death of the branch or of the tree. The disease is found on apples in moister areas, is worse on some soils than others and can be affected by the rootstock and vigour of the tree. It particularly affects certain old varieties such as 'James Grieve', 'Cox's Orange Pippin' and 'Worcester Pearmain', but is rarely seen on 'Bramley's Seedling', 'Sturmer Pippin' and 'Blenheim Orange'.

Gibberella zeae, with blackish-blue ascocarps, bears some resemblance to *Nectria*. It infects cereals, killing the young plants, and also appears as a pinkish, mycelium bearing sickle-shaped, multicellular conidia on the ears in wet weather. This conidial phase (the anamorph) is given the name *Fusarium graminearum*. The fungus is sometimes troublesome on maize in the USA and occurs with varying severity on wheat and barley. Its most interesting feature is the production of mycotoxins (fungal toxins that affect animals and/or humans) such as zearalenone in the infected grain. The first spectacular mycotoxins to be identified were the aflatoxins formed in ground nuts contaminated with the saprophytic fungus *Aspergillus flavus* and its relatives. In the initial outbreak in 1960 they were responsible for the death of 100,000 turkey poults from 'Turkey X disease', then of unknown origin. The aflatoxin B1 is now regarded as one of the most carcinogenic substances known. From the time of the first outbreak, attempts were made to associate sickness in animals, and to a lesser

Fig. 5.7 Canker of apple caused by *Nectria galligena* at Aberlady, East Lothian, Scotland. (D.S. Ingram.)

extent in humans, with the contamination of feed grains by fungi. The process of identifying the different syndromes and relating them to specific fungal contamination was a slow one, but a clear connection was made quite early between grain contaminated with *Gibberella zeae* and a number of conditions, most particularly infertility in pigs.

Zeranol, a synthetic oestrogen with growth-promoting properties, was produced from zearalenone and used extensively in the 1970s and 1980s on castrated beef animals to promote muscle growth, but the use of such compounds is now prohibited.

Another *Gibberella* species, *G. fujikuroi*, attacks cereals in subtropical areas. Morphologically identical forms are found on tropical fruit, particularly banana (*Musa* spp.) and pineapple (*Ananas comosus*). The anamorph is known as *Fusarium moniliforme*. In rice (*Oryza sativa*) some plants are killed but others, at least initially, grow taller than their neighbours. Work by P.W. Brian on the cause of this bakanae (foolish seedling) disease showed that the overgrowth was caused by material secreted by the fungus, later identified as gibberellic acid. This turned out to be one of a series of related terpenoids, the gibberellins, many of which are important plant hormones. Gibberellic acid is now used commercially to induce fruit set without pollination in the production of seedless grapes, for example, and in brewing to speed up and synchronise the germination of barley during the malting process.

Fig. 5.8 Sphacelia (anamorph) stage of ergot (*Claviceps purpurea*): a droplet of honeydew containing conidia on an infected flower of sea lyme grass (*Elymus arenarius*) at Gullane, East Lothian, Scotland (D.S. Ingram.)

Two members of the related family Clavicipitaceae demand our attention. The first, *Epichloe typhina*, cause of choke disease of a range of grasses (Plate 4), is especially conspicuous on cocksfoot (*Dactylis glomerata*). The fungus forms a creamy sheath, later turning golden yellow to orange, around the basal parts of the leaves and stems of infected plants. Single-celled conidia form on the surface of this fungal sheath (technically a stroma), which then develops within itself a number of flask-shaped ascocarps containing asci, each with eight filamentous septate ascospores. The openings of these ascocarps can be seen as minute pits all over the surface of the stroma. Flowering of the grasses is impeded. The mycelium grows systematically throughout the plant and is transmitted in seeds or by vegetative propagation. The role of the ascospores in infecting flowers varies in different genera and requires further investigation. Infected plants are said to be toxic to animals.

There are many similarities between *Epichloe typhina* and our second example, *Claviceps purpurea*, the cause of ergot disease of cereals and grasses (Plate 3 and Figs. 5.8 and 5.9). The outstanding characteristic of ergot is the production of a large elongated and ovoid sclerotium (Fig. 5.9) incorporating the infected ovary. The massed surface hyphae of sclerotia are impregnated with dark chemicals that protect the internal hyphae from desiccation and damage by sunlight. Sclerotia fall into the soil in the autumn and survive for long periods, just as seeds do. They often require a cold spell to induce germination. This helps to ensure that germination does not occur until the spring, when new hosts are available for infection. Upon germination they produce stalked structures with pinkish-violet globose heads, within which flask-shaped ascocarps are formed, their openings being visible as pits all over the heads (Plate 3). The ascospores are filamentous and after discharge are caught by the protruding stigmas of the grass flower, where they germinate and invade the ovary. The first part of the flower to be infected is the style and style base, where the fungus multiplies to form a dirty white structure containing internal folds, on the surface of which are produced the unicellular conidia given the anamorph name *Sphacelia*; these conidia are secreted in a drop of sugary liquid (the honeydew) and spread by visiting insects (Fig. 5.8). The rest of the ovary is infected by the fungus growing over its surface and invading from the base upwards, at the same time stimulating the ovary to grow very much larger, to many times the size of healthy seed. As the resulting ergot matures the surface hardens and darkens to a violet-black colour. The ergots protruding from the individual florets can easily be seen by the naturalist out for a walk before they are shaken free and fall to the ground.

Ergot disease occurs very widely on grass species, but is particularly obvious in the large-flowered grasses that grow on the coast, such as sea lyme (*Elymus arenarius*) and marram (*Ammophila arenaria*). At one time the disease was commonly found in cereals, most frequently on rye (*Secale cereale*). It also occurred on barley, durum and Rivet wheat (*Triticum turgidum*) and some North American hard spring wheats, but was much less common on other wheats. One might guess that rye

Fig. 5.9 Ergots of *Claviceps purpurea* on meadow grass (*Poa* sp.) at Gullane, East Lothian, Scotland. (Debbie White, Royal Botanic Garden Edinburgh.)

is particularly susceptible because, unlike other cereals, it is cross-pollinated and exposes the stigmas by opening the flowers for long periods.

There is evidence that *Claviceps purpurea* exists in a number of *formae speciales*, each of which attacks a particular group of grasses. Sometimes, however, the absence of cross-infection can be attributed to other causes; for example, the glumes of wheat (bracts enclosing the flowers) open to reveal the stigmata for a very short period each day, and if for reasons of local climate there is no dis-charge of the ascospores or *Sphacelia* spores are unavailable there will be no cross-infection.

Ergots have been used with varying success in folk medicine for several cen-turies, but began to be used with some degree of precision in orthodox medi-cine from the beginning of the nineteenth century. They contain a number of alkaloids based on lysergic acid. Indeed, LSD (d-lysergic acid diethylamide) itself was originally isolated from ergots. Alkaloids present in ergots and used for medicinal purposes include ergometrine and ergotamine, and the fungus is also rich in other substances, including histamine and ergosterol. In medi-cine ergots from rye are used for pharmaceutical preparations, and in eastern Europe and Russia the disease is deliberately 'cultivated' on a commercial scale. While there is no doubt that ergots from all sources contain pharmaco-logically active materials, the evidence for the strict comparability of ergots derived from different species is lacking. Nor do we know for certain whether any variation between the ergots from different grass species is due to differ-ences in the fungus or differences in the grasses on which the fungus is growing.

Ergots for pharmaceutical use must be harvested at full maturity in dry weather, before the crop on which they are growing is ripe, and then stored in strictly dry conditions, since ergots that have been damp for any length of time lose their potency. Ergot extracts were originally used for two main purposes, hastening childbirth and controlling haemorrhage afterwards. There is no doubt that in skilled hands they were useful, but knowledge of the severe symp-toms and death caused by accidentally eating ergots introduced a note of cau-tion into their use and caused their replacement by other substances wherev-er possible. The pharmacological basis for their use lay in their effects on smooth muscle contraction and constriction of the smaller blood vessels. Ergometrine tartrate has been used in recent times to control excessive bleed-ing after childbirth. Ergotamine, which also acts as a vasoconstrictor, reducing the diameter of abnormally dilated blood vessels in the brain, has become a boon to migraine sufferers because it alleviates the pulsating headache that is a characteristic of the condition.

Ergots can cause the disease of ergotism if eaten accidentally but, although this has been recognised for many centuries and plausible accounts exist from about the tenth century, convincing circumstantial detail is only available from about the eighteenth century. Ergotism was not common in Britain but there is a case of ergotism recorded in England, in 1762, at Wattisham near Bury St Edmunds, where the family of a farm labourer was afflicted after eating bread made from discoloured grain of Rivet wheat. We have found no reports of it affecting humans in the USA, but there are numerous reports from mainland Europe, especially France and Germany.

The disease appeared there in two forms. 'Gangrenous ergotism' began with a vague lassitude and pains in the lumbar region or limbs, particularly in the calf muscle of the leg. This was followed by swelling in the feet or hands or in whole limbs, and a sensation of violent burning alternating with a feeling of icy cold. The burning sensation gave rise to an early name for the disease, 'Ignis Sacer'. Later the name 'St Anthony's fire' was coined after miraculous cures were experienced at the church of La Motte au Bois in France, where the bones of St Anthony were said to rest, and the monastic order of St Anthony became closely linked with the treatment of the disease. The stage of alternating heat and cold was followed by the shrivelling and gangrenous decay of the affected area, which might then be shed spontaneously. The disease was horrific. In the initial stages the burning sensations could be so intense as to cause the patients to leave their beds and seek relief in the open air, after which the icy cold caused them to seek relief by immersion in warm water, and a repeat of the cycle of misery. Although in mild cases only the finger nails or solitary fingers or toes would be shed, in severe cases whole limbs might be lost. If death did not intervene those afflicted might recover and live for many years, but if they ate more ergot they could exhibit symptoms for a second time.

'Convulsive ergotism' presented a very wide range of varying symptoms, but was distinguished from the gangrenous form by the absence of loss of appendages and limbs. A very common early symptom was a numbness of the extremities and the feeling, known as formication, that ants (some said mice!) were running about under and on the skin surface. This gave way to spasms of the limb muscles, which resulted in the fingers being contracted to give the appearance of an eagle's beak and limbs being fixed with the thighs drawn forwards and the legs below the knee bent backwards, or the feet forwards and the toes backwards. There were similar contortions of the arms. These contractures often could not be overcome even by a strong man, although patients asked for this service and appeared to find relief when it was successfully offered. Other symptoms included convulsions, psychological problems, dementia, maniacal excitement and loss of sight. Patients could recover from mild attacks, but often died or were left with mental and physical impairment in severe attacks.

The two forms of ergotism were geographically separated by the River Rhine, the gangrenous form occurring in France and the convulsive form in Germany. Barger makes a convincing case for the gangrenous form being a direct result of the poisonous principles of ergots and for the convulsive form being due to an interaction between the ergot poison and vitamin A deficiency. The areas in both France and Germany where ergotism occurred were areas of extreme poverty where soil conditions favoured the growing of rye, a plant that survives in poor, acid and sandy conditions. In France, in Barger's view, these were also areas where some dairy products were available, whereas in Germany the land was quite unsuitable for any form of dairying. Hence in Germany the people were susceptible to vitamin deficiency in near starvation conditions, the very conditions under which grains of dubious quality, including ergots, might be eaten.

Modern knowledge has all but eliminated ergotism in humans, but there was a mild outbreak in Britain in 1928 from rye grown in Yorkshire, and a more recent outbreak in France in 1951 has also been documented. In animals ergotism can cause loss of limbs and abortion, and in the USA the form of *Claviceps* attacking *Paspalum* spp. has particularly severe effects on cattle, sheep and horses.

The heavy brigade – the Sclerotiniaceae

The Sclerotiniaceae, unusual in the order Leotiales (Table 1.1, p.21) in the development of resting sclerotia, are characterised by the production in the teleomorph stages of saucer-shaped ascocarps, the apothecia. These consist of disks of fungal tissue on the open surface of which asci are formed in great numbers, packed shoulder to shoulder with elongated supporting hyphae called paraphyses. In the Sclerotiniaceae the apothecia are usually stalked to raise the disks to the soil surface from the buried sclerotia that produce them (Fig. 5.4d).

Sclerotia are produced by the anamorph stage and are frequently quite large, being between 5 and 10 mm in diameter. Others are much smaller and yet others form a stroma, incorporating host tissue and extending as a thin plate in the leaf or stem. At first no connection was made between the various teleomorph and anamorph stages, but then Anton de Bary demonstrated that the very common *Botrytis cinerea* was the anamorph of an apothecial fungus which he named *Sclerotinia fuckeliana*. *Botrytis cinerea*, now *Botryotinia fuckeliana*, can be found in temperate climates if wild herbaceous vegetation is examined. When the plants are parted the beautiful conidiophores of the fungus will frequently be found on shaded leaf or stem tissue. It can also be seen in great masses on young plants (Plate 3), the dead stems of tomato plants at the end of the growing season or at the flower end of old cucumbers and courgettes (*Cucurbita pipo*) from where it may grow on to rot the whole fruit. In wine-producing areas it occurs on grapes at and after maturity and is the 'noble rot' that is said to remove the excess water from the grapes and concentrate the sugars to produce the shrivelled berries from which sweet dessert wines are made.

Botryotinia fuckeliana also parasitises the delicate tissues of the petals of roses (Fig. 5.10), sweet peas (*Lathyrus odoratus*) and other flowers, producing, after a spell of damp weather, circular bleached or darkened spots on the coloured background. These are infection sites derived from individual spores. To begin

Fig. 5.10 Lesions of *Botryotinia fuckeliana* on senescent rose petals in Aberlady, East Lothian, Scotland. (D.S. Ingram.)

with the fungus is restricted to a limited area, but it grows out into the surrounding tissue as it ages. In roses it may spread from the outer petals into the whole bud, which then collapses as a wet mass. On strawberries and raspberries it forms restricted infection sites in the green tissue of the sepals and young berries, later spreading to the ripening tissues at maturity. With other related species such as *Sclerotinia trifoliorum* on clover (*Trifolium* spp.) and *S. sclerotiorum* on a range of horticultural crops, a similar pattern may be observed.

Taxonomy never stands still, and the simple relationship between *Botrytis* and *Sclerotinia* was soon modified to make a range of teleomorph genera and associated, widely varying, anamorphs. These are fascinating organisms, in many of which the pathogenic biology has still to be worked out. An attempt is made to summarise their morphology in Table 5.1, and other points of interest in Table 5.2.

Table 5.1 Some common genera of the Sclerotiniaceae.

Teleomorph	Anamorph	Sclerotia or stromata formed
Botryotinia	*Botrytis*	Sclerotia
Ciboria	None known	Stromata in tissue
Ciborinia	None known	Stromata in tissue with black sclerotia at host surface
Gloeotinia	*Endoconidium* (pink and slimy)	Stromata as mummified grass seeds
Monilinia	'Monilia' chains of conidia	Mummified infected pome and prune fruits become giant 'sclerotia'
Myrioclerotinia	*Myrioconium*	Black sclerotia
Ovulinia	*Ovulitis*	Black sclerotia
Sclerotinia	None known	Black sclerotia
Septotinia	*Septotis*	Stromata
Stromatinia	None known	Stromata in host tissues, black microsclerotia on surface

Table 5.2 Some interesting members of the Sclerotiniaceae.

Botryotinia
Apart from the common *Botryotinia fuckeliana* there are many specialised species that infect narcissus, tulip, iris (*Iris* spp.) and onion.
Ciboria
Infects oak (*Quercus* spp.) and other broadleaved trees; large apothecia emerge from blackened acorns in oak litter in the autumn, caused by *C. batschiana*. Others develop on fallen catkins of various trees and on fruits of sedges (*Carex* spp.) and rushes (*Juncus* spp.).
Ciborinia
C. candolleana causes brown spots on living oak leaves before fruiting on the fallen leaves. Other species infect poplar (*Populus* spp.) and willow (*Salix* spp.) in the USA and *Camellia* spp. in Asia.
Gloeotinia
G. granigena causes blind seed disease of rye grass (*Lolium* spp.). Apothecia are produced from the mummified seed in spring.

cont.

Monilinia

The conidial pustules on fruit and mummified fruits are common on apples (*Malus* spp.), plums (*Prunus* spp.) and wild members of the family Rosaceae. Two species occur on bilberries and their relatives (*Vaccinium* spp.).

Ovulinia

Causes spots on azalea petals with *Ovulitis* conidia visible. Apothecia have not been found with certainty in Britain.

Sclerotinia

S. trifoliorum is very destructive of clovers, especially *Trifolium pratense*. Apothecia are produced in the autumn; latent infection sites are formed before spread in the host tissues. *S. sclerotiorum* is probably similar but infects many hosts. Grows and infects from sclerotia which send out spreading hyphae and hyphal strands. *S. homoeocarpa* causes dollar spot disease of fine lawns.

Septotinia

Causes poplar leaf blotch in USA; this may lead to defoliation.

Stromatinia

The cause of dry rot of cultivated gladiolus (*Gladiolus* spp.). Stromata and tiny sclerotia on leaf bases. Surface of corm with small dry sunken lesions which progress in store and often kill the corms.

Sclerotium

Sclerotium cepivorum attacks cultivated and wild onions and other *Allium* spp. (Another fungus attributed to this genus in the old literature, as *S. rolfsii*, which is very destructive to many crop plants in warm damp climates, is in fact not related to the Sclerotiniaceae but is an anamorph of the genus *Corticium* in the Basidiomycota.)

Not all sclerotial ascomycetes can be associated with apothecia. *Sclerotium cepivorum*, the cause of white rot of onions, produces sclerotia that resemble those of related species, but has not itself been associated with apothecia. It attacks onions, garlic (*Allium sativum*) and to a lesser extent leeks (*A. ampeloprasum*) in cooler climates. The first symptoms are yellowing and dieback. The roots and base of the growing bulb become covered with a white mycelium, and numerous small black sclerotia are formed within the scales. If the infection has not proceeded far at harvest the decay of the onion continues in storage. The disease can be devastating, and control is currently by soil treatments to kill off the sclerotia, with varying degrees of success, and by crop rotation, to avoid the fungus. However, the ability of the sclerotia to survive for at least ten years in the soil reduces the value of rotation as a control measure. Strenuous efforts have been made to find some means of biological control. The sclerotia are prevented from germinating in natural soil by a mycostasis (inhibition of growth) induced by the saprophytic soil flora, but this suspended animation is broken if the sclerotia are placed in pure water. It is also broken by the presence of onion roots in natural, unsterile soil. The stimulus can also be provided by the alkyl sulphides, chemicals that give onion and its relatives their distinctive flavour; these apparently have no direct effect on the mycostasis but enable the germinating sclerotia to tolerate it.

Other interesting fungi with an enclosing stromatic structure reminiscent of the Sclerotiniaceae are the Rhytismataceae, characterised by dark closed fruiting structures that form on leaves and stems. There they overwinter, breaking open in spring to display an apothecium disk bearing asci. The Rhytismataceae includes an interesting group of conifer pathogens, the needle cast fungi.

One, *Lophodermium pinastri*, is mentioned in Chapter 7 as behaving as a latent pathogen in the green leaves of Scots pine (*Pinus sylvestris*). After the leaves die and fall the ascocarps appear during the winter in dead plant material before opening as apothecia to release their spores in spring. *L. seditiosum*, formerly confused with this fungus, is a serious pathogen of Scots pine and apparently has only a limited latent period, actively killing leaves, on which it then fruits.

Also included here is the common tar spot fungus of sycamore leaves, *Rhytisma acerinum* (Fig. 5.11). The heavily pigmented, shiny, melanised lesions do not kill the infected leaves. However, following leaf fall in the autumn the fungus survives the winter on the dead tissues. In spring the lesions break open through convoluted lips (Fig. 5.12) to reveal a massive green disk-shaped apothecium. The spores from the asci in the disk are shot out and carried by wind to the young leaves, where each infection begins as a chlorotic area which becomes a tar spot. Associated with the maturing ascocarps is the production

Fig. 5.11 (left) Lesions of *Rhytisma acerinum*, cause of tar spot of sycamore (*Acer pseudoplatanus*) on a young tree in Thetford Forest, Norfolk, England. (D.S. Ingram.)

Fig. 5.12 (below) Lesions of *Rhytisma acerinum*, cause of tar spot, on fallen, overwintered leaves of sycamore (*Acer pseudoplatanus*), opening to reveal the apothecia. (D.S. Ingram.)

of minute spores called microconidia, which are believed to have a sexual function. In the days when coal was the main industrial and domestic fuel, sycamore trees in cities were free of tar spot disease, the pathogen being killed on the leaf surface by the sulphur pollutants from smoke. With the institution of clean air zones and a move away from coal as a fuel, tar spot has returned to city trees. In years when extended wet weather accompanies the period of spore discharge of this fungus, leaves become very heavily spotted and shrivel, making the trees unsightly and causing early leaf fall.

In summary

This broad survey of the pathogenic Ascomycota demonstrates how a group of fungi has broken its dependence on water by producing non-motile spores and achieving a close relationship with plants. Different species have adopted necrotrophic, hemibiotrophic and biotrophic modes of nutrition, enabling the group as a whole to occupy a wide range of ecological niches. Moreover, a wide range of growth strategies is deployed, including extensive biotrophic growth beneath the cuticle, destructive ramifications of necrotrophic hyphae through the tissues, and latency. Later we shall encounter yet more strategies in two specialised groups of Ascomycota: firstly those that have evolved to inhabit the vascular system of the host (Chapter 6); and secondly the powdery mildews (Chapter 8), biotrophs that live on the surface of the plant, tapping the living cells of the epidermis by means of complex haustoria.

The greatest number of organisms that cause plant disease are either members of the phylum Ascomycota or anamorphs related to them. And yet, of the 47 orders currently considered to make up the phylum, only about 15 include plant pathogenic species. Apart from a few specialised animal parasites the rest are saprophytes. Many of these are host delimited, however, being able to grow only on one particular dead host species. This leads us to ask if many of these species share with their pathogenic relatives the ability to establish a restricted infection in particular plant species, remaining poised until the death of the host from other causes, when they grow out as saprophytes in the dead tissue. Like necrotrophs they would then have an advantage in being able to colonise the senescent material in advance of competitors.

The free spirits: the mitosporic fungi

Throughout this chapter we have frequently referred to the anamorph stages of the various members of the Ascomycota described. In addition, there are vast numbers of other fungal anamorphs that have not yet been associated with a teleomorph; some possibly never will be. They are classified for convenience as the 'Fungi imperfecti' (sometimes Deuteromycetes or mitosporic fungi) with two sub-groups, the Hyphomycetes and the Coelomycetes.

Among the Hyphomycetes, which produce conidia on conidiophores arising direct from the hyphae, are the causal organisms of a number of well-known plant diseases – *Fusarium oxysporum* and *Verticillium albo-atrum*, for example, are among the most important plant pathogens known (see Chapter 6). Other well-known disease organisms in this group are the *Alternaria* spp.: *A. brassicae* and *A. brassicicola*, for example, cause leaf spots on brassicas, and *A. solani* causes early blight of potato and tomato, destroying large areas of the infected leaves. These fungi are recognised by their dark, pear-shaped, multi-celled conidia, often arranged in chains. *Ramularia pratensis* and *R. rubella* cause

Fig. 5.13
Lesions of
Ramularia sp.
on a leaf of
dock (*Rumex*
sp.) near
Temple,
Midlothian,
Scotland.
(D.S. Ingram.)

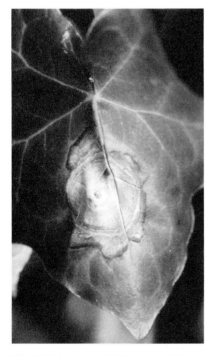

Fig. 5.14 A lesion of *Phyllosticta hedericola* on ivy (*Hedera* sp.) in Edinburgh, Scotland. (D.S. Ingram.)

characteristic purple-edged spots on the leaves of sorrels and docks (*Rumex* spp.) (Fig. 5.13), with the white conidia being produced on the lower surfaces of the lesions.

Members of the Coelomycetes, in contrast, form their conidia in pycnidia or acervuli. The pycnidia are dark-walled structures embedded in the host tissue. They are nearly spherical, with a domed cap at the leaf surface; the conidia are released through a central pore. The acervuli are mats of fungal hyphae on which conidia are borne in a mass, initially within the host tissue.

Pycnidial genera include *Phoma* and *Septoria*, necrotrophic leaf-spotting pathogens of a great many plant species. The conidia are single-celled in *Phoma* and elongated with septa in *Septoria*. Familiar pathogenic species are *P. exigua*, which infects potato, its f.sp. *foveata* causing destructive lesions on tubers in store, and *S. apiicola*, which causes a seed-borne blight of celery (*Apium graveolens*). Sometimes the leaf spots show concentric rings of necrotic tissue, as with *Coniothyrium hellebori* on Christmas rose (*Helleborus* spp.) (Plate 3) and *Phyllosticta hedericola* on ivy (*Hedera* spp.) (Fig. 5.14). Such lesions are often called target spots and are useful for diagnostic purposes but the mechanism for their production is not fully understood. Observations to determine whether the zones are related to periods of light or darkness, temperature changes or an interaction between the pathogen and host might be very rewarding. Where pycnidial species have been linked to teleomorphs the genera of the Ascomycota involved have included *Mycosphaerella, Didymella* and *Leptosphaeria.*

Acervular forms include the pathogenic genus *Colletotrichum.* This produces sunken sharply defined lesions on leaves and pods, referred to as anthracnose. This symptom is well seen with *C. lindemuthianum,* cause of anthracnose of French bean and other phaseolus beans. The lesions on pods spread to infect the seeds, which then become the main source of infection from one generation of the host to another.

Although the taxonomy of the Hyphomycetes and Coelomycetes has been well studied, the biology of the diseases they cause, especially on wild plants, has not, and there is an open field of investigation here for the amateur naturalist.

6

Wilt

The plant's vascular system

We turn now to a diverse group of specialised pathogens adapted to grow in the water-transporting system of their hosts. Just as humans suffer from afflictions of the arteries, veins and lymph glands, parasitic invasions of the conducting system in the plant may lead to disease and death, collectively known as wilt. The comparison cannot be taken too far, however, for the conducting system of plants is very different in structure and function from that of humans.

In the plant the water and dissolved salts from the soil and some metabolites from the roots are carried, generally in an upward direction, in the xylem tissue. This is largely composed of elongated cells strengthened by the polymer lignin and joined end to end by perforated end walls. Perforated plates also occur in the side walls and connect these vessels laterally. There are also less precisely defined, but very important, lateral connections through pits where a permeable membrane forms the floor of the pit, matched by a similar structure in the wall of the adjacent cell. Where lignified cells have solid end walls and are only connected through pits they are described as tracheids. The functioning xylem in flowering plants consists of the dead cells of the vessels and accompanying tracheids, together with lignified fibres. The whole mass is surrounded by living cells and is in contact with them through the pits. In some less advanced plants such as conifers there are no vessels, only tracheids, which are packed close together and connected laterally by valvular structures, called bordered pits.

Water molecules adhere to one another with great tenacity, making it possible for the columns of water in the xylem to be pulled up through the plant by the negative pressure generated as water is lost by transpiration through the stomatal pores of the leaf.

Close to the xylem and usually external to it there is another transport system, the phloem. This consists of elongated cells with cellulose walls and some part of the cell cytoplasm remaining. Each has a living companion cell. The phloem vessels carry complex organic materials such as sugars and hormones from the leaves and young stems to other areas of the plant body. The energy for this process is provided by the metabolic processes of the phloem cells themselves. The xylem and phloem tissues are arranged in discrete bundles in herbaceous plants, but in woody plants the xylem is massed together to form a solid column with a thin layer of phloem surrounding it.

The xylem and phloem tissues are linked and interpenetrated by radial plates of living cells, the rays. The two tissues are separated by a layer of dividing cells, the cambium, which produces xylem cells on its inner face and phloem cells on its outer face.

What causes wilt?

Strictly speaking, wilt diseases are caused by fungi or bacteria that invade the xylem vessels and grow in them, perturbing the water transportation system and leading ultimately to an irrecoverable loss of internal pressure (turgor) in the cells of the leaves and young stems. These droop, turn yellow and eventually die.

Some misconceptions had to be overcome before it was possible to think clearly about the cause of these symptoms. For example, it was postulated on false evidence that wilt fungi, on gaining entry to the vessels of the plant, produce materials that cause excessive transpiration and consequent wilt. Much evidence to the contrary has now been obtained and it is clear that the water loss from the plant is in fact reduced by the invasion of the pathogen.

Also, it was thought that the vessels of the plant, being inert and without cytoplasm, were not capable of normal tissue reactions to invaders. Indeed, early workers thought of the environment inside the vessels as approximating to that in the moisture surrounding the particles of soil where the wilt fungi grew as saprophytes. The consequence was the postulate that the pathogens were safely cocooned in an artificial soil solution in the xylem, and that their metabolites alone were responsible for the wilt symptoms. In a sense this is true, but further study of the xylem showed that the living cells could be activated through the pit connections, and that host tissue reactions to the presence of the fungus could lead to the release of materials that contributed to the wilt.

The nature of wilt is still a favourite topic for argument among plant pathologists. The following are just some of the explanations that have been advanced in recent years to explain the symptoms of wilt diseases (Fig. 6.1).

(i) Pathogen damage to the roots cutting off the water supply to the shoots;

(ii) blocking of the xylem vessels by pathogen cells or spores, or by gums produced by the pathogen or released from host cell walls by enzyme activity;

(iii) the formation of tyloses – balloons of membrane-bound cytoplasm from adjacent living cells projecting into the xylem vessels through pits weakened by enzyme activity;

(iv) embolism – gas bubbles released or drawn into damaged xylem vessels, thus breaking the water columns;

(v) crushing of the smaller vessels by hormone-induced expansion of adjacent cells;

(vi) production by the pathogen of toxins that poison and affect the functioning of leaf cells.

In fact there is virtually no evidence of injury to the roots or of a pathological reduction of their functional area when infected by true wilt fungi. Microscopic analysis shows little evidence of any massive growth of the fungi in the xylem. The normal situation appears to be a limited growth of hyphae in the vessels, with the release of small bud spores that spread in the vessels but are to some degree restrained by the perforated end plates. However, there is some evidence from experiments with lucerne infected artificially with *Verticillium albo-atrum* that the leaves wilt sequentially, from bottom to top of the plant, in step with the spread of the fungus in the vascular system.

The evidence for blockage of the vessels by materials that might be the products of the dissolution of the pit walls by pectic enzymes is put in doubt by the fact that some fungi which lack the capacity to produce the major pectic enzymes can nevertheless induce some of the symptoms of wilt. Other structures and materials that can be found by microscopic examination of infected plants, such as tyloses, emboli (gas bubbles), pathogen-produced gums, crushed vessels and phenolic materials undoubtedly play some part in the wilt syndrome and may help to impair the transpiration flow, especially in the

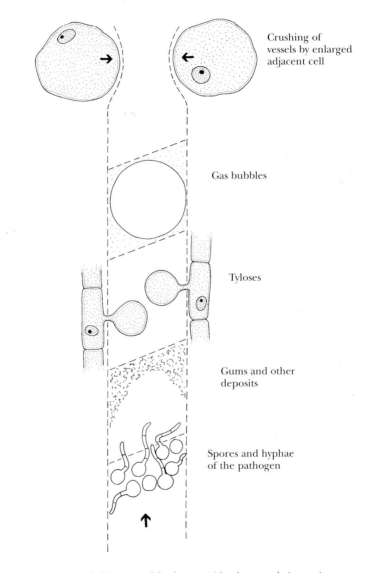

Crushing of
vessels by enlarged
adjacent cell

Gas bubbles

Tyloses

Gums and other
deposits

Spores and hyphae
of the pathogen

Fig. 6.1 Diagram of a xylem vessel with some of the factors said to have a role in causing symptoms of wilt diseases. (Drawn by Mary Bates.)

leaves. It is not clear, however, whether these are materials produced by the plant in response to any form of invasion, perhaps in an attempt to limit spread of the invader, or whether they are specific components of the wilt syndrome.

Finally, the evidence for the production of wilt-inducing toxins by the fungi themselves began with the demonstration that two substances, lycomarasmin and fusaric acid, could be produced in culture by the *Fusarium* spp. responsible for tomato wilt, and that these could be introduced into the stems of healthy cuttings to produce wilt. But the production of material in the artificial conditions of culture is not evidence for its production in the plant and it has proved difficult to identify such materials in infected plants. Nevertheless, evidence is now emerging to implicate pathotoxins in some major wilt diseases, such as Dutch elm disease caused by *Ophiostoma* (*Ceratocystis*) *ulmi*. Moreover, there is good evidence that overproduction of hormones plays a role in causing some wilt symptoms, including the downward bending of leaves (epinasty), root production by stem cells and senescence.

The situation seems to be that all the factors postulated as being the cause of wilt combine variously in different wilt diseases to produce the same terminal symptoms of irrecoverably drooping leaves.

Some suggest that the plant's defence reactions (see Chapter 3) play a significant part in the development of symptoms. Wilt fungi, they say, are probably recognised only slowly by the plant, and are therefore capable of spreading a considerable distance in the xylem, as far as the ultimate leaflets, before the defence reactions are fully activated. It is possible that it is the delayed deployment of the defence reactions by living cells associated with the xylem, including the formation of tyloses and substances that block vessels and limit pathogen growth, that finally causes the blockage of the smaller vessels and the development of wilt.

A sidelight on this theory arises from some very ingenious experiments in which tomato plants were simultaneously infected with inoculum of *Fusarium oxysporum* f.sp. *lycopersici* (a wilt-inducing pathogen of tomato) and *F. oxysporum* f.sp. *pisi* (a wilt-inducing pathogen of pea) in differing proportions. Full wilt developed with f.sp. *lycopersici* inoculum and no wilt with the f.sp. *pisi* inoculum. Mixtures showed marked reduction of the symptoms and at 4:1 and above of f.sp. *pisi* to f.sp. *lycopersici* the latter's pathogenicity was suppressed. The effect was not obtained with heat-killed spores of f.sp. *pisi*. This suggests that the control of pathogenicity can be exercised in the vessels of the tomato plant in the presence of living pathogens.

Classic wilt diseases

Classic wilt (or true wilt) symptoms may be produced by a variety of pathogenic organisms: occasionally by mycoplasmas such as the phloem yellows of American elms (*Ulmus americana*) in the USA; by bacteria such as *Pseudomonas solanacearum*, cause of brown rot of potatoes and the moko disease of bananas, and by a variety of fungi, of which *Fusarium oxysporum* and *Verticillium albo-atrum* are the most important. The classic symptoms of wilt were first identified in tomatoes in 1895 by George Massee, a mycologist at Kew, and were attributed to attack by both species which were not clearly separated. Both *Fusarium* and *Verticillium* are imperfect fungi and are probably the anamorphic phases of members of the Ascomycota, since morphologically similar forms can be linked to genera in the Hypocreales (see Table 1.1, p.21).

Plate 1 Phytophthora infestans, cause of late blight of potato

(a) Leaf lesion with a necrotic centre; note the white 'haze' of sporangiophores and sporangia on the green tissue around the margin of the lesion.

(b) Infected tuber; note the foxy red discolouration of the infected cells.

(c) Potato plot with a primary focus of infection in the top left hand corner.

(d) The same plot ten days later; spores have spread from the primary focus to infect most of the plants.

Plate 2 Zoosporic pathogens

(a) *Pythium ultimum* causing damping-off in a pot of lettuce (*Lactuca sativa*) seedlings.

(b) *Phytophthora* sp. (probably *cactorum*) causing partial death of Japanese maple (*Acer palmatum*) at Benmore, Argyll.

(c) *Bremia lactucae* causing downy mildew of lettuce; note that the infected tissues between the veins bear white spores and are quite green.

(d) *Albugo candida* causing white blister on shepherd's purse (*Capsella bursa-pastoris*) at Gullane, East Lothian; note the masses of white spores and the distortion of the infected flower stalks.

Plate 3 Pathogens spread by conidia or ascospores

(a) *Botryotinia fuckeliana* (= *Botrytis cinerea*) causing grey mould on *Kalanchoe blossfeldiana*; note the grey 'haze' of conidiophores and conidia on the rotted tissues.

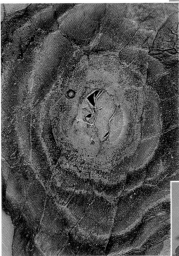

(b) *Coniothyrium hellebori* causing target spot on a leaf of Christmas rose (*Helleborus niger*).

(c) *Taphrina deformans* causing 'peach' leaf curl on almond (*Prunus amygdalus*); note the distortion and red coloration of the infected leaves.

(d) An ergot of *Claviceps purpurea*, collected from sea lyme grass (*Elymus arenarius*), germinating to produce drum-stick perithecial stromas.

Plate 4 Pathogens spread by conidia or ascospores continued

(a) *Epichloe typhina* causing choke of fescue (*Festuca* sp.) near Kilmartin, Argyll.

(b) *Venturia inaequalis* causing scab of crab apple (*Malus* sp.) in Aberlady, East Lothian.

(d) Galleries of elm bark beetle

(c) Early signs of disease on one of the elms (*Ulmus* sp.) along the 'Backs' in Cambridge in 1975. Within weeks this magnificent tree was dead.

Plate 5 Powdery Mildews

(a) *Erysiphe pisi* on pea (*Pisum sativum*) in Edinburgh; note the whitish, powdery covering of hyphae and conidia on the green leaves and stems.

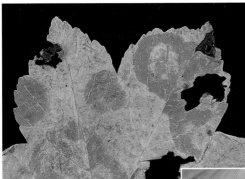

(b) Green islands induced by colonies of *Uncinula bicornis* on sycamore (*Acer pseudoplatanus*).

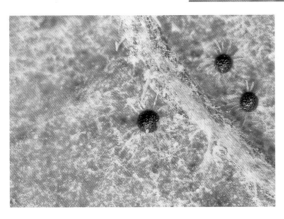

(c) *Sphaerotheca mors-uvae* on gooseberry (*Ribes uva-crispa*); note the masses of brownish hyphae and conidia on the surface of the green fruit and leaves.

(d) Magnified ascocarps (cleistothecia) and surface hyphae of rhododendron powdery mildew ('*Microsphaera* type').

Plate 6 Rusts

(a) Honey-coloured, sweet-smelling pycnia of *Puccinia punctiformis* (creeping thistle rust) on elongated, paler green leaves of *Cirsium arvense*.

(b) Orange aecia of *Puccinia* sp. (probably *caricina*) on the surface of nettle (*Urtica dioica*); note the distortion of the infected tissues.

(c) Lesions with concentric rings of urediospores of *Puccinia graminis* f. sp. *tritici* (black stem rust) on wheat.

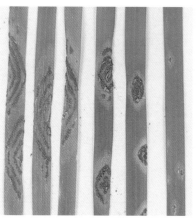

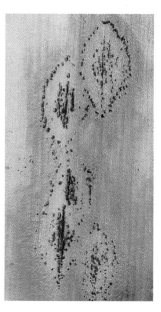

(d) Teliospores of *Uromyces muscari*, together with green islands, on English bluebell or wild hyacinth (*Hyacinthoides non-scripta*).

Plate 7 Rusts continued and Smuts

(a) Telial horns of *Gymnosporangium cornutum* on *Juniperus communis* near Walkerburn, Peeblesshire.

(b) Pointed aecia of *G. cornutum* on a leaflet of rowan (*Sorbus aucuparia*), the alternate host.

(c) Black ustilospores of *Ustilago violacea* on the anthers of red campion (*Silene dioica*).

(d) White bud spores (= conidia) produced by budding of hyphae of *Urocystis primulicola* infecting the anthers of primrose (*Primula vulgaris*).

Plate 8 Wood-rots

(a) *Heterobasidion annosum* causing death of young pine trees (*Pinus* sp.) being shown to students by the late John Rishbeth in Thetford Forest, Norfolk.

(b) Basidiocarps of *H. annosum* pushing up through Sitka spruce leaf litter at Juniper Bank, Peeblesshire.

(c) *Meripilus giganteus* basidiocarps growing from infected roots of beech (*Fagus sylvatica*) in Cambridge.

(d) *Ganoderma applanatum* basidiocarps growing from a mature beech tree; note the masses of rusty brown basidiospores.

The diseases produced by *Fusarium* and *Verticillium* are hard to distinguish by visual symptoms alone, but with *Fusarium* the first signs are seen in the young upper leaves, where the veins become translucent, followed by downward bending (epinasty) of some lower leaves. As the disease takes hold all the leaves lose turgor, soften and droop, and no recovery results when the plants are watered. In plants infected by *Verticillium* and other pathogens there may be a distinct succession of leaf symptoms, with yellowing and wilting developing from the bottom to the top of the plant.

The genus *Fusarium* is easily recognised in artificial culture by its production of two types of spore: the single-celled microconidia, which are produced on small conidiophores, either very widely over the colony surface or in well marked areas where the conidiophores are grouped, and the macroconidia, which are elongate, multicellular and crescent-shaped, and are produced most often on groups of small conidiophores (Fig. 6.2). As the colonies age many of the *Fusarium* species produce pigments, some of which change colour across the pH spectrum. In addition, some species, especially *F. oxysporum*, are sensitive to light, producing microconidia and diffusible red-purple (napthaquinone) pigments in the dark and macroconidia and orange carotenoid pigments in the light.

The difference between the production of microconidia and macroconidia appears to lie in the nature of the wall of the spores as they are budded off from the conidiophores. The microconidia round off rapidly to produce a chain of separate cells, while the macrospore wall remains intact and elongates without rounding off into separate cells. It is not difficult to imagine the problems of the taxonomist faced with such environmentally-induced variation. *F. oxysporum* also produces thick-walled asexual resting spores, called chlamydospores (Fig. 6.2), by conversion of sections of hyphae in the mycelium, and these are very important for the survival of the organism in adverse conditions.

The genus *Verticillium* is not so varied morphologically. Two very similar species, *V. albo-atrum* and *V. dahliae*, are the causes of many wilts in horticulture and agriculture. Both produce colourless conidiophores, like tiny, slender elongated hock bottles, in whorls. The spores collect in drops of mucilage at

Fig. 6.2 Conidia of *Fusarium oxysporum*: (a) macroconidia (approximately 30 μm long); (b) microconidia (approximately 10–15 μm long); and (c) chlamydospores with thick walls (approximately 10 μm diameter). (Drawn by Mary Bates after the late E.W. Buxton.)

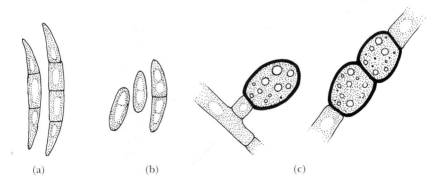

(a)   (b)   (c)

the tips of the conidiophores. These can be seen in cultures or can be induced to form by placing slices of infected stem in damp conditions for a day or so. The conidiophore with its mass of conidia makes a very elegant object under the microscope. *V. dahliae* differs in that it produces tiny sclerotia (called microsclerotia) and therefore has greater powers of survival in the soil than *V. albo-atrum*. The two species also differ physiologically: *V. albo-atrum* causes disease at temperatures up to 24°C while *V. dahliae* will cause disease at temperatures as high as 30°C.

Fusarium and *Verticillium* spp. are soil fungi with the capacity to persist in the soil, both in plant debris and as separate resting structures in the absence of a host. There is also some evidence for limited growth in dying root cells. As a root grows it sloughs off cells from the growing apex; just behind the apex ephemeral root hairs are formed, function and die; and further back still the outer cortex begins to collapse, leaving an inner impervious layer around the vascular tissues and an outer layer of dying cells. All these dying cells are potential food sources for facultative parasites, facilitating the saprotrophic survival of the pathogen, although our actual knowledge of their relative importance is small. Where a soil becomes infested by a pathogenic wilt fungus it remains so for many years, and susceptible plants grown in such infested soil invariably become wilted and die. The problem has been largely overcome in the glasshouse industry by the development of partially sterilised composts, artificial composts and hydroponic growing methods.

Fusarium oxysporum causes disease in a great many different hosts. The fungus attacking each host is a *forma specialis*: for example, f.sp. *lycopersici* – tomato; f.sp. *pisi* – peas; f.sp. *cubense* – banana; and f.sp. *vasinfectum* – cotton (*Gossypium* spp.). Altogether more than 75 common forms have been identified. The pathogenic forms can be isolated from diseased hosts and from soil where the hosts have grown. Non-pathogenic forms (or at least forms to which no pathogenicity can be assigned) can be isolated from almost all soils.

The life cycle of *Fusarium*

We do not know the whole life history of *Fusarium oxysporum* but, piecing together the information we have, it seems that chlamydospores are found in most soils and are stimulated to germinate in the presence of the appropriate host. In the wild they may germinate and enter the roots of a wide variety of hosts, where they set up a low level pathogenic relationship, or they may germinate and grow on root debris.

Although *Fusarium* appears to lack a sexual phase in its life cycle, we know that the hyphae of different strains are able to fuse and exchange haploid nuclei which can coexist in one mycelium. Nuclei of different origin may fuse within the mycelium and exchange chromosomal material as occurs in sexual reproduction. The normal haploid state is then restored by sequential loss of chromosomes. This 'parasexual' recombination (see p.29) generates continual variation in morphology and pathogenicity. Thus a soil population of *F. oxysporum* can produce more specialised forms to exploit the presence of a particular crop, either because the appropriate nuclei already exist in the soil population in small numbers, or because they appear in the genetic recombination of the parasexual cycle. This is a wonderful example of a passive lifestyle that promotes an active parasitism.

Is wilt a misnomer?

We have already described the classic wilt symptoms such as are induced in tomato by *Fusarium* and *Verticillium* spp., but when we dig a little deeper the picture is not quite so simple. Across the range of hosts the *formae speciales* of *F. oxysporum* show a wide variety of symptoms, only some of which may be attributed to true wilt. For example, a new *forma specialis* of *F. oxysporum* has appeared which causes 'foot rot' in various intensive growing systems, including hydroponic systems. The fungus enters the roots and lower stem and causes a decay of the cortex and subsequent collapse of the plant, but does not invade the xylem and does not cause wilt. A small genetic shift has produced a root rot from a wilt; perhaps the shift can take place in the opposite direction too?

When f.sp. *pisi* attacks peas the main symptoms are yellowing of the leaves and weakening of the plant. In some attacks a separate organism, *F. solani*, accompanies the *F. oxysporum* to cause a foot rot, and the yellowing symptoms are intensified. This double infection is referred to as St John's disease, because it regularly appears on or about the 24 June, St John's day.

In attacks by f.sp. *gladioli*, cause of gladiolus yellows, the typical symptoms are yellow striping of the leaves and early death of the plant, combined with a discoloration of the vascular system in the corm. Other strains of the fungus invade the corm through the roots without inducing yellowing symptoms. Then, during winter storage, a rot develops in the corm base or on the corm surface, in both cases commencing in the vascular tissue. Some cultivars of gladiolus exhibit only yellowing symptoms, while others exhibit only corm rot. Moreover, to complicate things still further, there is evidence that environmental factors, particularly temperature, are important in controlling the expression of the yellows symptoms. So we have a three way interaction between variation in the fungus, variation in the host and variation in the environment, which determines the degree of infection and the nature of the symptoms expressed. At one extreme the gladiolus yellows symptoms can be equated with wilt, but at the other the rotting phase is a long way from what one would expect of a typical wilt.

The problem of defining wilt symptoms appears again in the Panama disease of bananas caused by *F. oxysporum* f.sp. *cubense*. The fungus is found in the xylem vessels, which turn brown following invasion, and the symptoms first show as yellow patches on the lower leaves, which then spread to the upper leaves, sometimes accompanied by death of the leaves.

This is one of the most significant diseases of the world. In the early part of the twentieth century it caused major economic loss in the Caribbean area, with socioeconomic consequences as devastating as those caused by potato blight in western Europe. The banana has a long history and spread widely in the tropics before it became the subject of large scale cultivation in the twentieth century. However, wherever it was introduced it took *F. oxysporum* f.sp. *cubense* with it, and as the world trade in bananas developed, the disease became important in individual plantations, which quickly became saturated with the fungus and had to be abandoned. A series of Caribbean islands developed a banana industry and then lost it as the fungus spread.

The cultivated banana is a triploid (i.e. it has three sets of chromosomes instead of the normal two) and is a selection of the wild *Musa acuminata*. It occurs as a number of varieties, but the high quality, highly wilt-susceptible

'Gros Michel' became almost universally planted over the period of greatest expansion; not only was it a fruit of quality but its ripening could be controlled, making shipping relatively straightforward. When it was eventually replaced by the wilt-resistant 'Cavendish' variety (which occurs in a number of forms) new storage, shipping and ripening procedures had to be established. The 'Cavendish' variety is resistant to the strain of the pathogen that infects 'Gros Michel', and although it is attacked by another strain under subtropical conditions that strain has not yet spread to the tropics. Breeding programmes to produce resistant bananas are beset with the difficulty of working with a triploid that does not usually produce seed, and of restoring seedlessness at the end of a programme of crossing with seeded forms. One (very hard) seed in a commercial banana could have unfortunate consequences for the teeth of its consumer!

We could continue listing the various forms of *Fusarium oxysporum*, all of which induce a wide range of symptoms in the plants they attack. Some cause classic wilting, some are associated with quite disparate fungi and induce symptoms that are a result of the double infection, some lead to yellowing of the leaves, some induce these symptoms together with tissue rotting and some cause a straightforward foot rot, all originating from special forms of the same fungus. It is thus impossible for anyone except a greatly experienced pathologist to diagnose a wilt disease on symptoms alone. Confirmation, however, is easily obtained by culturing the organism from discoloured vascular tissue. It is also clear that wilt is a term that accurately describes only one particular manifestation of the effects of invasion by *F. oxysporum*.

The special features of *Verticillium* wilt

Compared with *Fusarium*, *Verticillium* is apparently a less complicated organism. *V. albo-atrum* is found most commonly in temperate regions, whereas *V. dahliae*, which grows at higher temperatures, is the commoner in warmer climates (*F. oxysporum* also infects at high soil temperatures). *Verticillium* shows less variation in general than *Fusarium*.

In temperate regions there is a limited number of important diseases attributable to *V. albo-atrum*. The *Verticillium* disease in tomatoes, which has similarities to *Fusarium* wilt of the same host, has been mentioned already. Another is the hop wilt, which is very damaging because once the soil of a hop 'garden' is infested the pathogen cannot be eliminated without a major fumigation programme. The disease on hops is caused by two forms of the pathogen: a so-called fluctuating strain, which causes mild symptoms, and a virulent form causing progressive wilt, which kills out plants and destroys hop plantations. The symptoms start as yellowing of the leaves and lead to eventual necrosis and death of the shoots. Resistant or tolerant varieties of the crop can be bred, but because of the esoteric qualities of hops in the art of the brewing industry the replacement of established varieties is not easy.

V. albo-atrum and *V. dahliae* also appear as sporadic pathogens of small trees in gardens. Maples and their relatives (*Acer* spp.) are particularly susceptible; the whole tree may wilt and collapse in a very short time, or single branches may die while the rest of the plant recovers.

V. dahliae is the species important in tropical and subtropical areas. Because of its ability to produce microsclerotia it is much more persistent in infested soil than *V. albo-atrum*. The latter disappears under a grass or fallow regime after two or three years, while *V. dahliae* remains for many years. *V. dahliae* is a cause of wilt in many tropical crops, including cotton, sunflower (*Helianthus* spp.) and tobacco. Isolates preferentially infect the host species from which they were obtained, causing wilt and also yellowing and necrosis of the leaves. This host specificity is not, however, clearly defined. Experimental work has shown that the microsclerotia can germinate and colonise portions of the roots of many wild species or crops other than that from which the isolate was obtained. They do not appear to extend beyond the dying cortex of such infected plants, but grow well enough to produce new microsclerotia and renew the stock of resting propagules in the soil. Some strains appear to produce no visible symptoms whatsoever in the plants they infect and are said to be latent, but how this is achieved is not known (see Chapter 7 for a discussion of latency). Others cause clear wilt symptoms to develop. Anomalies of the pathogen such as these make the development of a general theory of pathogenicity for *Verticillium* very challenging.

Resistance to infection

It might be thought that the many variations in the patterns of infection and the range of ecological behaviour, including saprotrophism, latency, yellows and full wilt shown by *Fusarium* and *Verticillium* species would preclude the use of resistance in disease control. This is not the case. Probably the first development of resistance against wilt disease was by the famous American plant pathologist L.R. Jones who, when faced with a yellows disease of cabbage caused by *F. oxysporum* f.sp. *conglutinans*, identified individuals in the cabbage population with resistance controlled by multiple genes. These were then bred into new varieties. Later J.C. Walker developed inbreeding techniques by bud pollination and isolated single dominant genes which conferred complete resistance in brassicas.

In tomatoes, resistance to *F. oxysporum* under complex genetic control was selected from commercial populations. It proved useful but rather imprecise and its expression was clearly influenced by environmental conditions, especially temperature. Single gene resistance was then incorporated from *Lycopersicon pimpinellifolium*, the currant tomato, and proved highly effective. It remained stable for many years. Although the original resistant varieties have now become susceptible to new races of the pathogen, stable resistance to these races has been incorporated into new varieties. A similar tale can be told about the control of *Verticillium albo-atrum* and *V. dahliae* in tomatoes.

Gardeners may remember that annual asters (*Callistephus chinensis*) were once exceedingly susceptible to attack by *Fusarium oxysporum* (aster wilt), but that seed of resistant selections has been on sale in catalogues for some years.

The stability of the single gene resistance to wilt, in contrast with the unstable nature of single gene resistance to potato blight and the cereal rusts, may be due to the fact that the wilt fungi are soil organisms. As such they are restricted in space and perhaps are less able to distribute genetic variants carrying new pathogenicity factors over a wide area, thereby slowing the process of change.

Wilt organisms in the soil

Here we have fungi that survive in the soil as a mixture of saprophytic, low grade pathogenic and highly pathogenic forms, usually in the form of chlamydospores. Populations of nuclei specifying these characteristics multiply in the population in response to the availability of suitable host material. In certain instances this will be a host that can be invaded and parasitised, as occurred when a new strain of Panama disease appeared in Florida in a situation where no importation of the pathogen had taken place. In addition, new genes for virulence and avirulence in the fungi are constantly evolving, matched by a genetic flux of evolving genes for susceptibility and resistance in the potential hosts. Thus, when we observe plasticity in the parasitism of the wilt fungi we are seeing an expression of the ability of the fungus to store, deploy and elaborate new genetic forms.

Another strand to the pathogenicity of *Fusarium* spp. in soils is the observation that some soils appear to be able to suppress the development of pathogenic forms of *F. oxysporum* and others seem to encourage them. These are referred to as suppressive and conducive soils respectively. Although the behaviour of *Fusarium* spp. in the soil is affected by the physical nature of the soil, the suppressive effect seems to be determined by biological factors. Experiments have shown that the suppressiveness can be transferred to other soils. Also, non-pathogenic forms of *F. oxysporum* in the soil do not appear to be affected by the suppressive soil in the same way as pathogenic forms, suggesting that the competition for sites within the soil between the non-pathogenic and the pathogenic forms may be important. A practical consequence of this situation is that large areas of land have been identified for tomatoes in France and California, and for cotton in India, where *Fusarium* wilt is unlikely to become a problem.

The Great wilt – Dutch elm disease: the first epidemic

Unlike *Fusarium* and *Verticillium*, *Ophiostoma* (*Ceratocystis*) *ulmi**, cause of Dutch elm disease and perhaps the most spectacular of the classic wilt diseases (Plate 4), has a well-marked sexual stage. Infected trees appeared in western Europe at the end of the First World War, first in France and then in neighbouring countries in quick succession. The pathogen was first isolated and identified in Holland in 1920. It appeared in Britain in 1927, at which time it was also spreading south to Italy and east as far as Uzbekistan, which it reached by 1939. The fungus first appeared in the USA in 1930, in Ohio, but did not immediately establish itself there. Over the next few years it was observed at various points in the USA and in Canada, always associated with timber yards where elms for veneer production had been imported. It finally established a secure footing in New York State in 1933, from where it spread across the continent during the next fifty years, wherever native elms were to be found or exotic elms had been planted.

The reason for the notoriety of Dutch elm disease must be sought in the relationship between man and the elm tree (*Ulmus* spp.). There are many traditional uses of elm wood, such as the making of coffins and wheelbarrows, because of its resistance to splitting and decay. In addition, highly decorative veneers may be cut from the large burrs that grow on the trunks. The bark was

*Taxonomists differ in their treatment of these genera. Some distinguish *Ophiostoma* by its conidia, externally produced on the hypha-like conidiophores (as in *Pesotum*), from *Ceratocystis*, with its conidia produced internally within the terminal cells of the hyphae.

used by the American Indians to construct shelters, and slippery elm bark from
U. rubra in North America is used in alternative medicine. There is some indication that the leaves of elms were used for feeding animals in times of shortage of fodder, and that the inner bark of elms was eaten by human beings in
times of famine. It was the use of elms as planting material, however, for
hedges, for windbreaks and as specimen trees in man-made landscapes, that
brought Dutch elm disease into prominence. The English elm (*U. procera*) is a
well marked variant (probably a clone) derived from the complex European
field elm (*U. carpinifolia*), and was valued for its part in forming hedges and as
mature trees to give shade around fields, in parkland and in towns. It was such
a characteristic feature of the English countryside and country town that its
impairment by disease was a threat to the English visual landscape heritage.
The same can be said of other forms of elm in mainland Europe.

In America, the American elm (*U. americana*) was similarly widely used for
street planting and as a shade tree in gardens, and when these elms disap-
peared their loss was deeply felt. The situation in the mid-west was particularly severe, since the elm yellows, a disease caused by a phytoplasma (a very primitive organism related to bacteria) had already killed some trees, creating
favourable sites in which the bark beetles, which carried *Ophiostoma*
(*Ceratocystis*) *ulmi,* were able to breed. They were thus present in large numbers
to spread Dutch elm disease when it came on the scene. Not only was it heart-
breaking to see whole streets lined by dead and dying trees, but the financial
problems of removing and replacing them were acute.

The relatively resistant wych elm (*Ulmus glabra*) is the commoner tree in
northern Britain and northern Europe, where the spread of Dutch elm disease
has been slower. It forms a hybrid with *U. carpinifolia,* the elegant *U.* x *hollandica,* which is extensively planted in mainland Europe.

The impact of the disease during this first epidemic was different in Europe
and America. The European elms, although consisting of a number of separate genetic groups, showed little variation within these groups because most
of them formed suckers that were used for hedging or creating larger plantings. Thus one or a limited number of genetic stocks was multiplied over large
areas. In Europe the impact of the disease was severe enough, for trees large
and small died, but many trees showed infection and did not die. It was quite
common to see mature elms with symptoms on one section of the crown which
did not apparently progress, the tree remaining alive. The numerical data on
the losses are difficult to interpret, partly because of irregular and inconsistent
records and partly because the epidemic spans the second world war, when
trees of all sorts were cut down for war purposes with few records kept.
Nevertheless it is possible to say that the disease was worse in mainland Europe
than in Britain; that in Britain on the whole the English elm was more susceptible in the field than the wych elm; and that the Huntingdon elm, a clone of
U. x *hollandica* named 'Vegeta', showed considerable resistance. T.R. Pearce,
who studied the problem for the UK Forestry Commission, estimated that
between 1927 and 1960, 10–20% of the British elm population was killed, but
that by 1960 new infections were becoming much less frequent, suggesting that
a relatively mild epidemic was coming to an end.

The American elms are genetically homogeneous and much more susceptible, so the progress of the epidemic in terms of infection and death followed
a steeper curve than in Europe. But the behaviour of the disease varied across

the United States and Canada; on the eastern seaboard it could be equated with that in Europe but in Illinois, probably because of the presence of dead trees as hosts for the bark beetles, the infection and death rates were much higher.

The process of infection and spread

Dutch elm disease is spread by bark beetles (Scolytidae). Three of these are important in Europe: *Scolytus scolytus*, *S. laevis* and S. *multistriatus*. Two are important in America: *S. multistriatus* (introduced from Europe in 1900) and *Hylurgopinus rufipes*, the native American bark beetle. In all these species the adult female lays her eggs in galleries which she excavates in the soft, living region of the bark, the general pattern being one main gallery with shorter side galleries (Plate 4). Weakened or dying trees are favoured. The larvae that hatch from the eggs burrow sideways into the phloem, feeding on it and creating tertiary short, narrow galleries.

In infected trees these galleries become lined with the mycelium of *Ophiostoma* (*Ceratocystis*) *ulmi*, which produces anamorphic reproductive structures placed in the genus *Pesotum*. They consist of bundles of conidiophores, from the apex of which single celled conidia are produced in a drop of sticky fluid. The bundles look like small black hairs to the naked eye. Occasionally ascocarps also form as dark, long-necked perithecia containing asci, which normally break down so that the ascospores extrude in a sticky fluid from the tops of the long necks. The emerging adult beetles become covered with the conidia and ascospores of the fungus. When they fly to feed on the young green shoots before flying off to mate and reproduce they inadvertently transfer fungal spores to the xylem of the new host tree and set up an infection there. Spread of the infection within the tree is rapid in an upward direction; spores artificially introduced at a lower level have moved as much as 10 m in an upward direction in three hours. In Europe the largest bark beetle, *Scolytus scolytus*, appears to be the most successful vector, but in the USA in warm moist conditions the small beetle *S. multistriatus* is a highly effective carrier.

Once the fungus has made initial entry into the xylem it grows and spreads until it has made sufficient inroads into the main trunk, when wilting follows. The disease is a classic wilt, with discoloration of the xylem and plugging of the vessels with gums, tyloses and gas emboli which break the water column. A toxin, cerato-ulmin, is also produced. Just as in *Fusarium* spp., toxins and enzymes have been isolated and used to bring about wilt symptoms artificially, but as in that fungus our knowledge of the process is not secure. However, there is clear evidence that the toxin cerato-ulmin is produced much more abundantly in the aggressive form of the fungus we discuss next than in the strain that caused the first epidemic. The wilt symptoms begin with a patch of yellow in the crown of the tree, followed by a bronzing of neighbouring leaves. There may be no further progress that year but in the following season the whole crown wilts in midsummer and the leaves rapidly turn brown and gradually disappear from the branches, leaving the bare skeleton of the tree. Where resistance is present, for example in some of the Asian elms (e.g. *Ulmus pumila*), the fungus spreads for a certain distance in the vessels and then appears to be walled in and unable to spread further.

The second epidemic – devastation on a grand scale in Europe

By the 1950s in Europe and some ten years later in the USA, Dutch elm disease seemed to have run its course. In the American mid-west, most of the available trees had been attacked and killed, so the epidemic was declining because of this reduction in host numbers. That can hardly have been the case in eastern America and Europe. Perhaps the slowing of the epidemic spread in these cases can be attributed to the so-called 'd-factors' discovered in the fungal cytoplasm, virus-like parasites of *Ophiostoma* (*Ceratocystis*) *ulmi* that debilitate the fungus.

Whatever the reason for the decline, the comfortable conclusion was reached by plant pathologists studying Dutch elm disease that they could prevent its further spread. The reduction in vigour of the pathogen brought about by the natural accumulation of d-factors in the fungus, combined with the spectrum of resistance in the genus *Ulmu*s – from the well marked resistance of the Asian elms through the tolerance of some clones of the European elms to the marked susceptibility of the American elms – offered at least two routes to effective control of the disease.

No sooner had this conclusion been reached, however, than a new epidemic began. Opinion is divided as to whether it started in the USA or in central Europe, but it was soon widespread in both continents, its spread hastened by beetles in the timber used as packing material in international trade. It was more virulent to the European elms than the previous pathogen and as virulent on the American elms as the early attack. In Britain, where the fungus of the first epidemic had removed comparatively few trees and invaded new territory at a slow pace, the new fungus took out huge swathes of hedgerow and parkland elms (as much as 60%) and suddenly transformed the appearance of much of the southern English countryside (Plate 4). In northern Britain and in Scotland, for example in the Tweed valley, its effects were just as severe but because of the greater diversity of trees in the landscape the visual impact was less. It is still spreading as we write.

Those who have studied the new 'strain' of the fungus have decided it is sufficiently different to be called a new species, *Ophiostoma novo-ulmi*; a similar fungus has been found in the western Himalayas, but not apparently causing epidemic disease.

Has Dutch elm disease a past?

One always wonders about the history of new epidemic diseases. Did they always exist in a mild form in particular parts of the world and then become epidemic when they were inadvertently moved to a new geographic area and met a new population of the susceptible host? That is what is believed to have happened with Dutch elm disease. Because pollen of elm is preserved in peats and sediments it has been possible to study the varying populations of elms over the Palaeolithic and later ages. The elm pollen preserved in deposits shows a marked decline at about 3,000 BC, the end of the Atlantic period. There are many possible reasons for this, such as Neolithic farming or climatic change, but some writers have associated the decline with an earlier epidemic of Dutch elm disease; a tempting theory that requires much more evidence to back it up.

Blue stain

Other members of the generic complex *Ophiostoma/Ceratocystis* cause a variety of diseases, including one or two species associated with the blue stain of coniferous wood, and because of their habitat in the vascular tissue these are of interest in the study of wilt. *C. coerulescens* and *C. ips*, which are closely similar to *C. ulmi* in the ascus-bearing stage, produce conidia inside conidiophores which are very like normal hyphae. This anamorph stage is given the name *Chalara*. The fungi attack recently felled coniferous timber and grow in the still living tissue that intersperses the woody material. From there the hyphae spread into neighbouring woody tissue through simple or bordered pits, or sometimes through bore holes in the cell wall. The fungi are colourless in the early colonisation of the wood but later become brown. An optical effect of the brown hyphae seen through the translucent wood cells gives a blue colour to the eye.

These blue stain fungi are relatively harmless to the wood, although they do weaken it slightly, but buyers become suspicious of the wood's structural strength and it becomes worthless. Infection is promoted by the activity of ambrosia beetles and bark beetles and by rain-washed spores. Rapidly dried wood or wood immersed in water is not liable to infection, and treatment with antifungal chemicals prevents the entry of the fungus.

It does not take much imagination to see that some genetic change leading to slight perturbation of the timescale of infection and the details of the growth of the fungi near to and in the tracheids would transform these fungi into wilt-producing organisms. Who knows what the future may hold?

Notes for naturalists

It is not possible to list for the amateur naturalists wild hosts which would consistently show classic wilt. However, *Fusarium* and *Verticillium* spp. and related fungi are largely associated with cultivated plants and can be observed in many plants in the vegetable garden (e.g. tomato (*Lycopersicon esculentum*), pea (*Pisum sativum*)) and herbaceous border (e.g. lupin (*Lupinus* spp.), pyrethrum (*Tanacetum coccineum*)). But the Dutch elm disease cannot be missed. Look for the galleries of bark beetle by pulling off the bark of dead trees. With a microscope or lens you should be able to see the fungal material. Also look for brown staining in the outer wood of felled trees and for classic wilt symptoms on elm suckers growing up from the stumps of apparently dead trees.

Simulated wilt – a red herring

We all die from circulatory failure in the end! But this failure may not be the primary cause of our demise. So it is with plants. A number of diseases cause wilt-like symptoms but do not invade the xylem to produce a true wilt. One of the most spectacular of these is the foot rot of *Piper nigrum*, the source of the black and white pepper in our kitchens. Pepper plants are evergreen tropical climbers, which are trained up tripods to form rounded, elongated pyramids. *Phytophthora capsici* f.sp. *piperis*, the fungal pathogen, may attack the leaves of the plant, where it forms small lesions causing little damage. It also attacks the roots and many plants may be attacked at approximately the same time as the *Phytophthora* spreads in surface water. The first obvious symptoms are a yellowing and wilting of the leaves, which fall off rapidly after a few days leaving a portion of the plantation derelict. Collapse of further plants follows. The systemic collapse accompanied by yellowed and wilted leaves suggests a classic wilt, but the basic cause is a foot rot; the fungus destroys the stem and root bases, and by killing the non-specialised (parenchymatous) cells that bind the xylem vessels together impairs the function of both xylem and phloem, leading to death of the plant. We have here a slow underground spread of the foot rot, unseen except by a close observer, which manifests itself suddenly as a spectacular above ground false wilt.

Another disease that presents misleading symptoms is the chestnut blight, a notorious disease which all but wiped out the chestnut forests of the United States from Maine down to Alabama, and west to Michigan and Missouri. Sweet chestnut (*Castanea dentata*) is a very important deciduous constituent of the American forests, greatly valued for its decorative wood, bark for tanning and nuts for eating. The introduction of the fungus *Cryphonectria* (formerly *Endothia*) *parasitica* from China or Japan in 1906 saw the native American chestnut virtually wiped out over a 50 year period in which an estimated 3.5 billion trees were destroyed. In the same area today there are no chestnut trees of any size, although stool shoots from killed trees grew up before being quickly destroyed in their turn. In Europe the disease occurred in Italy in 1938 and spread over southern Europe in a fashion similar to the epidemic in the USA, but it travelled more slowly and was less virulent on the European chestnut (*Castanea sativa*).

The pathogen is a member of the Ascomycota, with an anamorph in the genus *Endothiella*. It invades through wounds and grows in and kills the living phloem, the cambium and the newly formed sapwood. The consequence can be of two kinds: where the cambium is killed over a discrete area, a smooth sunken lesion is formed; where the fungus kills the phloem but does not everywhere penetrate to the cambium, new wood and phloem is produced which causes raised areas to form under the bark, eventually leading to splitting of the bark and the production of pycnidia in the cracks. Meanwhile the fungus grows into and kills the phloem or cambium at the edges of the lesion, and over time a small branch will be ringed and killed. Larger branches and main trunks are killed by the coalescence of a number of lesions (cankers) and the ringing of the trunk, a process which takes from one to ten years. When ringing is complete the crown of the branch or tree can no longer obtain water and nutrients, and the consequent yellowing and browning of the leaves, which can be seen from a distance, simulates the symptoms of some wilt diseases such as Dutch elm disease. But the chestnut blight fungus does not grow and spread in the xylem.

An interesting feature of the chestnut blight has been the discovery of spontaneous reductions in virulence (hypovirulence), first in the European epidemic and later in the USA. Where it occurs trees recover, and the hypovirulence can be passed on by inoculating the hypovirulent pathogen into a lesion already infected by a normal virulent form. There is evidence that this hypovirulence is carried in the fungal cytoplasm and is caused by a form of transmissible RNA, which is in fact a virus or viroid disease of the fungus (see Chapter 12).

7

A Secret Life

There is a whole world of microorganisms that invade the plant, but cause little or no obvious damage; indeed, some are beneficial to their hosts. Many do not produce spores in or on the plant tissues, and are scarcely visible to the casual observer. These organisms are all around us as we walk in the country, over the grass or under the trees and we are not aware of them. They may be likened to a fifth column, infiltrating and living behind enemy lines without being recognised and, in some of the cases we describe, acting as 'sleepers' and remaining hidden for months or years before being activated by some change in the balance between host and parasite. In the course of their evolution such parasites have either lost the characteristics that enable the invaded plant to recognise them, or have developed active mechanisms for confusing or destroying the recognition system. Alternatively they may have evolved to remain in a state of suspended animation until the host's senescence renders its resistance mechanism ineffective. Research workers are only slowly beginning to study this problem.

There are three different and unrelated groups of microorganisms known to have the capacity to invade a host without eliciting a visible disease reaction: the endophytes, the mycorrhizas and the latent pathogens.

The endophytes

In writing about this group one can appreciate the difficulties of the theologian writing about the unseen. Endophytes, fungal parasites which have been most studied on grasses, live in the tissue of the host, growing between the cells of the leaves, stems and roots, drawing nourishment from them but producing no haustoria. Eventually they grow into the developing seed, causing new plants to become infected, and the cycle is continued. They cause no symptoms easily visible to the naked eye and are thus entirely unseen except by the microscopist seeking their presence. As far as is known, such fungi reproduce in nature only by being carried in the seeds of the host plant, since no sporulation has been seen on living tissue, and there is no certain evidence for sporulation on dead tissue in the wild.

However, surface-sterilised leaves placed on nutrient agar in the laboratory will in many cases produce mycelium bearing asexual conidia. It is these artificially produced spores that allow us to attribute the endophytic fungi to the anamorph genera *Acremonium*, *Acremoniella*, *Gliocladium* and *Phialophora*. The fact that the common pathogen of grasses *Epichloe typhina* (see p.114 and Plate 4) has an endophytic phase and an *Acremonium* anamorph suggests that the endophytes may be related to this fungus and can be classified along with it in the Ascomycota. The ergot fungus *Claviceps* (p.115) and *Balansia*, its relative on tropical grasses, both in the Clavicipitaceae, are also similar to some endophyte fungi in structure and form.

The endophytes are widespread in the grasses of the tropics and subtropics.

They are also important in agricultural grasses such as tall fescue (*Festuca arundinacea*), particularly in the USA and occasionally in Britain, and in perennial rye grass (*Lolium perenne*), particularly in New Zealand. There is good evidence to suggest that such endophyte-infected grasses may sometimes have a deleterious effect on the health of the animals eating them. In the USA it is accepted that ingestion of infected tall fescue by cattle causes a fescue toxicosis leading to poor condition, weight loss and a reduction in milk yield. In addition, affected beasts may develop fescue foot, characterised by lameness and a dry scaly gangrene of the foot (somewhat reminiscent of some symptoms of ergotism in humans), which may lead to the loss of the hoof. Endophyte-free seed of tall fescue is now produced, and it has been found that the productivity of cattle enterprises is greatly increased on pastures free of infection, with an annual saving of many millions of dollars.

The effect of endophyte-infected perennial rye grass is quite different, although equally serious. In New Zealand sheep and other grazing animals suffer tremors and shaking on such pastures. In its most severe manifestation they develop muscular spasms and fall, but quickly recover. The danger is that affected animals sometimes injure themselves or drown by falling into watercourses; in particular seasons deaths are frequent from these causes. There is good presumptive evidence that endophyte-infected grass is responsible for these symptoms, but a strict causal relationship is harder to establish.

In Europe it has been suggested that the rye grass endophyte and similar fungi in other pasture grasses may be responsible for a variety of little understood clinical conditions in grazing animals, such as the well characterised but as yet unattributed 'grass sickness' disease of horses. This causes the sudden death of horses at pasture or, if they survive the attack, leaves them in a debilitated state. Efforts to incriminate the endophyte have been further encouraged by the discovery that the infected leaves produce a variety of poisonous alkaloids. Work is in progress to establish strains of the endophyte that do not produce the toxin in pasture grasses. Turf grasses benefit from the presence of toxins that reduce, for example, insect attack.

The relationship between these endophytes and their hosts is clearly biotrophic, and it is also believed to be mutualistic – that is to say, each partner gains from the other. The advantage to the fungus in terms of food, shelter and dissemination is obvious; the advantage to the green plant almost certainly lies in the protection it is afforded against grazing animals, including leaf-eating insects. By occupying potential infection sites the fungus may also protect the host against root pathogens like *Gaeumannomyces graminis* (take-all). There are some claims that infected plants grow taller than normal under the influence of endophyte-induced changes in the plant's hormone balance, but clear evidence for this is lacking.

The mycorrhizas

Although the degree of mutual dependence of host and parasite is still a matter of speculation in the endophytic fungi, the symbiosis is well understood in the mycorrhizal fungi, which are virtually ubiquitous in the plant kingdom, and year by year our understanding is growing of the important part these mutualistic symbioses play in the ecology of their plant communities. A number of clearly delineated groups may be recognised (see p.147).

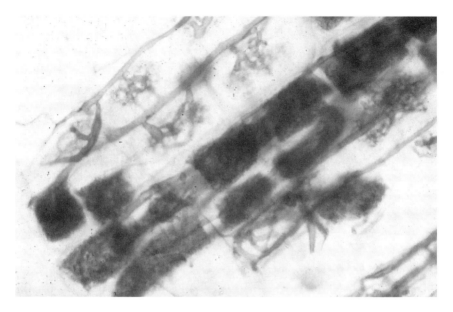

Fig. 7.1 Vesicular arbuscular mycorrhiza: intercellular hyphae with arbuscles in the root cells of cowpea (*Vigna unguiculata*). (Light microscope photograph of a stained, squashed root supplied by P.A. Mason, Institute of Terrestrial Ecology, Edinburgh.)

1. Endomycorrhizas: the vesicular-arbuscular (VA) mycorrhizas

These, the commonest mycorrhizal forms, grow within the host cells and are called endomycorrhizas. These are as hard to recognise with the naked eye as the endophytic parasites. They are best looked for with the microscope in pieces of root taken from herbaceous perennials, grasses, shrubs from the forest understorey and trees which are not dominant litter-forming species, both in temperate zones and the wet tropics. They are also found in the prothalli (the haploid plantlets which bear the sexual structures) of some ferns and fern allies and in the sporophyte (the adult fern plant that develops on the prothallus following sexual reproduction and, ultimately, bears the asexual spores) roots of others. Thin sections, suitably cleared and stained, are seen to contain a coarse intercellular mycelium without septa. This bears the characteristic arbuscules, haustorium-like structures with numerous slender, dichotomous branches which penetrate the living cells (Fig. 7.1). Swollen hyphal tips or hyphal segments rich in stored lipids, the vesicles, are also formed either within the living cells or in the spaces between them. It is the possession of these two characteristic structures that gives the fungi their name. The broad hyphae also grow out for considerable distances into the soil surrounding the roots. These may produce chlamydospores, by a ballooning out of the hyphal tip, which then develop a thickened wall. The chlamydospores can survive in the soil for many years. Infection of new plants takes place from the soil, either from the chlamydospores or from other living or possibly dead roots.

Vesicular-arbuscular mycorrhizal fungi appear to have very wide host ranges. In a situation in which the relationship benefits both partners by extending the range of niches available to each, it is probable that evolutionary selection has been in favour of compatibility rather than incompatibility, which would not lead to the high levels of specificity discussed in Chapters 2 and 3.

Vesicular-arbuscular mycorrhizal fungi eluded researchers for many years, but are now thought to belong to a very specialised group within the Zygomycota, the Endogonaceae. Their sexual fruiting bodies resemble the zygospores of other more familiar members of the Zygomycota (see p.25), and have been found attached to hyphae emerging from infected roots and also free in the soil.

It is very easy to find the large chlamydospores of vesicular-arbuscular mycorrhizal fungi. Take a tablespoonful of soil from natural grassland and stir it vigorously in a glass of water. Allow time for it to settle and for the scum to rise to the surface. Then with a piece of filter paper or absorbent kitchen paper skim some of the scum off the surface. Examine the paper with a low power microscope, lit from above, and look for the golden brown to black chlamydospores, often with a fragment of hypha attached. Some of the spores, such as those of *Glomus* spp., will be 150 μm across, as large as some of the soil particles; others, like those of *Sclerocystis* spp., will occur as clusters of smaller spores and resemble raspberries.

2. Ectomycorrhizas

Another very important form of mycorrhiza is the ectomycorrhizal association between fungi of the Basidiomycota (and sometimes the Ascomycota) and the roots of litter-forming trees, for example coniferous trees and broadleaved trees such as beech and oak. The fungi in this partnership do not enter and grow in the cells to any marked extent, hence the prefix 'ecto'. Instead they grow extensively between the cells of the root cortex, forming plates of fungal tissue. When these infected roots are cut in transverse section fungal tissue appears to surround most of the cells of the cortex. In stained preparations this appears as a net, the so-called 'Hartig net', named after Robert Hartig, an important innovator in forest pathology. Usually a thick sheath of fungal tissue also forms over the outside of the root (Fig. 7.2). From such infected roots fun-

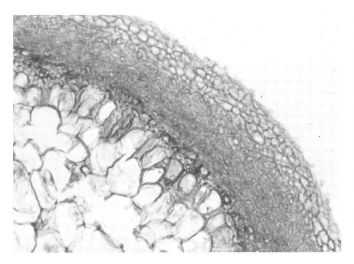

Fig. 7.2 Section of a beech root (*Fagus sylvatica*) colonised by an ectomycorrhizal fungus. A two-layered mycelial sheath can be seen around the outside, with the hyphae of the Hartig net between the outer cells of the root cortex. (H.J. Hudson, University of Cambridge.)

gal hyphae permeate the litter and soil.

Further understanding of these structures involves a digression into the morphology of tree roots. No two tree root systems are exactly the same, but they can be understood by reference to the structure of one of the simplest forms, that of the Scots pine. Here the new main roots are long and produce for most of their length short branch roots, about half a centimetre long. At intervals a short root will be replaced by a long root. When infection takes place from the soil, either from spores or from living mycelium growing from other roots, the cortex of the elongating root becomes infected and a Hartig net is formed. Both long and short roots can be infected, and once the long root is infected it is almost certain that the short roots that emerge through the infected cortex will also be infected. Short roots of the pine that become mycorrhizal fork (the technical term is 'branch dichotomously') and as they age they do so more than once (Fig. 7.3); because the oldest mycorrhizal roots are at the base

Fig. 7.3 Branching of roots infected by ectomycorrhizal fungi: (a) pine (*Pinus* sp.) infected by *Lactarius* sp; (b) birch (*Betula* sp.) infected by *Amanita muscaria*. (Photographs supplied by P.A. Mason, Institute of Terrestrial Ecology, Edinburgh.)

Fig 7.3a

Fig 7.3b

of the long pine root there is a progression along this long root, from tip to base, of mycorrhizal roots of increasing complexity of branching. In some host-fungus combinations this dichotomous branching is so marked that 'balls' of branched material (knollen mycorrhiza) are formed.

Another form of branching may also be induced by infection, in which branches arise on the laterals at acute or right angles (Fig. 7.3). The resulting racemose mycorrhizal root clusters resemble small bushes or tiny Christmas trees. It can be seen in such species as beech, birch, Norway spruce (*Picea abies*) and Douglas fir (*Pseudotsuga menziesii*). Some uninfected short roots close to the infected areas also branch to resemble mycorrhizal roots. In the uninfected roots of Scots pine all the cortical cells collapse and die to within a short distance of the root apex in both long and short roots, leaving the endodermal cells to delimit the root surface. In the infected roots, by contrast, the cortex remains alive for much of the growing season so that the living mycorrhizal short roots may persist from one season to the next.

The fungi mainly responsible for these ectotrophic mycorrhizas are the common members of the Basidiomycota: the toadstools, boletuses and gastromycetes found in woods (Fig. 7.4). However, the truffles and false truffles which are members of the Ascomycota may also form mycorrhizas, and the recent successful attempts to cultivate truffles by inoculating oak seedlings and growing them on to tree size (fast becoming an industry in New Zealand) derive directly from an increase in knowledge about the fungi that cause myc-

Fig. 7.4 Toadstools (basidiocarps) of two ectomycorrhizal members of the Basidiomycota: (a) *Amanita muscari* growing near beech (*Fagus sylvatica*) in Thetford Forest, Norfolk, England; (b) *Leccinum versipelle* growing near birch (*Betula* sp.) near Edinburgh, Scotland. (Photograph (a), D.S. Ingram; photograph (b), P.A. Mason, Institute of Terrestrial Ecology, Edinburgh.)

Fig. 7.4a   *Fig. 7.4b*

orrhizas. We know that some are very closely tied to one host species. For example the boletuses *Suillus grevillei* and *S. aeruginascens* infect only larches (*Larix decidua, L. leptolepis* and their hybrid). Other boletuses, *S. luteus* and *S. granulatus*, infect Scots pine and yet others, *Leccinum scabrum* and *L. versipelle*, associate with birches. We also know that in some of these trees there is a succession of mycorrhizal fungi as the trees age. Moreover, even where the main mycorrhizal partners are known, other mycorrhizal fungi may appear on a less regular basis. Much field observation is required to determine the distribution and ecology of these fungi, and the keen amateur could play a useful part in noting which fungal fruit bodies associate with which tree species.

In the Australian *Eucalyptus* spp., ectotrophic mycorrhizas are sometimes formed by members of the Basidiomycota but principally by members of the Ascomycota, and sometimes, very surprisingly, by *Glomus* spp., which normally form vesicular-arbuscular mycorrhizas.

In reading more widely about mycorrhizas reference may be found to ectendomycorrhizal roots of forest trees. These are difficult to define accurately, but in addition to forming a Hartig net the hyphae appear to invade the cortical cells. Whether they are specifically different from the ectotrophic forms or are a growth stage of these is not clear.

3. Orchid mycorrhizas

The mycorrhizas of orchids are associated with members of the Basidiomycota which are normally parasitic, particularly the anamorphic genus *Rhizoctonia*. This can be related to such teleomorph genera of the Basidiomycota as *Corticium* and *Thanatephorus*, which cause necrotrophic diseases on a range of other plants. The basidiomycete fungus *Armillaria mellea*, a cause of rots in trees (see p.227), also forms mycorrhizas with some orchid species (*Gastrodia* spp. and *Galeola* spp.).

The orchid mycorrhizas grow inside the host cells and are therefore endophytes. Two forms of infection take place in the interaction between the fungus and its partner. In the first the fungus infects the germinating seed, invades the first swollen plantlet, called the protocorm, and eventually enters the tuberous base of the adult. Indeed, the orchid seed cannot germinate in the absence of mycorrhizal fungi unless it is provided with an artificial supply of sugars. In the second type of infection the fleshy roots that develop as the plant increases in size are invaded. With both these types of infection the fungi penetrate the living cells, where they form coiled hyphal structures called pelotons. After a period these are digested by the plant. Invasion followed by digestion can take place several times.

How these fungal partners are prevented from becoming aggressively pathogenic necrotrophs, as they are in their non-mycorrhizal state, is not understood. One possibility is that the orchid partner actively suppresses the production of pathogenicity factors. Another is that the benign parasitism is an adaptation of the fungus, enabling it to occupy a niche that would not otherwise be available.

4. Ericoid mycorrhizas

Another specialised and relatively circumscribed mycorrhizal association is found in the heathland plants, heather or ling (*Calluna vulgaris*) and species of the related genera *Erica* (heaths), *Vaccinium* (including the whortleberries, bilberries and blueberries), *Epacris* (the Australian heath) and many species of *Rhododendron*. The plants produce fine roots around which fungi, now identified in heather as *Hymenoscyphus ericae* (a member of the Ascomycota), develop copiously. The root cortex, which is often only one cell thick, is heavily colonised, as are the spaces between the cells. The fungus penetrates the living cells, where it forms characteristic hyphal coils in large numbers; the fungus is said to make up more than 80% of the weight of the active roots. Eventually the coils and the cells containing them degenerate together.

In the related plant genera *Arctostaphylos* (bearberry) and *Arbutus* (strawberry tree), also members of the Ericaceae, the mycorrhizas – here called the arbutoid type – take on a different form more closely resembling the ectotrophic mycorrhizas of forest trees, with a Hartig net of hyphae being formed around the cells. However, the outer cortical cells are penetrated by hyphae which again form coils that are eventually digested. These infections are both endotrophic and ectotrophic.

The herbaceous members of the Ericaceae, *Pyrola* spp. (the wintergreens), sometimes separated as the Pyrolaceae, and *Monotropa* spp. (the bird's nests) sometimes separated in the Monotropaceae, form a range of mycorrhizal associations from ericoid to ectomycorrhizal. Plants in the saprophytic genus *Monotropa*, being without chlorophyll like the saprophytic orchids such as *Neottia nidus-avis* (bird's nest orchid) and *Corallorhiza* (coral-root), are totally dependent on a pre-formed source of carbohydrate provided by the agency of the mycorrhizal fungus which is infecting both the roots of the saprophytic plant and those of neighbouring trees carrying on normal photosynthesis.

A recent study of *Monotropa* and some related genera has illuminated our view of specificity in the mycorrhizal fungi. Throughout this chapter the wide range of parasitic behaviour by the fungi in mycorrhizal relationships has been emphasised, although the evidence for these assertions is not complete. A group of Californian pathologists have turned their attention to *Monotropa* and related genera in the conifer forests of west North America. These plants are so rare that repeat behaviour, or different behaviour, in two or more sites gives credence to the observations. Since it is too laborious to extract and identify the fungus by eye or to culture it, modern techniques for the identification of fungal DNA were used. It was shown that *Sarcodes sanguinea*, a *Monotropa* relative, associated with a wide range of fungal species. However, at the 31 sites studied, the related *Monotropa hypopithys* associated only with *Rhizopogon* spp, and *M. uniflora* only with fungi in the Russulaceae. Another relative, *Pterospora andromeda*, associated with only one fungal species, *Rhizopogon subcaerulescens*. Moreover, although this fungus formed mycorrhizas with some roots of nearby trees, it was only one of many fungi that made up the tree mycorrhizal population. So we see here in a related group of plants examples of specificity of association with mycorrhizal fungi varying from high to low. This gives us a more accurate picture of the behaviour of fungi on plant roots in the wild, but we still have to explain the mechanisms that control such a wide range of interactions.

The ecological importance of mycorrhizas

After nearly a century of study, during which all sorts of remarkable properties were attributed to mycorrhizal fungi, we now know that their importance lies in their almost universal presence. Most plants are mycorrhizal, the exceptions mainly being the ruderal (weedy) species that rapidly colonise disturbed ground and plants of wet areas and open water. Otherwise, most wild plants obtain their nutrients from the soil by means of both roots and fungi. There is nothing absolute about this because in nutrient-rich soils many of these plants can function without fungi and may even expel existing mycorrhizal infections, but in nature the delimitation and distribution of plant communities appears to be determined, at least in part, by the nature of the mycorrhizas associated with them. We still lack a complete tally of the mycorrhizal status of all British wild and cultivated plants, and this is an area where the amateur can be of great assistance.

The ecological importance of mycorrhizal fungi has been very clearly described by David Read. As a preliminary to discussing his hypotheses it is useful to look at the normal saprophytic fungal population of the soil. Here we see that different carbohydrate substrates added to the soil stimulate the growth of different fungi. Thus the Zygomycota are encouraged by the addition of simple sugars and on that account are often called the sugar fungi. Some members of the Ascomycota and some of the Basidiomycota are encouraged by the addition of cellulose; and the most successful degraders of cellulose and lignin are the Basidiomycota. It is a fair presumption that the ability to degrade these substances, which form a series from simple carbohydrate to more complex carbohydrate, has been accompanied by the evolution of enzymes with the capability of attacking them.

Read suggests that the types of mycorrhiza we have outlined can be associated with distinct ecological groupings of plants – the mineral soil plants, the broadleaved trees of temperate climates, the coniferous forests and the heathlands – groupings which bear ecological relationships with one another in a latitudinal as well as in an altitudinal fashion.

The fungi involved in vesicular-arbuscular mycorrhizas are probably related to the Zygomycota or sugar fungi. Their nutritional status is not yet certain, although the few experiments done to test their growth in laboratory cultures show an ability to use simple sugars. They are associated with plants on mineral soils rich in calcium and other minerals required for growth, where mineralisation of plant remains and perhaps nitrogen fixation by bacteria lead to rapid depletion of soil organic material. In these soils Read believes that phosphorus becomes the limiting nutrient because of the low solubility of its salts. The function of the vesicular-arbuscular mycorrhizas appears to be to concentrate this element by tapping a much larger volume of the mineral soil than the plant's roots could tap alone and then to transfer it to the plant. There is some evidence for the uptake and transfer of other minerals from the soil to the plant, but the evidence for the transfer of nitrogen salts, which are major plant nutrients, through this mycorrhizal system, remains scant. The plants must get their nitrogen by some means, but whether this is by uptake by normal roots or by mycorrhizal roots remains to be determined. The experimental difficulties in the way of resolving this question are very great, because vesicular-arbuscular mycorrhizal fungi are very difficult to grow in culture. What is clear is that the transfer of carbohydrate in the form of simple sugars is always from the host to the fungus, thus conferring a competitive advantage on the fungus as it explores the mineral soil in which few free carbohydrates occur.

The vesicular-arbuscular mycorrhizal fungi apparently have a long evolutionary history, for fungi indistinguishable from them have been found in sec-

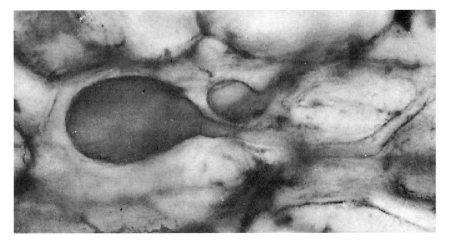

Fig. 7.5 Fossil intercellular hyphae and vesicles, perhaps of an endomycorrhizal fungus, in a thin section of plant material in Rhynie Chert (approximately 400 million years old). (D.S. Ingram.)

tions of the underground stems of fossil plants from the Rhynie Chert (Fig. 7.5) of the Devonian Period (400 million years ago), when plants were first invading the land, and also in coal measure plants from the Carboniferous Period (300 million years ago). It has been suggested that the earliest land plants, which did not have roots, were able to colonise the land from the water only because of the presence of mycorrhizal fungi in their underground stems acting as roots.

At the other extreme, the heathlands of cooler climates form on soils that are already short of plant nutrients. This situation is made worse by the acid litter deposited by the heathland plants which accumulates on the soil surface, facilitating the leaching of minerals from the upper layers of the soil and their deposit lower down in an often impermeable, iron-rich layer (a process called podzolisation). In particular these soils lack the nitrogen salts that all plants require for synthesis of the amino acids and proteins they need for growth.

It is only recently that the fungus associated with the roots of certain of these plants, identified as *Hymenoscyphus ericae* (a small ascomycete), has been isolated into culture and studied in detail. This and other similar mycorrhizal fungi can decompose the acid and highly recalcitrant heath leaf litter and dead roots from earlier growing plants. They are especially noteworthy for their ability to split the ring structure characteristic of phenolic molecules, thus making them able to break down complex phenol-rich polymers such as lignin to release simple sugars essential for fungal growth. They also possess the enzymes to break down cellulose into simple sugars. Together, cellulose and lignin make up the main constituents of the litter. What is left includes the cell proteins, which the fungus is able to break down into soluble amino acids for transport into the growing green plant. Put simply, the ericoid mycorrhizas are able to extract sugars for their own use and to provide nitrogen for their partners in an ecological situation where nitrogen is in such short supply as to limit plant growth. Also, because of their wide-ranging exploration of the litter-soil interface, they are able to provide phosphorus salts where these are in short supply.

While it may not be possible to say that heather cannot flourish without its mycorrhizal partner, it is abundantly clear that it could not occupy its particular ecological niche on difficult soils under harsh climatic conditions without its companion fungus to provide nutrients from the acid litter that accumulates under the canopy.

The orchid mycorrhizas, which are formed by basidiomycetes, are well understood. In the first place, the mycorrhizal fungi are significant in the germination of orchid seeds. Orchids produce very many tiny seeds consisting of only a few cells rich in starch granules. Sown on soil or on an agar gel in the absence of the mycorrhizal partner the seeds fail to germinate, whereas in the presence of the fungus they soon begin to swell and grow. The first sign of the interaction between seed and fungus is often the disappearance of the starch granules from the plant cells. Then the cells multiply, drawing on the sugars released from the starch, and eventually form a small mass of tissue called the protocorm. Further development depends on the nature of the orchid, but in many British orchids the protocorm swells to form a tuber. The fungus becomes concentrated in the basal layers of this fleshy mass, where it forms loops in the cells. At this stage the orchid is still not green and there is ample evidence that it is dependent on the inward carriage of carbohydrates by the fungus, which is digesting polymerised carbohydrates in the soil litter. The carbohydrates are in turn released to the plant cells by digestion of the fungal loops. Of course, materials other than carbohydrates are imported by the fungus, including minerals, nitrogenous compounds and vitamins. During growth most orchids become green and less dependent on the fungus for a supply of carbohydrates, although these other substances continue to be important to it. The role of the fungus can be simulated in the laboratory by sowing the orchid seeds on an agar gel containing sucrose, vitamins and materials such as yeast extract, when successful germination can be obtained aseptically.

In other orchids the non-green phase is permanent, and these species are totally dependent on their mycorrhizal fungi for carbohydrates and many other materials required for growth. We have already mentioned the small chlorophyll-less saprophytic ground orchids of Europe, *Neottia* and *Corallorhiza*; in tropical forests there are others, such as *Galeola hydra*, which form large climbing vines called lianas, all the time dependent on the presence of the fungus for nutrition. There is no doubt that carbohydrate from the non-living material in the soil is passed to the plant, but there is also evidence that in some saprophytic orchids, as in *Monotropa*, mycorrhizal associations shared with adjacent trees lead to transfer of carbohydrates manufactured by the tree to the saprophyte.

The ectomycorrhizal fungi are associated with trees growing on soils that are nutrient-deficient. These depleted soils not only present a challenge for the extraction of minerals by the trees, but also favour the accumulation of leaf litter, which breaks down less quickly than on mineral and base-rich soils. The active mycorrhizal roots grow particularly in the layer between the disintegrated litter and the litter composed of whole leaves. Such situations are found in coniferous and broadleaved forest, and in mineral-depleted soils such as those of tropical rainforests. Evidence now shows that the ectomycorrhizal fungi, supplied with plant carbohydrates by the host roots, are able to grow into the litter and secrete enzymes that help to break down the organic material present, mobilising nutrients, particularly nitrogen in the form of ammonia, which can be transported to and used by the growing trees. They are also able to supply phosphorus and copper salts in soils where these materials are in short supply.

Curiously, we have little information on the carbohydrate metabolism of these ectomycorrhizal associations. It is known from radioisotope studies that the mycorrhizal fungi depend on the host for carbohydrates, and shading experiments show that the development of mycorrhizal roots is closely related to the amount of light received by the host plant. It is also known that the fungi involved have the capacity to degrade cellulose and lignin in the litter, although perhaps to a lesser extent than the saprophytic members of the Basidiomycota, which are also involved in litter decay. There is plenty of evidence that the ectomycorrhizal fungi receive carbohydrate from the host plant, but no evidence of any flow of carbohydrate from the litter to the host.

There is another aspect of mycorrhizal associations that transforms our view of their importance. In any natural plant community, whatever the type of mycorrhiza present, we can now imagine large groups of plants interconnected by a vast network of hyphae. What this means in terms of the distribution of nutrients has still to be determined. Is it a self-help society that provides equal nutrients across the whole community? Is it a stable community of fungi? Or is it, as some evidence from the ectomycorrhizal fungi suggests, an evolving community with changes in the species mix according to age and soil quality, and if so why? There is much still to find out, particularly as regards the mycorrhizal Basidiomycota; careful monitoring of the changes in populations of these fungi could with time reveal a great deal.

The benefit of mycorrhizas to agriculture, horticulture and forestry

We know that the ectotrophic mycorrhizal fungi of forest trees can be spread through the air by spores, hyphal fragments and so on, and that this happens whenever young trees are planted within a reasonable distance of other forest trees. Indeed, in view of the difficulty of keeping young seedlings free from mycorrhizal infection when they are grown in the middle of a small town, it is probable that the infection can be carried considerable distances, although there is as yet no conclusive evidence on this point. In any event, in normal afforestation all trees will be mycorrhizal, or at least have the opportunity to be mycorrhizal, either because inoculum soon reaches them or because as young trees transplanted from forest nurseries they start life with mycorrhizal roots. It is therefore difficult to answer questions about the value of mycorrhizas to the tree or about the possible advantages they confer, since there are few uninfected trees for comparison. On nutrient-rich soils plants probably grow well in the absence of mycorrhizas. On poor soils where litter accumulates and where nitrogen and phosphorus salts are in short supply the presence of the mycorrhizal fungus increases the ability of the tree root to explore the soil and litter, and transport these compounds to the plant. There is also some evidence that the fungal sheath on the mycorrhizal root can store phosphorus and nitrogen salts and release them at times when they cannot easily be removed from the soil. Thus it no longer seems sensible to talk of value or advantage; the fact is that the tree would be incomplete without its mycorrhizal partner, which allows it to function efficiently in nutrient-poor soils.

There remain the exceptional circumstances where new forests are started from seed in isolated sites on prairies, steppes, islands or tropical highlands where there is no possibility of mycorrhizal inoculum reaching them. Here the importation of mycorrhizal inoculum has been shown to be advantageous to the establishment and growth of the trees.

Another interesting feature concerns the role of mycorrhizas in the imported trees which are such an important part of modern British forestry. Where did the fungi come from which form the mycorrhizal partners of the Douglas fir, the sitka spruce and the lodgepole pine (*Pinus contorta*)? They may have been brought in on potted plants at first introduction, but there is no evidence for this. It is more likely that the trees were able to form associations with the native mycorrhizal partners of indigenous trees. If so, this is one more indication of the wide host range of these fungi.

The vesicular-arbuscular mycorrhizas appear to be so universal that we must think of them as plant organs that aid the efficiency of the plant in a natural environment. In agriculture and horticulture, however, where mineral nutrients are supplied in artificial form, mycorrhizal associations do not seem to form. Mycorrhizas may thus be less important in conventional agriculture and horticulture than in forestry. In low input sustainable agricultural and horticultural systems, however, they are likely to be essential to the production of good yields.

The latent pathogens

Latency is hard to define. At first sight it seems obvious that it must refer to pathogens that lie quiescent in or on the plant, waiting for an opportunity to invade, parasitise and produce disease symptoms visible to the naked eye. But a little reflection suggests that every pathogen-host interaction divides into phases: deposition of inoculum; establishment of infection; tissue invasion without overt damage (establishment of the infection site); tissue damage resulting in symptoms, and, in the case of the fungi, sporulation. Each of these phases may be longer or shorter than the others, and greatly extended duration of the earlier ones may be regarded as latency.

The extreme expression of latency is shown by some viruses (see Chapters 1 and 12). The swollen shoot virus of cacao (*Theobroma cacao*), for example, exists in a number of strains. Some of these have been found in wild forest trees related to cacao, and in such cases symptom expression is muted or there are no visible symptoms at all; the virus is thus said to be latent in these hosts. When transferred to *T. cacao*, some of the virus strains produce marked symptoms and some are so mild as to be virtually symptom-free; the latter are again said to be latent. Similarly, in the potato, the paracrinkle virus is latent in the variety 'King Edward', i.e. it is carried without symptoms, but can cause severe symptoms in other varieties. Also the potato virus X can exist in such mild forms that in the USA it was once called the 'healthy potato virus'.

The situation with fungal pathogens is less clear cut. It is usual to talk of any situation where pathogens delay entry or, after entry have an extended symptomless phase, as showing a 'latent' phase, and in extreme cases such parasitic relationships would be referred to as 'latent infections'. The difficulty in using these terms, however, is well illustrated by the fungus *Fusarium oxysporum*, which causes wilt in a number of plants (see Chapter 6). A well-known form attacks vining peas (*Pisum sativum*) (peas grown for freezing). The fungus here infects the seedling through the root and grows in the vascular tissue until the plant achieves its full height, when suddenly the leaves wilt and the plant collapses. No one normally thinks of calling this a latent disease, partly because from an early stage a clean cut across the vascular tissue will reveal a brownish-red discoloration; however, the period when the plant is apparently uninfected when viewed externally could possibly qualify as a latent phase.

It is only in cases where the fungus seems to disappear for a large part of the infection cycle, reappearing at an appropriate point such as ripening of the fruit or senescing of the leaves, that we can with confidence talk of a latent disease. One such is *Monilinia fructigena*, which attacks the apricot fruit (*Prunus armeniaca*) when it is young and green. In spring, overwintering mummified fruits of the apricot scatter conidia and ascospores into the air around these young fruits. When the spores land they germinate, and the germ tubes enter the sub-stomatal cavity and sometimes the cells of the stomata themselves. The presence of the fungus stimulates the tissues of the young fruit to produce a cork barrier and phenolic material in the cells surrounding the infections, which are prevented from proceeding further and remain latent until the fruits mature and ripen. At this point, when the ability of the host cells to produce the resistance factors keeping the pathogen in check declines, the hyphae break out of their prisons and grow beneath the cuticle or epidermis into the healthy ripening tissues beyond the reaction areas. When full colonisation takes place the fruits discolour and rot. Eventually the hyphae sporulate on the surface of the fruits and the conidia infect other ripe fruits through wounds caused by wasps or pecking birds.

We mentioned the behaviour of *Sclerotinia trifoliorum* and *Botryotinia fuckeliana* (*Botrytis cinerea*) in Chapter 5. In the diseases they cause, an initial invasion of green tissue by the germinating spore is held in abeyance by a tissue reaction. The infection sites so formed remain latent until the foliage or fruit matures, presumably again accompanied by physiological changes making the plant no longer able to keep the pathogen in check. The fungus then becomes aggressive and destroys the surrounding tissues (Plate 3). The nature of the changes in the tissues which allow the fungus to become aggressive are not yet fully understood.

Harking back to the military analogy used earlier, we see here a possible comparison with a 'sleeper' in the secret service. The fungus enters host territory and then remains 'perched' just within the plant tissues until the host environment changes with ripening or maturation. To achieve this synchrony between host and fungus, the discharge of spores has to be critically timed to ensure that infection of the fruit or leaves takes place at the appropriate stage of their growth.

There is a practical value in understanding such life cycles; control can be focused on the initial infection process at a time when the pathogen is especially vulnerable to sprays that control its growth, long before the characteristic symptoms are expressed as a rot.

A slightly different latency mechanism, in which the fungus is contained before true infection has taken place, is shown by *Colletotrichum musae* and *C. gloesporioides*. Both fungi, which are the anamorphs of *Glomerella* spp. in the Ascomycota, cause a fruit rot of bananas (*Musa* x *paradisiaca*). The spores are produced on the ageing leaves of the banana plant, then dispersed by rain and wind to land on young green fruits and leaves, where they germinate to produce appressoria on the fruit surface or in the stomatal depressions. From these appressoria, germ tubes grow out, penetrate the cuticle and grow a little beneath it before becoming latent. The pathogen remains in this position until the individual bananas grow to full size and ripen naturally, or are harvested green and ripened artificially. At this point, as the fungistatic substances present on the surface of the green banana disappear and the flesh changes and becomes rich in sugars, the latent fungi are activated, invade the skin and penetrate the flesh. Superficial lesions on ripe bananas are not uncommon, but the level of control obtained in the plantations is such as to ensure that severe infections are comparatively rare. It should be mentioned in passing that although the main infection is by this latent route, infection can take place directly, and sometimes with severe effects, when the fruit or the stems on which the bunches are borne are wounded or bruised.

Colletotrichum gloesporioides is also a serious disease of, among other cultivated plants, the avocado (*Persea americana*). Work in Israel has shown that when spores land on the surface of the fruit they appear to produce germ tubes which dissolve a passage through the cuticle to the surface of the epidermal cells, where an appressorium is formed and latency ensues. A form of the fungus in Australia behaves slightly differently, producing appressoria on the cuticle surface, with extensions into the cuticular wax to a point just short of the epidermal cell wall, where it remains latent. As ripening of the fruit occurs an extension of the appressorium puts out a fine infection peg, which infects the fruit by penetrating the epidermis.

The parasites we have just discussed are active necrotrophs that have adopted a strategy of latent infection to facilitate their parasitism, at a stage when the host is near to natural senescence. The ecological advantage of this strategy is that the pathogens are able to become established in the host before death, giving them a considerable advantage over purely saprophytic competitors which can only invade the rich food source after the fruit has senesced.

Another group of fungi that has evolved the strategy of latency, the needle cast fungi of conifers, are as yet little understood. Some in the past have been considered as weak parasites or natural saprotrophs which occasionally attacked hosts already dying or weakened from other causes. If we consider them instead as latent fungi that normally sporulate on plant tissues after the death of the plant, they become much more comprehensible. Although this idea was first suggested in a study of *Hypodermella nervisequia* on common silver fir (*Abies alba*) in 1912, it was ignored for many years, but it now appears that most of the needle cast fungi are probably latent and can commonly be found on shed needles on the forest floor. Needles of evergreen conifers are usually shed on a regular basis after two to five years from their first appearance, and at that point the lens-shaped ascocarps appear (Fig. 7.6). These contain asci and discharge ascospores of varying shapes, often threadlike, in the spring, when the evidence suggests that they infect the young developing leaves. They appear to colonise the sub-stomatal chambers and the adjacent mesophyll, where they remain quiescent and latent for the several years the leaves are on the trees. If the leaves are weakened by some environmental effect on the host they may produce ascocarps on the dying leaves, but as yet we have no evidence that in their natural host they are active parasites.

Apart from these specialised forms it is now believed that a great many of the dark-coloured Ascomycota – which produce their ascocarps in spring on the dead stems of herbaceous perennials and the dead leaves of trees or shrubs – probably infect the young tissues of the plant by means of ascospores derived

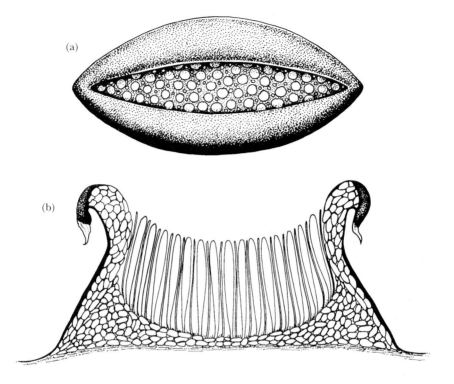

Fig. 7.6 Lophodermium sp., cause of needle-cast of conifers: (a) surface view of an ascocarp beginning to open; (b) section of an ascocarp fully expanded in moist conditions to expose the asci. (Illustrations by Mary Bates, based on drawings and photographs by M. Wilson and N.F. Robertson, University of Edinburgh.)

from ascocarps on dead material from the previous year, setting up latent infections. When the host dies down in winter these fungi may grow out and sporulate in the dead tissues. We can point to *Leptosphaeria acuta* on the stems of nettles (*Urtica dioica*), *Cochliobolus sativus* and *Gibberella zeae* on cereals, and a range of leaf-inhabiting forms on fallen leaves of trees, but as yet there is only circumstantial evidence to confirm this view. There is no doubt that there is a great field for amateur investigation in field studies of these and similar fungi, to determine the annual cycle of host and pathogen.

8

Life on the Surface
The Powdery Mildews

The surface of a leaf is an inhospitable place by any standard. It is subject to extremes of temperature, both high and low, from hour to hour and day to day. Moisture is supplied irregularly; drenching rain one moment may be followed by desiccation in the heat of a summer afternoon the next. All the time it is unprotected from the damaging components of sunshine. Food is in short supply, being available only by the parsimonious leakage of nutrients through the cuticle from the cells beneath, or by chance windfalls of pollen and dust deposited by wind and rain. And as if this were not enough, it may be bombarded by a variety of poisons – sulphur dioxide and other pollutants from cities and factories or salt spray carried on the wind around the coast. Like many hostile places, however, this bleak world is teeming with life.

The phylloplane (the surface of the leaf) and the other above-ground surfaces of the plant – stems, buds and flowers – are inhabited by a whole range of fungi and bacteria, mostly saprophytes, all fully adapted to cope with every environmental extreme and all competing for space and food. No group is more successful in this battle than the ubiquitous powdery mildew fungi, obligate biotrophic members of the order Erysiphales within the Ascomycota. They are important pathogens of a wide range of cultivated and wild plants.

The word 'mildew' has been in the English language a long time, being derived from *mildéauw*, an Old English word meaning honeydew. It now refers to any fungal growth on any surfaces, living or dead. Thus clothes and leather goods become mildewed by saprophytic *Penicillium* or *Aspergillus* spp. in damp climates, and sheets become mildewed by dark coloured saprophytes if left damp too long. Plants too may show white fungal parasitic growths, referred to as 'downy mildew' if caused by parasitic members of the CHROMISTA (see p.91) and as 'powdery mildew' if caused by members of the Erysiphales. The superficial grey-white vegetative mycelium, conidiophores and conidia of the Erysiphales usually make infected plants obvious, looking as if they have been partially whitewashed or dusted with talcum powder (Plate 5a). High summer is the time to see these powdery mildews at their best, both in the garden and in the field, especially if the weather has been dry.

Life on the leaf surface

The hyphae of the powdery mildews are confined to the surface of the plant, with all the stresses that this implies, and they maintain themselves on this hostile environment principally by sending haustoria into the epidermal cells of the host to tap the food and water available there. Some have also developed the capability of germinating under drier conditions than those which suit other parasites. For example the conidia of *Erysiphe polygoni*, now restricted to isolates infecting Polygonaceae, but formerly considered to be a pathogen of many wild and cultivated species (the problem of naming powdery mildews will be dealt with below) not only germinate without the presence of free water but can even germinate in a relatively dry atmosphere, and the same is true of the grapevine mildew, *Uncinula necator*. On the other hand *Sphaerotheca pannosa* (rose mildew) (Fig. 8.1) and *Podosphaera leucotricha* (apple mildew) need a saturated atmosphere for germination, whilst other powdery mildews appear to behave somewhere between these two extremes.

There are many observations that suggest that having germinated, many powdery mildews are able to grow in dry conditions, and are thus xerophytes. Observations on the powdery mildew of rubber trees (*Oidium heveae*) suggest that rain is required to start an epidemic but that the epidemic subsequently develops most rapidly if a period of fine weather follows rather than a period of continuous rain. Many other powdery mildews appear to flourish and spread in warm dry weather, such as *Uncinula necator* on vines (*Vitis vinifera*) mentioned above, and *Sphaerotheca fuliginea* on members of the cucumber family. But one can lay too much emphasis on this, for there are also observations to suggest that fine weather during the day needs to be accompanied by dews at night for the spread of species like *Podosphaeria leucotricha* on apple. It is easy to demonstrate the damage that free water can do to the conidia of powdery mildews, but it is also possible to observe a reduction in the spread of these fungi when the weather becomes very hot. Also there are biological factors such as the nutrition of the host, its age and its stage of development, which influence the ease of infection by different species. It is clear that we are dealing with a group of fungi, some of which have xerophytic features but certainly not all.

The different genera of the Erysiphales are identified by the details of their asexual and sexual reproductive processes. All have the typical life cycle of the Ascomycota (see p.26).

Fig. 8.1 Sphaerotheca pannosa causing powdery mildew of rose (cultivar 'Albertine') in Cambridge, England. Note the mycelium covering almost the entire surface of each leaf and the distortion of the infected leaves. (D.S. Ingram.)

The life cycle

The conidia, the non-motile asexual spores, are budded off in succession from a short, often flask-shaped, mother cell (the conidiophore) attached to a hypha on the leaf surface (Fig. 8.2). In this process the single nucleus of the mother cell divides repeatedly to provide one haploid nucleus to each conidium. The mature conidia, which contain large reserves of water, are resistant to drying out over short periods and are easily detached, to be dispersed on the wind to new hosts.

Fig. 8.2 Conidia of *Erysiphe graminis*, cause of powdery mildew of grasses and cereals (Gramineae); note the flask-shaped mother cell. The oldest conidium, at the tip of the chain, is approximately 28 μm long. (Illustration by Mary Bates.)

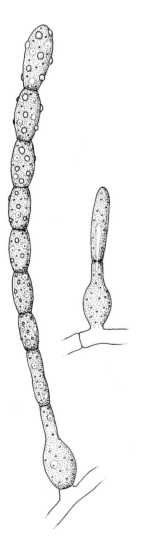

There are small differences between the conidial stages of different species, and these can be used for identification purposes. Sometimes the individual spores remain attached to one another to form long visible chains, as in *Erysiphe graminis* (powdery mildew of grasses and cereals) (Fig. 8.2); indeed, there is evidence from spore traps that this fungus is often distributed as unbroken chains of conidia. Sometimes conidia fall off one by one as they are formed and do not form chains, as in *E. polygoni*. The mother cell may be swollen and barrel-shaped (*E. graminis*), or scarcely distinguishable from the young conidia (*Sphaerotheca* spp.). The fully developed conidia themselves vary in shape from egg-like (*E. graminis*) to pointed or dumbbell-shaped (*Phyllactinia* spp.). The presence or absence of vacuoles or fibrosin bodies (bundles of elongated fibres of unknown nature) and oil droplets are also distinguishing characteristics.

The sexual bodies, the ascocarps, are roughly spherical cleistothecia with no neck or natural opening, containing irregularly-shaped sac-like asci (Fig. 8.3). The number of asci varies according to genus: one large one in the case of *Sphaerotheca* and *Podosphaera*, several in *Erysiphe*, *Uncinula*, *Microsphaera* and *Phyllactinia*. Melanin, a dark fatty substance that impregnates the walls, helps to prevent water loss and may also protect the developing ascospores from the harmful components of sunlight. Ascocarps are very resistant to environmental extremes, enabling the fungus to survive over winter.

Being almost 1 mm in diameter, the ascocarps are usually visible to the naked eye as black or dark brown specks, rather like specks of dust, on heavily infected plants (Plate 5).

Close examination with a lens or low power microscope reveals that the ascocarps are frequently ornamented with grotesque branches or appendages that grow out from the surface, the different shapes and branching patterns of

Fig. 8.3 Diagrammatic longitudinal sections (not to scale) of ascocarps of powdery mildew genera to illustrate ascus shape and number, and the form of the appendages: (a) *Erysiphe*; (b) *Sphaerotheca*; (c) *Microsphaera*; (d) *Podosphaera*; (e) *Uncinula*. (Drawn by Mary Bates after Ellis and Ellis (1997).)

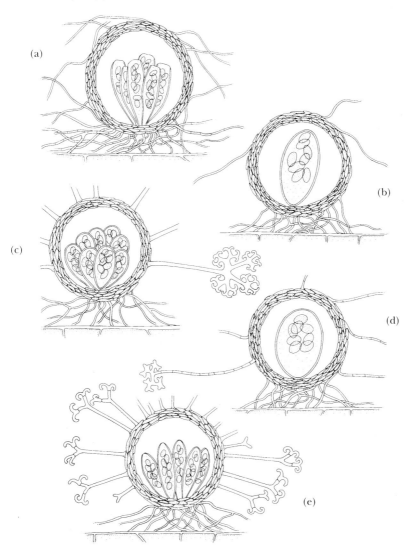

these structures being characteristic of different genera. In *Microsphaera alphi-toides* (powdery mildew of oak) (Fig. 8.5), for example, they branch repeatedly at the tip to form patterns resembling a cluster of old-fashioned H-shaped television aerials; in *Uncinula necator* they are hooked at the tip; whilst in *Podosphaera leucotricha* they are boringly straight. Their function is not known with any certainty, although it is possible they have a role in dispersal.

The cleistothecia open irregularly to reveal the asci, in some cases by an equatorial split which removes one half leaving the asci protruding from the matrix in the other, like eggs in a nest.

Tapping the aquifer

In describing the ecological niche that the mildews occupy some clues to their behaviour have emerged. Individual conidia or ascospores land on the leaf surface, germinate and rapidly send out a germ tube. In some species, such as *Erisyphe graminis*, the first formed germ tube is very short, and by attaching itself to the leaf surface appears to act as an anchor for the spore. Then a second germ tube emerges from the other end of the spore, grows and swells to form an appressorium, which also sticks to the leaf surface. Finally an infection peg emerges from the underside of the appressorium and penetrates the underlying cuticle and cell wall. In other species there is a single germ tube with an appressorium at its tip. If the appressorium were not securely attached to the surface of the leaf the hydrostatic pressure within the host cell would make penetration impossible. The entry of the infection peg is facilitated by the secretion of enzymes such as cutinases, and perhaps cellulases, which soften the cuticle of the leaf surface and the underlying cell wall to allow the peg to grow into the cell.

It is this rapid penetration of the host cell, to tap its reserves of water and nutrients, that allows the fungus to flourish in its inhospitable environment. On entering the cell the infection peg comes into contact with the plasmalemma. Immediately changes begin to take place. The swollen infection peg grows to become a characteristic lobed or fingered haustorium, which occupies a large part of the cell. The plasmalemma of the host cell grows rapidly to accommodate this structure and becomes thicker than normal. It is inhibited from producing cell wall materials, and this prevents the haustorium from being walled off. Nevertheless, layers of material containing protein and lipids are laid down at the interface between the plasmalemma and the haustorium wall. This 'sheath matrix' may aid the exchange of nutrients and other substances between host and parasite. The haustoria are easily seen by stripping off the epidermis of an infected leaf and examining it with a microscope.

The timescale of infection is interesting. In *Erysiphe graminis* f.sp. *hordei*, the form that attacks barley, germination and the formation of an appressorium takes about 9–10 hours. Penetration occurs after 10–12 hours, and then the growth of the haustorium begins. The finger-like extensions of the functional haustorium in this species are found after 18 hours, and at that point a second hypha begins to grow out from the appressorium onto the surface of the leaf. This branches repeatedly, and an approximately circular colony forms. Secondary haustoria are formed after about 50–70 hours. Eventually approximately one haustorium is produced for every three hyphal 'cells' of the fungus and more than one haustorium may be found in each epidermal cell of the host.

Hyphal growth and branching usually leads eventually to the whole leaf surface being covered by a network of grey-white mycelium. Grey-white mycelium is characteristic of most powdery mildews, but in some the hyphae darken as they age, as with the American gooseberry mildew, *Sphaerotheca mors-uvae*, which attacks gooseberries in Britain (Plate 5).

Just to make things difficult, not all powdery mildews behave in this way. An exception is *Phyllactinia guttata*, which infects hazel (*Corylus avellana*) in Europe and other species in the USA and Asia (Fig. 8.6). The hyphae grow from the leaf surface, through the stomata and into the mesophyll. Another is *Leveillula taurica*, powdery mildew of xerophytic Mediterranean species and of crops grown under warm dry conditions, such as pepper (*Capsicum* spp.), cotton, and globe artichoke (*Cynara scolymus*); here the vegetative mycelium grows mostly within the leaf. Infection takes place through the stomata rather than directly through the cuticle, and the hyphae then sporulate on the leaf surface. Were the original mildews like this, or is this a secondary development in response to particular conditions? At present we do not know.

In the epiphytic forms, close examination of the neck of the haustorium where it passes through the plant cell wall shows a thickening resulting from the deposition of a ring of fatty, suberin-like material. It is tempting to think that this functions as a mechanism for isolating the haustorial wall from the wall of the external mycelium, to prevent evaporation or leakage. Further speculation demands more understanding of the qualities of the wall in the surface mycelium, especially if the mechanism for survival in the hostile environment of the leaf surface is to be understood.

Each haustorium contains a nucleus and is metabolically very active. Haustoria not only remove water, sugars and other nutrients from the host but also affect the host cells. In some instances, for example, chlorophyll is retained by infected cells after the normal onset of ageing, giving a green island effect (Plate 5b); this may be caused by the secretion by the fungus of plant hormones such as cytokinin. The fungus, presumably by way of the haustoria, also stimulates an increase in the rate of flow of nutritive material from the mesophyll of the leaf to the epidermal cells in contact with the fungus.

Powdery mildews have been valuable in studies of the mechanisms involved in the uptake of sugars by haustoria, most experiments being made with *Erysiphe pisi*, the powdery mildew of pea (Plate 5a). Two approaches have been used. In the first, haustorial complexes, each consisting of the haustorium itself, the surrounding host plasmalemma and the sheath matrix between this and haustorial wall, were isolated. Once isolated, the neck of the haustorium becomes plugged as a result of the wound responses of the fungus, so the only route into the haustorial complex is through the semipermeable host plasmalemma surrounding it. These isolated complexes were then immersed in various solutions containing radioactively-labelled sugars and the pattern of uptake monitored. In other experiments, thin sections of infected, living leaves were stained with chemicals that react with and highlight the molecular 'pumps' responsible for the transport of sugars across membranes. The results from these experiments (Fig. 8.4) suggest that the presence of a powdery mildew haustorium in a host epidermal cell leads to activation of the inward facing pumps on the plasmalemma of the cell. Thus the infected cell begins to accumulate sugars from surrounding, uninfected cells. The only part of the plasmalemma on which the pumps are not activated is that surrounding the

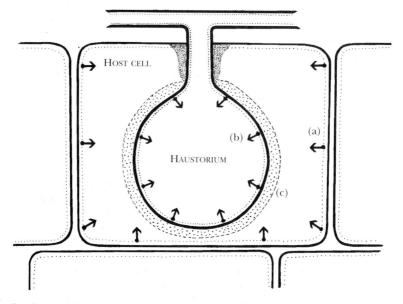

Fig. 8.4 Suggested location and orientation of molecular 'pumps' in a host epidermal cell containing an haustorium of an *Erysiphe* species. Note the inward facing pumps on the host cell membrane (a) and the inner haustorial membrane (b) (arrows), and the lack of pumps on the section of host cell membrane invaginated to accommodate the haustorium (c). (Illustration by Mary Bates, based on Manners and Gay (see Callow, 1987).)

haustorium. If these pumps were activated sugars would be pumped out of the haustorium because the section of the plasmalemma enclosing the haustorium is turned back on itself. As sugars build up in the infected cell a concentration gradient between the cytoplasm and the haustorium is established. This leads to simple diffusion of sugars across the host plasmalemma that encloses the haustorium, into the sheath matrix. From there they are pumped into the haustorium by another set of inward facing pumps on the membrane surrounding the fungal cytoplasm inside the haustorium itself.

The diversity of species and the diseases they cause

Powdery mildews on wild plants

Very few wild plants escape attack by mildews. Few of them have been studied in depth and the taxonomy of the fungi involved is unsatisfactory, for two reasons. Firstly, most occur only in the anamorph phase and where ascocarps are present they are usually hard to find, being confined to a particular part of the plant's growth cycle; many of the herbs of the European flora develop infection in midsummer, but the cleistothecia are not seen until late summer or autumn. Secondly, the taxonomy of the mildews is still evolving. Many of the forms on wild plants were assigned in earlier classifications to one or two species with wide host ranges such as *Erysiphe polygoni* and *E. cichoracearum*, but later investigations have separated particular groups from these. Thus *E. trifolii* has been distinguished from *E. polygoni* because it is confined to members of the clover and pea family (Leguminoseae) and has distinctive morphological features; similarly *E. fischeri* on *Senecio* spp. was originally part of *E. cichoracearum*, widely distributed on members of the daisy family (Compositae). However, the taxonomic literature that delineates the differences between these more closely defined species is only available in the major botanical centres, so it may be difficult for the amateur to make a precise identification even if the sexual stage is present, although Ellis and Ellis (see Bibliography) offer a guide to sound identification.

A different problem is highlighted by the species *E. polyphaga*. This apparently attacks plants as unrelated as members of the families Cucurbitaceae, Scrophulariaceae, Begoniaceae, Compositae and Crassulaceae. *E. cichoracearum* occurs on a number of distinct genera of the Compositae and *E. heraclei* attacks several species in widely separate genera in the Umbelliferae. The difficulty here is accommodating these biotrophs with wide host ranges in the model of pathogenicity developed in Chapters 2 and 3. If recognition is, as seems certain, such an important element in the delicate relationship built up between a biotrophic pathogen and its host, between parasitic gene(s) for virulence and avirulence and host gene(s) for susceptibility and resistance, then how can one genetic form of a pathogen attack many different hosts? The simplest explanation would be that the fungus either consisted of a mixture of appropriate pathogenic forms or was heterokaryotic (containing a mixture of nuclei of differing pathogenic potential), pathogenic races or forms being selected from the population of spores landing on the host surface. We do not have the experimental evidence to confirm or deny this hypothesis, but naturalists interested in plant diseases could go some way to solving the problem by studying the host range of some of the commoner mildews of wild plants. Inoculating them with single spores would be difficult, but serial inoculation of a single host to see if, after several passages on one host, it would transfer to another, would be simply done and of great interest.

There is no space to discuss in detail the many mildews found associated with wild plants. There are, however, three that are particularly worthy of notice. The oak mildew (*Microsphaera alphitoides*) (Fig. 8.5) is widely distributed in the north temperate zone and is seen every year on *Quercus robur* and *Q. petraea* in Europe. It is probably the only oak mildew in Britain; *Microsphaera extensa*, which may be part of the same taxonomic complex, has a similar distribution but is not recognised in Britain. Both are rare in the USA, but *M. lanestris* is widespread on oaks in North America and Asia. *M. alphitoides* is most easily seen on young shoots; their leaves are often distorted and look as if they have been painted with white-

Fig. 8.5 Microsphaera alphitoides, cause of powdery mildew of oak (*Quercus* sp.) in Aberlady, East Lothian, Scotland. (D.S. Ingram.)

wash. The disease was not recognised in Europe until 1907, and was thought to have been imported from the USA, but this is by no means certain. In its early years the cleistothecia, with their much branched appendages and several asci, were rarely seen and were not identified in Britain until 1946 (Fig. 8.3). Attempts have been made to relate their occurrence to physical factors such as high temperatures and drought, or to the later appearance of a sexually compatible strain. In terms of taxonomy, history and physiology there is much still to discover about the oak mildews.

The hazel mildew *Phyllactinia guttata* (= *P. corylea*, the nomenclature is confused) (Fig. 8.6) is probably widely distributed in the north temperate zone and is especially noticeable on young coppice shoots of hazel late in the year. Its immediate interest lies in the presence of some mycelium inside the leaf and in the apparently eccentric but superbly functional design of the cleistothecia (Fig. 8.7). Cleistothecia are found on the lower surface of the hazel leaf with the top facing downwards. The appendages are very distinct, being long and straight, pointed at the end and with a swollen base. In addition, on the surface of the cleistothecia facing downwards are several secretory structures like small shaving brushes which secrete a large blob of mucilage. In autumn, as each cleistothecium dries out, the irregular thickening of the appendage bases causes them to bend towards the leaf and lever the whole structure off the leaf surface, so that it falls to the ground. As it falls the appendages act like the flights of a shuttlecock, ensuring that the cleistothecium lands with the blob of mucilage facing downwards; this sticks it firmly to the surface of the leaf litter on the ground, where it remains all winter. In spring, as the new leaves of the hazel emerge, each cleistothecium splits around its equator and the upper half (originally the base, when attached to the underside of the leaf) hinges back, carrying with it the asci, each containing two ascospores. The ascospores are then shot out from the tip of each ascus to infect new leaves and begin the cycle again. Morphologically similar fungi apparently infect other plants such as chestnuts (*Castanea* spp.) and oaks (*Quercus* spp.)

The sycamore mildew *Uncinula bicornis* (*U. aceris*) commonly occurs in the north temperate zone on a variety of *Acer* spp.; in Britain it is found on field

Fig. 8.6 Phyllactinia guttata, powdery mildew of hazel (*Corylus avellana*) from Elibank, Peeblesshire, Scotland. (C. Prior, Royal Horticultural Society.)

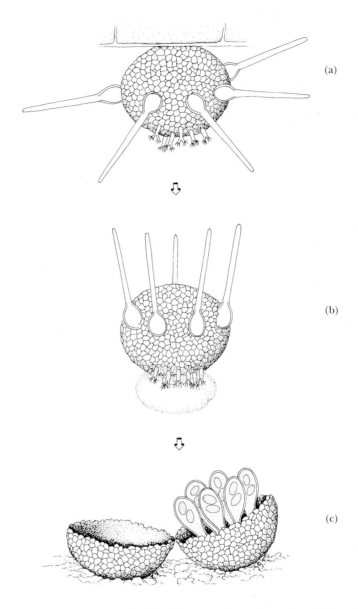

(a)

(b)

(c)

Fig. 8.7 Diagram of an ascocarp of *Phyllactinia guttata,* cause of powdery mildew of hazel (*Corylus* sp.). (a) An ascocarp attached by its base to the underside of the host leaf; note the secretory structures at the apex (pointing downwards) and the bulbous appendages. (b) The downward flight of the ascocarp; note that the appendages have bent to lever the ascocarp off the leaf, and act as flights; also note the blob of mucilage formed by the secretory structures. (c) The ascocarp attached to the ground surface; the base, with the asci attached to it, has hinged back ready for ascospore release. (Ilustration by Mary Bates, based on illustrations and descriptions by Webster (1980) and Ellis and Ellis (1997).)

maple (*A. campestre*) and sycamore (*A. pseudoplatanus*), and less commonly on Norway maple (*A. platanoides*). The mycelium and conidia occur in obvious patches late in the year, mostly on the lower surface but sometimes on both sides of the leaf, and spectacular green islands develop around the lesions as the leaves senesce in autumn (Plate 4b). The cleistothecia can often be found on the undersides of newly fallen leaves in the green island area, by which point no very obvious vegetative mycelium can be seen. They have beautiful appendages which branch dichotomously twice at the ends, with tips that curl over like ram's horns (Fig. 8.3).

Powdery mildews on cultivated plants

There are several other species that occur on both wild and cultivated plants. Where serious damage to the cultivated hosts has led to economic loss there has been intensive study of the disease. Examples include the powdery mildews of arable crops and their wild relatives, vines, hops, fruits and garden ornamentals.

Erysiphe graminis is probably the commonest of the powdery mildews, occurring on wild and cultivated grasses and on cereals (Fig. 8.8). The disease almost always develops on young plants, where the thick white powdery masses of fungal mycelium, bearing conidia, may be seen forming patches on leaves and stems. The host is not killed by the infection but, as the season advances and the disease extends, infected plants become increasingly debilitated and brown areas of dead and dying cells form around the margins of the lesions, although shaded leaves may develop green islands.

To all intents and purposes *E. graminis* looks the same whatever its host. Research has shown, however, that it exhibits a high degree of physiological specialisation with respect to host range. Firstly, as a species it is restricted to hosts within the grass family (Gramineae). Secondly, morphologically identical but physiologically distinct forms (*formae speciales*) are restricted to particular genera and species within that family. Within the *formae speciales* there may be further specialisation into races capable of attacking particular cultivars of the host, according to the combination of resistance genes carried by the host and the virulence/avirulence genes present in the pathogen (see Chapter 3).

It is difficult to estimate the crop losses caused by *Erysiphe graminis*, but they are large. Reliable figures for experiments comparing yields of sprayed and unsprayed barley show a loss in excess of 20% in unsprayed plots. The average estimated loss for all crops year on year is 5 to 6% for barley and 2 to 3% for wheat, but farmers know that when resistant cultivars are overtaken by a new pathogenic race the crop yield may be dramatically reduced and they make strenuous efforts to control the disease with sprays, which is expensive in itself.

E. graminis forms cleistothecia without clearly defined appendages on the older leaves and stubble of cereal plants in late summer (Fig. 8.8). These discharge and infect autumn-grown cereals and plants that have germinated from dropped seed, which carry the disease over to the next spring. Any cleistothecia that remain over winter are generally without fertile asci. We shall return to *E. graminis* later.

Of the several other members of the Erysiphales that cause disease in crops *E. betae* is the most studied. It attacks sugar beet crops everywhere, but is most damaging where the weather remains dry for long spells and where temperatures are high throughout the growing season. Thus it causes major problems in the southwestern USA, eastern Europe, North Africa and western Asia, but is not significant in temperate climates such as those of Britain and the eastern seaboard of the USA. Forms of the same fungus are to be found on wild species of *Beta* but not on other members of the beet family (Chenopodiaceae). Other common *Erysiphe* spp. occurring on crops include *E. pisi* on peas and other

Fig. 8.8 Erysiphe graminis, cause of powdery mildew of grasses and cereals (Gramineae): (a) surface mycelium and ascocarps on a leaf sheath of winter oat (*Avena sativa*) in Cambridge, England; (b) vertical section of an ascocarp showing the thickened wall and elongate asci, each with eight ascospores. (Photograph (a), D.S. Ingram; photograph (b) H.J. Hudson, University of Cambridge.)

Fig. 8.8a

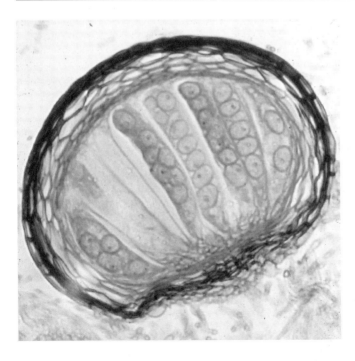

Fig. 8.8b

legumes (Plate 5a) and *E. cruciferarum* on members of the family Cruciferae including swedes, turnips and rape grown for oilseed and fodder. *E. cichoracearum*, as its name implies, attacks chicory (*Cichorium intybus*) and is also found on endive (*Cichorium endivia*), lettuce and globe artichokes (*Cynara scolymus*), becoming especially damaging on these and other crops in the daisy family in warmer climates; in British gardens it is a severe problem on *Aster* spp. (e.g. Michaelmas daisies).

Another major disease is the powdery mildew of vines, caused by *Uncinula necator*. This fungus appeared in epidemic proportions at about the same time as the potato blight (1842–5). It was first recognised in the United States and turned up in a glasshouse in Margate, England, in 1847, from whence it spread to the vineyards of Europe and threatened the wine industry, a disaster too awful to contemplate. It should not be confused with the downy mildew of vines, *Plasmopara viticola*, which was first imported from America in 1878 and which can also cause serious disease problems.

Uncinula necator, although capable of surviving very dry conditions, is not as dependent on warmth and dryness for its spread as are some other mildews. If left unchecked it soon covers all the leaves, debilitating the vines and affecting their growth and productivity. It also spreads to the fruits, stopping growth of part of the fruit wall, leading to bursting. The disease is often seen in Britain on ornamental vines, and is especially easy to spot on those with purple leaves.

The conidia are borne singly on the conidiophore, although in still conditions small chains may occasionally form. The sexual stage is rarely found and indeed was unknown in Europe until 1892; the cleistothecia have appendages with curved, forked ends. Ascospores are discharged in the spring, but by far the greatest source of overwintered fungus lies in buds infected with vegetative hyphae.

From thoughts of wine we move to beer. There are many similarities between the life cycle of the vine mildew and *Sphaerotheca humuli* (*S. macularis*), the powdery mildew of hops (*Humulus lupulus*). The characteristics of the cleistothecia in the genus *Sphaerotheca* are the long sinuous appendages and the presence of a single ascus. The disease has been recognised in British hop growing from about 1700, although the causal organism was not recognised until the late nineteenth century. It also occurs frequently on wild hops. The importance of the disease for the hop industry lies in the general debility of a severely affected crop. In addition, the pathogen often multiplies on the fruiting branches in late summer, which may lead to premature ripening of the 'cones', a reduction in the production of the flavouring principle known in the trade as alpha-acid and a consequent reduction in the appeal of the hops to buyers, who are greatly influenced by the appearance and smell of the cones in the hand.

There are a number of interesting aspects to this disease. It is widespread in Britain, mainland Europe and the eastern states of America. Hop-growing in the eastern USA, which suffered severely from epidemics of the disease in the early twentieth century, has probably for this reason been overtaken by the spread of the crop to the Pacific northwest, where the mildew does not yet seem to occur. Its significance worldwide has varied as other hop diseases such as downy mildew (*Pseudoperonospora humuli* – see p.95) and wilt (*Verticillium albo-atrum* – see p.132) have become important in their turn, but it is still recognised as a serious threat to production.

Survival of the pathogen in the absence of hop plants is in part by cleistothecia, which mature the eight ascospores in the single ascus over winter and discharge them in spring when the young foliage is emerging. The vegetative mycelium also persists over winter in external buds on the crowns of the hop stools. In the spring these give rise to heavily infected shoots, which initiate the spread of powdery mildew in the hop garden. Interestingly, the significance of both of these sources of inoculum has increased with the adoption of laboursaving methods in hop cultivation. In more labour-intensive days, infected buds were recognised and cut out; the cultivation of the soil then resulted in the burial of infected debris. Nowadays, with the replacement of cultivation by minimal tillage methods, infective debris with viable cleistothecia often accumulates around the bases of plants and this leads to an increase in infection.

Powdery mildews of fruits can be seen in almost any garden. *Podosphaera leucotricha* on apples has already been mentioned, its only method of survival over winter being within the dormant buds. It is more troublesome in some climates than in others, and in Britain is particularly severe, causing distortion and even death of heavily infected leaves. Mildew on soft fruit is caused by a number of different species. On strawberry the fungus is *Sphaerotheca macularis*, and on raspberry it is possibly a *forma specialis* of the same species. The most spectacular soft fruit infection, however, is that of the American gooseberry mildew (*Sphaerotheca mors-uvae*) on *Ribes uva-crispa* (Plate 5c). This fungus, which is indigenous on various *Ribes* spp. in the USA, was introduced into Europe via Northern Ireland in 1900. It attacks young shoots and leaves in the spring and then moves to the developing fruit, which is first covered with white sporulating mycelium, then becomes felted and black. This mycelium, despite its ugly appearance, can easily be rubbed off, demonstrating the surface nature of the mildew as well as the lack of structural damage to the host cells, as one would expect of a biotroph. Cleistothecia develop on leaves and stems and the fungus also overwinters in the terminal buds. The latest formed cleistothecia on shoots appear to be the most important for the release of ascospores in the spring. The fungus also attacks red, white and blackcurrants, especially the latter. There is in addition a European gooseberry mildew, caused by *Microsphaera grossulariae*, which occurs on leaves late in the season and is generally innocuous.

Powdery mildews are also widespread on ornamental plants grown in gardens, and in hot dry summers whole herbaceous borders may be afflicted. We highlight here just two that will easily attract the attention of the naturalist. The most striking is the rose mildew, *Sphaerotheca pannosa*. This occurs in a number of guises. Infection is most obvious and growth of the fungus most luxuriant on the shoots of wild roses (*Rosa* spp.) and on cultivars that bloom several times a year. In the old floribunda variety 'Frensham', for example, the flower stems and buds may be covered with a profuse growth of the fungus while the leaves on the rest of the plant are scarcely affected. In other cultivars the mycelium is not very obvious on the leaves or stems but, as the plant matures, patches of felt-like (pannose) mycelium consisting of a creamy layer of thickened hyphae appear. It is in these that the cleistothecia are formed. They lack appendages and contain a single large ascus with eight ascospores. The patches of pannose mycelium are often around the thorn bases or on the maturing fruit. Indeed, we can recognise certain genotypes in the wild rose population in Scotland by the striking pannose mycelium on the ripening 'hips'. Cleistothecia are known to overwinter and release spores in the spring, but again mycelium surviving in buds is responsible for much of the carry over of infection. The release of the conidia follows a diurnal rhythm probably conditioned by changes in relative humidity, with the greatest release and dispersal taking place at about noon on a rainless day, reducing in the afternoon and at night.

Our second example is powdery mildew of *Rhododendron*. This disease is instructive because it is at an early stage in its development. However, we have some way to go before we feel secure in our knowledge of it. The first problem it poses is one of identity. In the mid-1950s some diseased rhododendrons in the glasshouses of the Royal Botanic Garden, Edinburgh were found to be infected by a mildew in the anamorph (*Oidium*) phase. No sexual stage was present. As far as could be ascertained, the fungus responsible was exterminated. In 1955 powdery mildew was identified on rhododendrons in Australia, and again no sexual stage was present. No more mildew was reported in Britain until 1969, when it appeared once more on plants in the Edinburgh glasshouses, this time on *Rhododendron zoelleri*. These plants had been introduced from New Guinea the previous year, and the supposition was that the fungus had been introduced with its host, since careful examination suggested it was different from the earlier examples in Edinburgh and Australia.

The history since then is of the identification of at least two types of *Rhododendron* mildew: one similar to *Sphaerotheca pannosa* (the rose mildew), on glasshouse rhododendrons; and one similar to *Erysiphe cruciferarum* (a powdery mildew of the cabbage family), on outdoor rhododendrons (Fig. 8.9). This last

Fig. 8.9 Erysiphe cruciferarum-type on *Rhododendron stenophyllum* at the Royal Botanic Garden Edinburgh, Scotland. Note the chlorosis and distortion of the infected leaves. (S. Helfer, Royal Botanic Garden Edinburgh.)

has spread to nearly all *Rhododendron* collections in Britain, and probably the same species is also present in New Zealand; the situation in Australia and Japan awaits clarification. Powdery mildew of *Rhododendron* has been found in the USA, mostly of a different type and thought perhaps to relate to *Microsphaera azaleae*, although two other types have possibly been recognised there as well.

The reason for concern regarding the taxonomy of the pathogen is that we cannot be sure of the exact relationship between these various isolates, nor do we know whether their chance introduction to other countries might complicate the problem of control. Moreover, although the British isolates have been equated with *Sphaerotheca pannosa* and *Erysiphe cruciferarum*, this is on the flimsiest of morphological evidence and it is impossible at present to extrapolate any of our knowledge of these organisms to the situation in *Rhododendron* mildew.

The effects of infection are various. In some *Rhododendron* spp. shoots and leaves are obviously infected, often yellowing and dropping off the plant, and the typical mildew mycelium can be seen. In others the effects are less obvious and limited infections on the undersides of leaves produce dark lesions on the upper surface, often persisting in the older leaves and sometimes causing defoliation. The degree of damage varies from species to species, sometimes causing massive leaf loss and death of the plant, sometimes being hardly noticeable. *R. ponticum*, the pernicious weed of many natural ecosystems is, ironically, relatively resistant to infection.

Diversity within species

Probably the best documented variation within species of powdery mildews is that of *Erysiphe graminis* on cereals. The presence of *formae speciales* on different cereal and grass species has already been mentioned. Although there is occasional evidence of mildew crossing over from grass hosts to cereals, the specialised forms on cereals seem to be fairly strictly confined, with no crossover among wheat, oats and barley. This is well shown in New Zealand, where powdery mildew has always been present on wheat and barley, but did not appear on oats until at least 1973.

The recognition of variation within these specialised forms followed the development of cereal varieties in the early days of the twentieth century. Previously cereals had been grown as 'land races', morphologically similar genetic mixtures from particular parts of the country, recognised by a name. At that time mildew does not seem to have been a problem, perhaps because of the genetic variation among the hosts.

Advanced farmers have always selected especially fine ears of wheat or barley from these land races to create new varieties, but especially so since the emergence of Mendelian genetics at the beginning of the twentieth century. The cereals are largely self-pollinating, so over a number of generations of ear selection they breed true and a variety becomes 'fixed'. As the new varieties appeared so also did the science of agronomy and the mildew. Comparing these varieties in field plots, farmers saw mildew in a way that had not been possible before. The next step was to make crosses between varieties and to select new varieties from the progeny of these crosses. A whole spectrum of susceptibility and resistance to mildew was displayed by the new varieties. Nevertheless, in the early years of the twentieth century, although cereal mildew was recognised as a disease that could be troublesome, it was still thought of as a natural hazard that had to be endured. There was a mildew epidemic in Germany in 1903 but the pathologists had no remedy for it.

In barley (and a similar story can be told of wheat and oats) an attempt was made to incorporate mildew resistance in some of the new crosses after the first world war. In 1931 'Pflug's Intensive' was identified as a barley apparently completely resistant to mildew. It was a selection from a European land race and was used in crossing programmes to produce a number of the well-known barley varieties of that period. In succeeding decades, further sources of mildew resistance have been found, and since the Second World War the barley breeding programme has reflected the need to incorporate new sources of resistance in new varieties. These resistant varieties allow the identification of races of the mildew and in turn of genes in the host for resistance and susceptibility and in the fungus for virulence and avirulence (see Chapter 3). Resistance has been found in further selections from land races and also in exotic species like wild barley (*Hordeum spontaneum*). Crossing cultivated barleys with these exotic species is sometimes difficult, but an even greater problem is the array of 'wild' characteristics that are carried into the breeding line and have to be bred out by intensive selection. Different types of resistance are available, from near immunity to various levels of reduced susceptibility with small lesions and limited sporulation.

In barley most of the breeding has concentrated on genes for near immunity. The incorporation of these has generally led to a corresponding change in the fungus over a shorter or longer time, so that the resistance has broken down. We have here nothing more than a 'plant pathologist's merry-go-round' similar to that described for the rusts (see p.97). Before moving on to discuss the utility of these plant breeding efforts we should note that some plant breeders believe that this approach is wrong, and the best way to breed 'adapted' varieties is by discarding the most susceptible plants in any breeding line and gradually building up a genotype with a tolerance of, rather than a high level of resistance to, the pathogen.

However, since breeding programmes from the 1930s to the present day have relied on the incorporation of genes for resistance or immunity, what strategies can be adopted to make best use of them? One is to turn out new varieties incorporating new genes for resistance as fast as possible and to combine with them, in a 'belt and braces' approach, the development of new fungicides and spraying techniques. This approach, of course, is favoured by commercial interests.

Another approach is to sow varieties with different resistance genes in a patchwork of fields over a wide geographical area, to slow down the rate of epidemic spread of different forms of the pathogen and possibly prevent the build-up of new races. A similar strategy is to separate the areas for growing winter and spring barley. For example, when a build-up of mildew on winter barley posed a threat to the spring barleys in Scotland the advisory service sought to discourage the growing of winter barley in areas where the valuable, mildew-susceptible spring barley 'Golden Promise' was grown. But this did not succeed. Farmers require flexibility and the freedom to grow what seems to them to make commercial sense. The high yields of winter barley and its value in spreading farm work between autumn and spring far outweighed the problem of mildew.

Another strategy is to hark back to the land races of the past, where mildew was apparently not a problem. It is known that they contained a population of barley genotypes with differing levels of mildew resistance and susceptibility. Could they be simulated by growing mixtures of barley cultivars containing a variety of genes for resistance to present day mildew races? Such mixtures have indeed been made and are effective, but they pose problems in farming. Straight mixtures of cultivars certainly reduce the severity of mildew attacks and more doubtfully may even show an advantage in terms of the exploitation of above-ground space and below-ground soil fertility, because they are morphologically different. The problems arise, however, because of the very tight specifications of the end user. In brewing, for example, when two or more cultivars are present in a crop it is impossible to get the uniform endosperm quality or the uniform germination required for making good malt. And since the brewer already accepts or rejects the crop in terms of a set of somewhat subjective commercial criteria the farmer cannot afford to add any more disadvantages to the crop for sale.

It is true that it is possible to construct a 'multiline variety' which is reasonably uniform for morphological and commercial features, but which contains a population of different resistance factors for mildew. The problems in the use of such 'varieties' lie in the commercial strategies of the breeding companies and the legislative framework within which varieties are defined and released. When it is realised that improvements are continually being made in yield and commercial qualities, and that disease resistance has to be incorporated for other pathogens too, it is clear that the success of a cultivar depends on elements of chance, fashion and other indefinable factors. Nevertheless, barley mixtures designed to overcome the problems of disease can now be found in the catalogues of some seed merchants. It is good to know that the science behind resistance breeding is partly understood and that methods exist for the rational production of varieties that resist disease.

It is one thing to find evidence of pathogenic variation in fungi causing disease of cultivated plants, where the actions of plant breeders slowly reveal the range of pathogenic variation present. It is quite another matter to investigate this phenomenon in wild plants. However, in Israel, which is thought to be a centre of origin of cultivated barleys, not only are cultivated barleys infected by mildew but their wild relatives are also infected by genetically similar forms of the fungus. This has made possible a detailed investigation of the range of pathogenicity present in the wild populations. It is at least equal to that identified in cultivated barley. Some of the races in geographical or ecological isolation are more virulent on cultivated barleys than any previously known race. Conversely, there are genes in the wild *Hordeum* population for all degrees of resistance, including some for resistance to all the mildew races so far recognised on cultivated barleys.

Although this is an indication of the presence of two interacting genetic systems in wild host and pathogen combinations, there is a lingering worry that the presence of the population of mildews on cultivated barleys might be influencing the situation. This problem has been taken care of in studies of the interaction between the common groundsel (*Senecio vulgaris*) and its mildew *Erysiphe fischeri*, for no cultivated form of *S. vulgaris* exists. First a number of inbred lines of the *Senecio* were selected and tested for susceptibility to a number of genetically homogeneous isolates of the mildew. The lines were easily obtained by isolating single chains of conidia, picked up on a dry needle, as the primary inoculum (every spore in a chain is derived from the same mother cell, so all are identical). Once the inbred lines of the host and the pure lines of the pathogen were established, replicated trials of the reactions of the host to the pathogen were carried out, using detached leaf segments placed (in Petri dishes) on water agar containing the senescence-retarding chemical benzimidazole. In this simple set up the leaf segments were kept fresh and there was no inhibition of the growth of the mildew.

It was shown that a wide range of resistance and susceptibility to mildew existed in the host population, with an equally wide range of pathogenic and non-pathogenic forms among isolates from the wild population of the pathogen. For example, a sample of only 24 mildew isolates yielded 19 different races on the differential hosts employed. Isolates from Glasgow and from Wellesbourne in Warwickshire, England, showed a similar and partially overlapping heterogeneity. From this and other evidence provided by studying the behaviour of pathogens on wild plants it is possible to postulate that co-evolution of host and pathogens continues throughout the evolutionary life of plants.

Mildews and fungicides

Powdery mildews have played a significant role in the development of modern fungicides. It was shown in about 1850 that sulphur could be used to control vine mildew, then a serious threat to European wine production. It was used in various forms as a sulphur dust, as sulphur painted onto glasshouse heating pipes, and in mixtures with water, soap and so on to control a variety of powdery mildews. 'Eau Grison', or 'Bouille Versaillaise', was compounded from sulphur and lime boiled together and in time became the lime-sulphur widely used in orchards in Europe and the USA. It comprised a concentrated mixture of calcium polysulphides which broke down on dilution with water and exposure to the air to give a suspension of fine particles of sulphur along with other sulphur compounds. However, lime-sulphur had many disadvantages: the spraying regimes required the expensive transport of great quantities of water; it could under certain circumstances be toxic to fruit varieties; and in all circumstances it was toxic to certain 'sulphur-shy' varieties such as the apple 'Stirling Castle' and the gooseberry 'Leveller'. Even so, lime-suphur was increasingly a standard part of orchard management in the first half of the twentieth century. At the same time organic fungicides were being sought which would be easier to use and which might be used in lower volume sprays.

Meanwhile the breeding programme that was producing cereal varieties resistant to mildew also brought into commerce a number of high-yielding varieties which became spectacularly infected with mildew when the genetic resistance in the host was overcome by a change in the pathogen. Farmers sought to ameliorate this situation by the application of sprays, and although this was a difficult operation in fields of standing barley or wheat, new spraying equipment with long booms mounted high on the tractor and the development of sowing techniques that left 'tramlines' (unsown drills for tractor access) allowed the spraying of cereal crops without damage. This opened a new market to the manufacturers of sprays and even greater efforts were put into finding and developing new compounds. By 1960 non-phytotoxic derivatives of the hitherto phytotoxic dinitrophenols were being produced and these were very effective in controlling mildew. There followed a series of syntheses which produced other useful fungicides. To iterate them here would be tedious, but three groups are particularly noteworthy. The first, the 2-aminopyridines, were synthesised in the mid-1960s. These were systemic fungicides which could enter the xylem of the plant and diffuse upwards. Some of them proved to be particularly useful as soil drenches for the control of mildew in glasshouses on, for example, cucumber, one application keeping the plants disease-free until harvesting. Others could be used in similar fashion as a seed-dressing on cereals, as well as being valuable sprays; a seed application kept the young plant free of mildew for a while, but the protection was not sufficient to last for the whole life of the plant and was usually supplemented with subsequent sprays. The timing of the application of such sprays is critical, and constant monitoring is important for the correct spraying sequence. An interesting feature of these fungicides is their capacity to disrupt the behaviour of the mildew on the leaf. Instead of growing over the leaf surface it begins to grow up and away from it.

The other two synthetic systemic fungicides can be considered together. These are the benzimidazole and morpholin fungicides, which inhibit sterol synthesis, but at a different site in each case. The benzimidazoles are active against a range of organisms, including mildew, while the morpholines are particularly effective against mildew species.

Benomyl, which is one of the benzimidazole fungicides, came into use in the late 1960s and shortly afterwards adapted forms of the mildews were found which it could not control. This phenomenon of genetic adaptation to resist fungicides has a superficial similarity to the changes that take place when pathogenic variants of a fungus attack hitherto resistant host varieties. There is now ample evidence that pathogens develop resistance to a range of new organic fungicides soon after their release. Whether the nuclei controlling such resistance pre-exist in the population, or whether resistance develops as a consequence of new mutations, is not yet clear. The practical consequence is that it is necessary to develop new fungicides, to ring the changes with existing ones or to apply mixtures.

The way ahead lies in a fuller understanding of the nature of pathogenicity and resistance, so that spraying and the manipulation of resistance go hand in hand. The selection of resistant types with genes for slow colonisation by the fungus and for mature plant resistance, which seem less liable to be overtaken by a change in the fungus, may produce more stable varieties. The ideal would be a situation in which the utilisation of different types of resistant cultivars, alone or in mixtures, eliminated the need for commercial fungicides altogether. We recognise, however, that there are a great many vested commercial and political interests that might militate against the early development of such a situation.

The powdery mildews are a particularly interesting group for the naturalist to study. Easy to see and to collect, in the teleomorphic phase they are readily identifiable to the level of the genus. At the species level there is still much observational work to be done by the field naturalist, and as we have illustrated here there are many unsolved pathological problems which would respond to careful study with limited equipment.

9

A Treacherous, Mutable Tribe
The Rusts

The rust fungi may derive their common name from the wonderful colours of the various spore stages, which range from yellow-ochre and orange to dark brown, reminiscent of the range of colours generated by the oxidation of iron. They are members of the Uredinales (see p.28) in the Basidiomycota and are a widespread group of highly specialised biotrophic pathogens with a great capacity for genetic change. Moreover, they parallel some of the parasitic invertebrate animals such as the Protozoa and the Platyhelminthes in having complex life cycles with a number of different spore types produced in sequence, often on two unrelated hosts. The hyphae grow internally between the living cells of these hosts and normally produce haustoria to tap the internal tissue cells. Until 1959 the rusts were considered to be the epitome of physiological obligate parasitism (see p.46) and even now only a few species have been successfully cultured on artificial media (Fig. 2.5, p.47).

Rusts and crops

The rusts are so intimately tied in with the history of cultivation that this is where we shall start. On cereal hosts, where they are particularly conspicuous and a cause of much economic loss, yellow, orange, or brown pustules of asexual dispersal spores, the urediospores, are found on leaves and stems throughout the growing season. These are generally followed on the same plant by pustules of the dark brown over-wintering teliospores, in which the two haploid nuclei fuse and undergo meiosis, eventually entering the haploid basidiospores which infect an alternate, unrelated host (the word 'alternate' is used because the rust alternates between the cereal host and the barberry in sequence, as part of its life cycle) and give rise to pycniospores and aeciospores (see p.178 and Figs. 9.1 and 9.2). Severe infection of the cereal host leads to increased water loss, to diversion of food material from the developing grain and consequently to poor yields or even the death of the plant.

Epidemics of cereal disease were known in the ancient world and it has been argued, persuasively but without real evidence, that the Biblical seven lean years in Egypt and the Middle East which, with their consequences for Joseph and Moses, so profoundly influenced the history of the Jewish people, were the result of a rust epidemic. Various references in Greek and Roman writers are also taken to refer to the rusts, but it is not always easy from the written accounts to separate the effects from those of powdery mildews (see Chapter 8), smuts and bunts (see Chapter 10) and unfavourable droughts or drying winds.

The Romans were sufficiently troubled by crop failure, in part probably from the effects of cereal rusts, to believe that they were caused by a supernatural force – Robigus, the spirit of the diseased crop. They thought this spirit behaved like a fox with its brush on fire, running through the crop and setting it alight. To propitiate Robigus, an annual ceremonial procession for the lustration of the crops was instituted on 25 April, at about the time rust usually first appeared. This Festival of the Robigalia was also designed to propitiate Mars, the god of war and farming, and the dog star Sirius, which appeared in the sky at the same time. The procession left Rome by the Flaminian Gate and proceeded by way of the Milvian Bridge to a sanctuary at the fifth mile of the

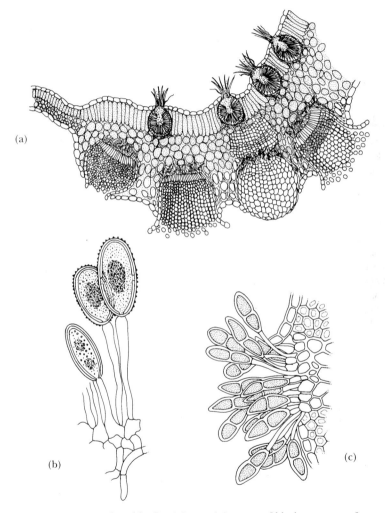

Fig. 9.1 Spore stages produced by *Puccinia graminis*, cause of black stem rust of grasses and wheat (Gramineae): (a) transverse section of a leaf of barberry (*Berberis vulgaris*) with pycnia producing pycniospores and receptive hyphae on the upper surface and cup-shaped aecia with aeciospores on the lower surface; (b) urediospores emerging from a wheat leaf; (c) a transverse section of a wheat leaf with a teliospore pustule containing typical 2-celled, thick walled teliospores. (Drawings (not to scale) by Mary Bates after A. De Bary (1887).)

Via Claudia. There a sheep and a dog, the latter as fox-like as possible, were sacrificed to propitiate Robigus and ensure freedom from disease in the growing season to follow. The strategy did not work! Later, in medieval Europe, barberry was thought to exert a malign influence on wheat by causing rust, so bushes were destroyed to prevent this. It was to be many centuries before this belief was proved correct, and barberry was shown to be the 'alternate' host of the black stem rust of cereals, *Puccinia graminis*.

Black stem rust is the classic rust disease (Plate 6c; Figs. 9.1 (previous page) and 9.2), with pustules of rusty brown urediospores being produced on the wheat leaves throughout the summer. Later, black teliospores form in elongated streak-like lesions, especially on the stem (Fig. 9.3), and later still, in spring, the bright orange aeciospores occur on the barberry (Fig. 9.4). Jethro Tull, the inventor of 'horse-hoeing husbandry' and a keen observer and commentator on the contemporary agricultural scene, says that in 1725 England suffered a rust year 'the like of which had never been seen'. Crop losses were recorded in Italy in 1766 and in Sweden in 1794, and further epidemics occurred in England in 1804 and in the 1820s. These epidemics followed particular weather conditions when the warmth and moisture were optimal for the spread of *P. graminis*, and were dependent on the presence of the alternate host, the barberry. Similar though less severe epidemics still occur.

In the USA the increase in wheat-growing after the eighteenth century followed the successive thrusts from the settled areas into the undeveloped frontier lands. There was a major expansion with the opening up of the prairie

Fig. 9.2 Stages in the long life cycles of rust fungi such as *Puccinia graminis* and *Puccinia poarum* with two hosts and several spore types.

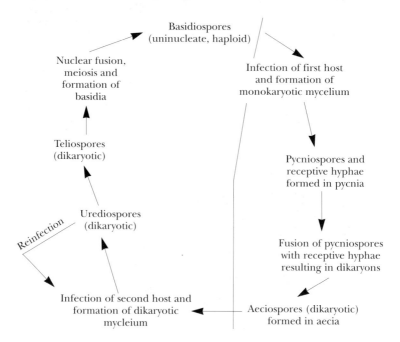

grassland and this cropping soon began to suffer from rust epidemics, which still continue. The figures quoted for current annual losses from black stem rust in the USA amount to many millions of dollars. Bald figures from the pioneering days hide the human tragedies that rust epidemics could inflict. Imagine a settler family that has braved distance, rough tracks, natural hazards and hostile native inhabitants to reach a new farm. The sod has been broken with great difficulty and a seed bed prepared for wheat. The produce from that crop lies between the family and poverty or starvation. An epidemic of black stem rust would destroy not only the crop but also that of neighbours; the whole area would be desolate and the outlook would be dire, for there would be no seed for the next year's crop. Fortunately the weather conditions conducive to such epidemics were not present every year. This rust, which grows best at higher temperatures, has now spread to all parts of the world where wheat is grown, but is especially serious in countries where the growing season is hot, notably the USA and Australia.

The barberry, the alternate host of black stem rust, is a European plant generally assumed to be native, although it probably originated, like the cereal grasses, in the Middle East. From there it spread throughout Europe with the Romans. There are very similar *Berberis* spp, such as *B. lycium* and *B. aristata,* in

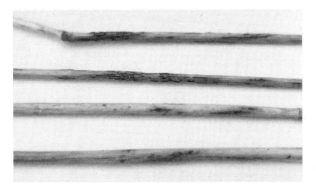

Fig. 9.3a (above) and *Fig. 9.3b* (below)

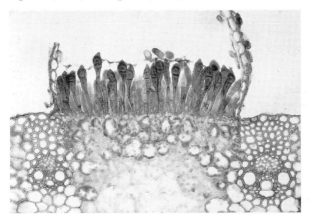

Fig. 9.3 Teliospores of *Puccinia graminis,* cause of black stem rust of grasses and wheat (Gramineae): (a) lesions on a grass stem in Thetford Forest, Norfolk, England; (b) transverse section through a lesion with teliospores emerging through the broken epidermis and cuticle. (Photograph (a), D.S. Ingram; photograph (b), H.J. Hudson, University of Cambridge.)

the Himalayan region of Asia but these, although infected by *Puccinia graminis*, seem to play little part in the epidemiology of cereal rusts in that area.

In Britain the barberry was planted as a hedge, its thorns providing an effective deterrent for keeping animals in and trespassers out and its fruits being used for conserves. This useful plant was transported to America with the early colonists, but when its connection with the wheat rust was established strenuous and mostly successful attempts were made to eradicate it, and it is still an offence to plant barberries in certain states. Nevertheless, the fungus also attacks other wild *Berberis* spp. in the United States, and in Europe it occasionally infects the young berries of *Mahonia aquifolium*, a related species of American origin; the berries, unlike the leaves, have not normally developed a thick cuticle when spores are available for infection. Chinese and Chilean barberries are generally not attacked.

Several other species of *Puccinia* attack cereals, as we shall see later, and almost all other important crop plants are affected by rust fungi. Strenuous efforts made since the beginning of the twentieth century to control these diseases, by eradicating the alternate hosts, using fungicides and breeding resistant cultivars, however, have much reduced their effects on agriculture. But all is not lost for the naturalist, for they occur in great numbers on both wild and garden plants.

Dipping a toe into rust diversity

First a disclaimer: rusts that we refer to as 'easy to find' or 'common' may not be found in a particular district and where they do occur they may be plentiful or rare in different seasons. Attempts have been made to understand the reasons for such variation, but so far with little success. There are too many potential variables: the fungus and the host are both evolving for virulence and resistance; the species mixture and the density of the host in any population or region varies with environmental factors; and the climate locally and seasonally varies too.

The best time to start looking for rusts is as soon as the green leaves begin to expand in the spring and early summer, when they are plentiful and easily recognised. Among the earliest in Britain and northwest Europe is *Uromyces muscari*, a rust with only one host (Plate 6). It forms characteristic large lens-shaped

Fig. 9.4 An aecium of *Puccinia graminis* on the underside of a leaf of barberry (*Berberis vulgaris*). (D.S. Ingram.)

lesions orientated lengthways on leaves of the (English) bluebell or wild hyacinth (*Hyacinthoides non-scripta*) and its relatives in the wild and in gardens. In each lesion concentric rings of dark brown teliospores will be found erupting through the surface of the living green leaf, each spore consisting of a single thick-walled cell borne on a stalk (the pedicel), a characteristic of the genus *Uromyces*. If infected leaves are placed in a polythene bag with a moist tissue and left in a dark cupboard for a couple of days they will begin to senesce and turn yellow, leaving a clear pale green island around each lesion (Plate 6).

Another common and easily seen spring rust with a single host is *U. ficariae* on the young leaves of the lesser celandine (*Ranunculus ficaria*) (Fig. 9.5). Pustules of dark brown, almost black teliospores are clearly visible on the infected green leaves. But, just to emphasise the complexity of rust life cycles, another quite separate rust, *U. dactylidis*, can be found on lesser celandine at the same time of year. This is a rust with a longer life cycle and two distinct hosts. The spores formed on *Ranunculus* are the final stage of the sexual phase of the life cycle, called aeciospores. They occur on the leaves in sunken, cup-like structures, the aecia. The urediospores and teliospores of this rust (the dispersal and resting spores respectively) are produced on the leaves of quite different alternate hosts, the grasses *Poa* spp. (meadow grasses), *Dactylis* spp. (cocksfoots) and *Festuca* spp (fescues).

Yet another common spring rust is *Puccinia violae*, which first appears as very definite aecia containing orange aeciospores on the leaves of a range of violets (*Viola* spp.), in spring. Later in the year the urediospores and teliospores are formed on the same plant.

In this short exploration we have seen examples of the three different kinds of rust life cycle: short-cycle and autoecious as in *Uromyces muscari* and *U. ficariae*, with a reduced number of spore stages all occurring on a single host species; long-cycle and heteroecious as in *U. dactylidis*, which, like *Puccinia graminis*, has several spore stages divided between two quite separate host species; and long-cycle and autoecious as in *P. violae*, with several spore stages all occurring on one host species. Before going further it might be helpful to look at a single life cycle in a little more detail, to clarify the relationships between the different spore types.

Fig. 9.5 Teliospores of *Uromyces ficariae* on leaves of *Ranunculus ficaria* in Edinburgh. (D.S. Ingram.)

Life cycles

Puccinia poarum is a good rust to begin with (Fig. 9.2): it is plentiful and easy to find in the spring, and has a long life cycle with stages on coltsfoot (*Tussilago farfara*) and on meadow grasses, especially *Poa pratensis* and *P. trivialis*, cocksfoot and fescue, *Festuca* spp.

There is no sign of rust on the coltsfoot when the early flowers shoot up from the bare earth in early spring, but as soon as the leaves appear and reach full size look for pale, circular, blister-like spots on some of them. Each spot, which is at first yellowish, then deepening and sometimes becoming red and purple in colour, represents an infection focus, probably derived from a single basidiospore. The source of these spores will become clear shortly.

In the earliest stages the upper surface of each pale spot comprises many tiny, yellow, flask-shaped structures embedded in the leaf surface called pycnia (an alternative name is spermagonia). In damp weather these ooze droplets of slime which on microscopic examination can be seen to be full of minute uninucleate spores, the pycniospores (or spermatia). In addition, hyphae, called receptive hyphae, which have one nucleus per cell, protrude from the mouth of the pycnia. The pycniospores are carried from plant to plant by insects. When a pycniospore from one plant is deposited on a receptive hypha of another it adheres to it and the two fuse, thus allowing the entry of a nucleus and the creation of a dikaryotic cell (see pp.27–29). This process is akin to fertilisation in normal sexual reproduction. Following the entry of the nucleus, hyphal cells deep in the leaf become dikaryotic, either by the migration of the introduced nuclei or by growth of hyphae from the fertilised cell.

Eventually, groups of dikaryotic hyphae begin to aggregate on the underside of the leaf immediately below the groups of pycnia, and produce beautiful clusters of tiny cups, the aecia (Fig. 9.6). Each cup is packed with rows of bright orange, squarish dikaryotic spores, the aeciospores. The oldest cups are in the centre of the cluster, with the younger ones to the outside.

By this time the young leaves of the alternate hosts, meadow grasses, are beginning to emerge. The aeciospores are carried to these on the wind, and there, if sufficient moisture is available, they germinate. The germ tube grows parallel to the longitudinal axis of the grass leaf, often along the slight depression where two epidermal cells adjoin one another. Eventually, over one of the stomata, it stops growing and forms an appressorium. From this a penetration hypha grows into the leaf and establishes an intercellular, dikaryotic mycelium that taps the living mesophyll cells with haustoria. Eventually aggregations of hyphae occur under the epidermis and from these the orange, single-celled and thin-walled urediospores are produced. These are the dispersal spores on the grass host. Each is dikaryotic. In due course the epidermis splits and the

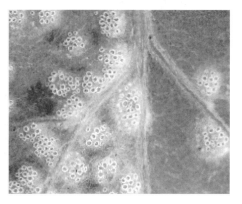

Fig. 9.6 Aecia of *Puccinia poarum* on the underside of a leaf of coltsfoot (*Tussilago farfara*). (D.S. Ingram.)

urediospores are discharged into the air. Throughout the summer they continue infecting other *Poa* plants, which in turn produce more urediospores which infect further *Poa* plants, and so on. Thus the number of infected plants in the population rapidly increases.

As the summer wears on the dikaryotic mycelium stops producing urediospores and begins to form instead dark brown, two-celled, thick-walled resting spores, the teliospores. Each of these again contains two nuclei in each cell. Later in the development of the teliospores the two nuclei fuse together to produce one diploid nucleus per cell containing the genetic information from both parents.

In the spring the teliospores germinate, but first each diploid nucleus undergoes a meiotic division to form four haploid nuclei, each containing genetic information from both parents. At about the same time a short hypha, the basidium, grows from each cell of the teliospore and buds off four basidiospores along its length, each containing one of the four haploid nuclei. This completes the sexual phase of the life cycle. The basidiospores are finally carried in air currents to the emerging coltsfoot leaves, which they infect, beginning the cycle again. In fact, *Puccinia poarum* is exceptional in having two complete generations in one year, with teliospores forming and germinating in midsummer, and after a further cycle forming in the autumn before overwintering.

Thus the complex life history of *P. poarum*, like that of *P. graminis*, is neatly integrated with the complementary life cycles of its two hosts, allowing reproduction and dissemination of the pathogen to take place throughout the spring and summer months. Moreover, a sexual phase, with consequent genetic mixing, is built in and ensures that the teliospores are furnished with new combinations of genetic information to cope with any changed environmental conditions that may have arisen during the resting period.

At the other extreme are *P. malvacearum* (Fig. 9.7, overpage), common on garden hollyhock (*Althaea rosea*) and its relatives in the mallow family (Malvaceae), including musk mallow (*Malva moschata*) and common mallow (*M. sylvestris*); and *Endophyllum sempervivi*, the rust of common houseleek (*Sempervivum arachnoideum*).

In *Puccinia malvacearum* the dark teliospores germinate where they are formed on the leaf to give four uninucleate basidiospores which infect new leaves of the same host and re-establish the binucleate mycelium, possibly by an exchange of nuclei between mycelia. In *Endophyllum sempervivi* the mycelium established by the germination of a basidiospore is perennial in the host, and from it are developed first pycnia and then aecia. The aeciospores function as teliospores, while retaining the appearance of aeciospores, giving rise to basidia which produce four basidiospores on long stalks, to repeat the cycle. The exact nature of the sexuality of these and other forms is not fully understood.

Between these extremes there are many variations, as we have seen. Where the climate is mild, rusts often propagate continuously by means of the nonsexual urediospores, without the intervention of teliospores or aeciospores. In contrast, in climates with a cold winter or a long dry spell, there seems value in having a resting teliospore stage, with or without an alternate host. It is suggested that in climates with a very short summer period, as on mountain tops or in higher latitudes, shortened life cycles ensure a rapid throughput of host

*Fig. 9.7 Puccinia
malvacearum* on mallow
(*Malva sylvestris*) in
Cambridge, England.
(D.S. Ingram.)

and fungus to overcome the early onset of winter. This may also be the case where the host has only a short period of vegetative growth, as with the spring-flowering species of temperate woodlands, for example.

There has been much speculation about the origin of the life cycles of the rusts. Which way did evolution go – from short to long life cycles or, as seems more logical, from long to short? And which came first, the heteroecious forms with two hosts or the autoecious forms with one host? The oldest rusts may be those which infect the most primitive plants: *Uredinopsis*, *Milesina* and *Hyalospora* spp., for example, which produce urediospores and teliospores on ferns and their aecia on various conifers. It has been suggested that this confirms heteroecious species as the original forms, with autoecious forms emerging later. These speculations are probably fruitless, however, for each rust will have evolved a life cycle appropriate to the climate, hosts and ecosystem with which it is currently successful. The evolutionary changes required may have been in more than one direction, as circumstances changed. The techniques of molecular biology are now making it possible to elucidate with relative accuracy the evolutionary origins of many organisms (see p.98) and may in due course throw more light on the vexed question of rust life cycles.

Modern understanding of genes and mutation, and of the nature of pathogenicity, gives us now an opportunity to begin to appreciate what changes are necessary to allow one biotroph to attack two unrelated hosts, developing at one and the same time the capacity to overcome two quite separate host defence mechanisms, a capacity that no other group of biotrophic fungi appears to have achieved. Indeed, a close acquaintance with the rusts and the ease with which 'new' hosts are attacked – witness the large number of grass species infected by the urediospores of *Puccinia graminis* derived from aecia on common barberry – suggests that a very simple step in the pathogenicity process is sufficient to allow the colonisation of a new host. Recently a gene locus has been identified at which a simple change is all that is needed to cause the rust to change hosts.

Many things are possible in the rusts, and it is the study of the different short and long-cycle rusts and their possible interrelationships which gives much of the fascination to these organisms. Indeed, close observation in the field may sometimes reveal evidence of the possible origin of short-cycled rusts when urediospores, or more spectacularly teliospores, are found partially or completely replacing the aeciospores on the aecial host. A good example of this plasticity has been seen in *Uromyces striatus* which is related to and may indeed be a part of the *U. pisi* complex. It forms its urediospores and teliospores on the legume hosts. The alternate host is the cypress spurge (*Euphorbia cyparissias*). In a population of *E. cyparissias* in Suffolk, one colony was found to have groups of shoots with well-marked teliospores and occasional urediospores replacing the aecia. These correspond to the short-cycled species referred to as *U. scutellatus*. Other shoots of the colony bore normal aecia and were shown by inoculation to infect peas and *Medicago* spp., on which they produced urediospores and teliospores and would be referred to as the heteroecious species *U. striatus*. This is evolution in the making.

Going deeper into the taxonomy of the rusts

The characteristics of the different spore types are central to the classification of the rusts, with the teliospore characteristics being especially important. There are 14 families recognised worldwide, but the temperate genera (Table 9.1) are often grouped into three families, as follows.

Table 9.1 The teliospore characteristics of some common genera of the uredinales (rusts).

Genus	Teliospore characteristics
Uromyces	Single-celled with a distinct stalk (pedicel) and a clearly defined apical pore.
Puccinia	Two-celled, occasionally 1, 3 or 4-celled with a pedicel which may be long or short; pore apical in top cell, sub-apical in lower cells.
Triphragmium	Three cells in a group, flattened laterally and verrucose, with a short pedicel; pore apical in each cell.
Phragmidium	Two to several cells in a row with thick verrucose walls and a long pedicel, often swollen at the basal end; two or more pores in each cell.
Gymnosporangium	Two elliptical or elongated cells with a long pedicel that becomes gelatinous when moistened; massed pedicels form 'horns' with teliospores all over the surface; usually two pores per cell.
Cronartium	Single-celled, arranged in chains, many chains being combined to form a cylindrical column that is horny when dry. Teliospores germinate *in situ*.
Coleosporium	Single-celled at first, later dividing into four cells in a line; sessile; formed in red waxy crusts, sometimes beneath the epidermis.
Melampsora	Many contiguous single cells arranged in a layer and adhering laterally; sessile; single indistinct apical pore.
Pucciniastrum	Two to four-celled, sessile, in a sub-epidermal layer; one pore in upper part of each cell.

The Pucciniaceae

In the Pucciniaceae the teliospores are stalked. In *Puccinia* spp. they are two-celled and in *Uromyces* spp. they are single-celled. A number of common species of *Puccinia* and *Uromyces* have already been referred to, but many other members of these genera may be encountered in the wild and in the garden. Indeed, the garden can be a good place to 'get one's eye in' before venturing further afield. *Puccinia antirrhini* (Fig. 9.8), for example, produces dark chocolate brown urediospores and sometimes teliospores on snapdragon (*Antirrhinum* spp.) all through the summer. *P. iridis* is found in some localities on irises (Iridaceae), where its brown urediospores and more rarely teliospores are produced in abundance on the leaves, the lesions often being surrounded with a yellowish halo, and *P. allii* produces orange urediospores on leeks and wild and cultivated onion. Two species of *Uromyces*, *U. appendiculatus* and *U. viciae-fabae*, infect runner beans (*Phaseolus vulgaris*) and broad beans (*Vicia faba*) respectively. White aecia as well as dark brown urediospores and teliospores are sometimes seen on runner beans. This rust was rare in Britain until 1952, since when it has become common and damaging in southern England. On broad beans the yellowish-white aecia are more common. However, *Puccinia menthae*, the mint rust, is the best for studying all spore stages on a single host. It grows within the living tissues of the underground stems of the mint plant (*Mentha spicata*) and may overwinter there. Then in spring large numbers of pale, almost dirty-looking orange aecia are formed on the swollen and distorted stems that grow up from the underground stems (Fig. 9.9). Later, pale brown urediospores are produced on the leaves, followed by much darker brown teliospores. These fall to the soil and infect emerging young shoots the following spring. Clearly there is more to mint sauce than is commonly thought!

Triphragmium, also in the Pucciniaceae, has three-celled teliospores. The most commonly encountered species is *T. ulmariae*, which produces bright

Fig. 9.8 Urediospores of *Puccinia antirrhini* on *Antirrhinum grossii* in Edinburgh, Scotland. (Debbie White, Royal Botanic Garden Edinburgh.)

orange uredinoid aecia and yellowish urediospores and brown teliospores on meadowsweet (*Filipendula ulmaria*). In *Phragmidium* the teliospores have several cells and all species are autoecious, occurring only on members of the rose family (Rosaceae). The most easily seen species is *P. violaceum*, on the bramble (*Rubus fruticosus*): the aecia form in reddish-brown lesions with a violet margin on stems and leaves. The urediospores that follow are bright yellow-orange and the teliospores chestnut-brown (Fig. 9.10, overpage).

The genus *Xenodochus,* another member of the Pucciniaceae, has teliospores comprising long chains of up to twenty dark brown cells. The only species occurring in Britain is *X. carbonarius,* the rust of great burnet (*Sanguisorba officinalis*). Last in our limited survey of this family comes *Gymnosporangium.* The mycelium is perennial within the juniper host (*Juniperus* spp.), causing infected branches to swell abnormally. The masses of two-celled teliospores form on these branches each spring. When dry they are small, rather hard structures, yellowish-brown in colour. When wetted, however, a remarkable change takes place. The long stalks of the teliospores, massed together, rapidly absorb water and are transformed into spectacular bright orange-yellow horn-like structures with teliospores all over the surface (Plate 7). The alternate host of *G. clavariiforme,* a species of *Gymnosporangium* common in Britain, is hawthorn (*Crataegus* spp.). The curious groups of cylindrical pointed aecia are clearly visible on the leaves of infected trees, as are those of *G. cornutum* on the leaves of rowan (*Sorbus* spp.), found more commonly in Scotland (Plate 7).

Fig. 9.9 Aecia of *Puccinia menthae* on the distorted lower stems of garden mint (*Mentha spicata*) in Cambridge, England. (D.S. Ingram.)

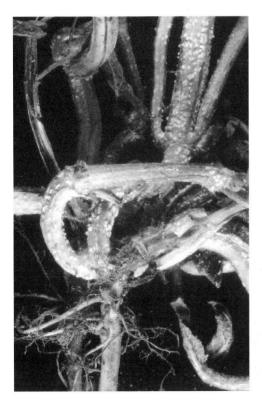

One of the first rusts to be grown in artificial culture in the 1950s was probably a *Gymnosporangium, G. juniperivirginianae,* a native of North America. Since rusts had hitherto been considered to be supreme examples of obligate parasitism the cultures were thought to be of contaminant saprotrophs, not the rust, and the findings were ridiculed. It was not until ten years later, when a group of Australian plant pathologists were able to grow a second rust (*Puccinia graminis*) in culture, that it was finally accepted that rusts could be grown away from their hosts, albeit with great difficulty.

Fig.9.10 Puccinia violaceum on leaves of bramble (*Rubus fruticosus*) in Edinburgh, Scotland. (Debbie White, Royal Botanic Garden Edinburgh.)

The Coleosporiaceae

The second family is the Coleosporiaceae, in which the teliospores have no stalk and are therefore referred to as being sessile. The uredio- and teliospores are formed in red to orange waxy crusts. The teliospores form one or two layers under the epidermis of the host and each divides internally to form a chain of four cells, each of which buds off a basidiospore. *Coleosporium tussilaginis*, a common species in Britain, forms teliospores on various hosts, including coltsfoot, ragworts and sow thistle (*Sonchus* spp.). The aecia are found on the needles of pines such as Scots pine.

The Melampsoraceae

Finally in this brief survey we come to the family Melampsoraceae. The unicellular teliospores are again sessile and often form a crust beneath the cuticle or epidermis. The best known species to plant pathologists is the autoecious *Melampsora lini*, the flax rust, since this was used by Flor in his classic experiments on the genetic basis of resistance to pathogens, from which emerged the gene-for-gene hypothesis (see p.68). In the northern hemisphere, from Europe through to North America, many heteroecious species of *Melampsora* and the related *Melampsoridium* are found in abundance on willow (Fig. 9.11), poplar and birch (Fig. 9.12), while their aecia are found on a variety of hosts, including larch. The bright yellow urediospores are produced in abundance in midsummer on the leaves of these trees. Anyone who grows the weeping form of the Kilmarnock willow (*Salix caprea*) will be well aware of the massive summer infection by *Melamspora caprearum* that causes the leaves to wither and drop.

Spectacular spores

Plant pathologists have long been aware that the spores of rusts have sculptured, often spiny walls (Fig. 9.13, overpage). It was not until the invention of the scanning electron microscope that the full complexity of this ornamentation was revealed; the three dimensional image shows the spore surface

Fig. 9.11 (above) *Melampsora larica-caprea* on *Salix caprea* at Juniper Bank, Peeblesshire, Scotland. (Debbie White, Royal Botanic Garden Edinburgh.)

Fig 9.12 (left) Green islands of *Melampsoridium betulinum* on leaves of birch (*Betula* sp), Cambridge, England. (D.S. Ingram.

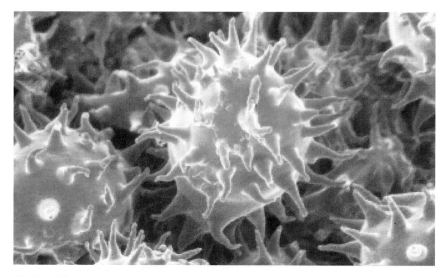

Fig. 9.13 The highly ornamented teliospores of *Puccinia prostii* (each is about 18 x 60 μm in size). (Scanning electron micrograph by S. Helfer, Royal Botanic Garden Edinburgh.)

magnified many thousands of times. Urediospores are almost without exception spiny (echinulate). Sometimes the conical spines, curved at the tip, are evenly distributed over the spore surface, but sometimes they are arranged in patterns such as spirals or lines. Occasionally they are set in circular depressions or are encircled by raised rings.

Teliospores show a much wider range of ornamentation and have been classified as: smooth (no ornamentation); echinulate; verrucose (warty); spinose (with long spines, sometimes branched); apically digitate (the apex has long finger-like projections); reticulate (covered with more or less rectangular depressions separated by ridges); striate (covered with linear depressions separated by ridges); rugose (irregularly sculptured); punctate (sharply pitted); and tomentose (covered with matted hairs). These categories do not correlate with particular families or genera.

Aeciospores are always ornamented, most being covered either with cylindrical knobs with flattened ends, or knobs comprising stacks of five to eight discs. Sometimes they are covered with a mixture of long spines and short projections, and occasionally with conical spines with hemispherical caps. Again, there is no correlation with any of the conventional taxonomic categories. Basidiospores usually have only slightly roughened surfaces, while pycniospores are usually smooth.

It is possible that these surface features of rust spores are important in aiding wind dispersal or adhesion to surfaces, especially those of the host, but we are not aware of any experimental evidence to confirm this.

It might next be asked why rust spores are usually such wonderful colours − yellow, orange, brown and almost black. We know that in common with many fungal spores the yellow, orange and pale brown aeciospores and urediospores contain carotenoid pigments dissolved in droplets of oil. It has been suggested that these act as a screen to protect the cell nucleus and its nucleic acids from the harmful effects of radiation, particularly ultraviolet radiation. The additional darker brown and black pigments of the thick-walled teliospores seem to give extra protection to spores which are longer lived. In spite of the supposed protective properties of these substances, however, the effect of exposure to bright daylight is to reduce the germination rate of urediospores of *Puccinia striiformis* to less than 1%, and of *P. graminis* to less than 5%.

Hormonal disturbance of the host

Biotrophs frequently cause hormonal disturbances of their hosts, resulting in green islands or distorted growth. With rusts, green islands can often be seen where infected leaves are shaded and have senesced prematurely, or can be induced by placing infected leaves in a dark moist atmosphere for a few days. Often, but not with every rust, the haloes of pale green cells surrounding the lesions stand out clearly against the yellowish background of the leaf as it senesces (Fig. 9.12 and Plate 6).

Where distorted growth occurs it usually takes the form of local swelling, but occasionally major developmental changes occur. The rusts of conifers in the genus *Cronartium*, for example, which infect the stems of pines in the aecial phase, affect the dividing cells between the xylem and phloem (the cambium), either killing them directly in the following year or causing increased wood deposition and swelling before killing the cambium, often girdling and killing the whole branch. Sometimes the swollen gall is a very prominent structure, as in the related rust *Peridermium cerebrum* (see below), where it resembles a small brain. The *Gymnosporangiums*, which attack junipers, behave in a similar way to *Cronartium*, with the degree of swelling and the rate of death of the cambium varying from species to species.

Fig. 9.14 An etiolated shoot of creeping thistle (*Cirsium arvense*) infected by *Puccinia punctiformis*. (D.S. Ingram.)

In Europe, one of the most intriguing changes in the growth of the host is induced in the common creeping thistle (*Cirsium arvense*) by *Puccinia punctiformis* (Plate 6; Fig. 9.14). The first sign of infection is the appearance of pale shoots among the normal growths of the perennial thistle in spring and early summer. The leaves are narrow and paler green than normal, and appear to sit more openly on the stems. These shoots are totally permeated by the mycelium of the rust and are said to be systemically infected. They are at first covered with pycnia, producing masses of pycniospores (Plate 6) with a sweet distinctive smell, reminiscent of privet (*Ligustrum* spp.) flowers on a warm day. Later, the shoots bearing the primary urediospores (which are produced instead of aeciospores) are also elongated and paler green than normal (Fig. 9.14). These effects seem to result from disturbance of a group of plant hormones called the gibberellins, but whether the rust induces the host to overproduce its own gibberellins or whether *P. punctiformis* itself is capable of gibberellin production is not known. Later, secondary urediospores and teliospores form discrete infections and do not lead to discoloration or growth changes in the thistle shoots.

An even more remarkable transformation of a host plant may be seen in the USA. *Arabis holboellii* and *A. drummondii*, when systemically infected by *Puccinia monoica*, produce counterfeit flowers on abnormal elongated shoots. The terminal leaves become expanded and covered with crowded bright yellow pycnia. The pseudo-flowers are visited by insects in much the same way as normal *Arabis* flowers, and these carry pycniospores to other similarly infected plants, where fertilisation occurs. Once the dikaryotic mycelium is established, the shoot reverts to its normal green colour.

World travellers – epidemic spread on a large scale

Worldwide the rusts probably pose the greatest threat to human existence of any plant disease. It is appropriate, therefore, that we should next look at this larger picture. Rusts attack all the major cereals of the world with the exception of rice (although rust has been found on rice in Sierra Leone it is not significant on that crop elsewhere) and every so often a particular rust becomes epidemic over a large geographical area, with catastrophic effects. Thus the well-documented major epidemic of black stem rust of wheat (*Puccinia graminis*) in the United States in 1916 was followed by a number of lesser epidemics

Fig. 9.15
Urediospore lesions of *Puccinia striiformis*, stripe or yellow rust, on leaves of wheat in Cambridge, England.
(D.S. Ingram.)

at intervals over the next 40 years, and then other major epidemics occurred over a four year period from 1950–54 with the emergence of Race 15B (see below). There was an unrelated outbreak in Scandinavia in 1951 and again in Western Australia in 1963–4. Those farmers with crops affected by rust disease faced a major disruption of their income and those who depended on the grain for subsistence faced starvation.

A number of other interrelated species of *Puccinia* also attack the cereals (Table 9.2). Probably they all originated on grasses and adapted to their particular cereal host as agriculture developed; indeed strains can commonly be found on grasses in the wild. From this table we can see that black stem rust (*P. graminis*), which flourishes under warm continental conditions, attacks a number of cereal species; that oats have in addition a specific and easily identifiable rust of their own (*P. coronata*); and that wheat, barley and rye are attacked by two other important rust diseases, the yellow (or stripe) rust *P. striiformis* (Fig. 9.15), which is favoured by cool conditions, and brown (or leaf) rust (*P. recondita*), which is intermediate in its climatic requirements. An interesting example of adaptation is shown by *P. recondita*, brown rust of wheat, barley and rye, which has adapted to different aecial hosts: in Portugal it has been found in geographically separate areas producing aecia on *Thalictrum* spp. in the family Ranunculaceae; in other areas it is reputed to produce aecia on *Anchusa* spp. and other members of the family Boraginaceae; while in Siberia it has been found on *Isopyrum fumarioides*, another member of the Ranunculaceae.

Table 9.2 Cereal rusts.

Host	Alternate host	Rust	Common name
wheat	*Berberis*	*Puccinia graminis* f.sp. *tritici*	black stem rust
wheat		*Puccinia striiformis* f.sp. *tritici*	yellow rust (stripe rust)
wheat	*Thalictrum* *Anchusa*	*Puccinia recondita* f.sp. *tritici*	brown rust (leaf rust)
barley	*Berberis*	*Puccinia graminis* f.sp. *hordei*	black stem rust
barley		*Puccinia striiformis* f.sp. *hordei*	yellow rust (stripe rust)
barley	*Ornithogalum*	*Puccinia hordei*	brown rust (leaf rust)
oats	*Berberis*	*Puccinia graminis* f.sp. *avenae*	black stem rust
oats	*Rhamnus*	*Puccinia coronata* f.sp. *avenae*	crown rust
rye	*Berberis*	*Puccinia graminis* f.sp. *secalis*	black stem rust
rye		*Puccinia striiformis* f.sp. *secalis*	yellow rust (stripe rust)
rye	*Anchusa*	*Puccinia recondita* f.sp. *secalis*	brown rust (leaf rust)

On the western seaboard of mainland Europe and in Britain the yellow rust, *Puccinia striiformis*, is the most important rust of wheat and barley. It has no known alternate host. In the first half of the twentieth century, yellow rust was common on wheat in the autumn, and the extent of the autumn/winter build up largely determined the number of spores available for the summer infection and whether it would reach epidemic proportions or not; the disease was worse in a temperate summer than a hot one. Various agricultural practices were used in an attempt to control the disease: sowing thinly, reducing the fertiliser input, grazing a winter-proud (excessively luxuriant) crop by sheep to reduce the winter leafage available for infection, but to little avail. The disease was at that time not of great consequence on barley because of the lack of spores during the winter months. The arrival of winter-hardy barleys brought the disease once more to the fore on this crop, but by then the knowledge of plant breeding techniques for the incorporation of genes for resistance to the disease and the arrival of techniques for chemical control reduced the risk of epidemic attack.

These methods brought their own problems. Mistakes made in the breeding of new wheat varieties led to a series of epidemics. The emergence of new pathogenic field races of yellow rust led to the spectacular failure of the 'Heine VII' wheat variety in the Netherlands in 1955, of the 'Probus' variety in Switzerland in 1961 and of the 'Cleo', 'Falcus' and 'Opal' varieties in Holland in the same year. In Britain two spectacular epidemics took place when resistance to yellow rust collapsed in 1966 in the wheat 'Rothwell Perdix' and in 1970 in the wheat 'Maris Beacon', due to the emergence of new pathogenic races. These were two of the most promising varieties of their generation and were widely planted, so their failure was the more keenly felt.

The brown rust *Puccinia recondita* usually attacks late in the summer in Britain, so it has negligible effects. It is, however, the most important cereal rust in Argentina and caused a major epidemic in Mexico in 1976–7. Currently it may well be the most serious wheat rust on a world basis.

Teliospores and aeciospores are not essential for the epidemic spread of cereal rusts. Thus in the Great Lakes area of North America urediospores from crops infected with *P. graminis* are carried south on the wind in autumn to infect volunteer wheat (plants growing from grain spilt accidentally) in Texas and Northern Mexico, where the pathogen overwinters. From this source the disease moves north again in early spring in sweeps, infecting the newly emerging wheat in sequence, as the climate allows. This has been referred to as the '*Puccinia* pathway' by the American plant pathologist E.C. Stakman. Other rusts such as *P. recondita* and *P. coronata*, as well as other pathogens such as *Bipolaris maydis* (see Fig. 2.4, p.45) and insect vectors of viruses (see Chapter 12) follow the same route.

In Europe there is a similar pathway, or even two. In the first, authenticated by identification of the same genetic variants along the route, urediospores of *P. graminis* overwintering in North Africa drift into southern Portugal in January and February, multiply and then move north as crops reach a stage suitable for infection. The pathway ends in Scandinavia in late summer. In years when the summer is warm and the wind direction appropriate to blow spores across the Channel there may be a sideways diversion to southern England. Another pathway is said to have its origin in Turkey and the Middle East, and to travel northwards through eastern Europe and Russia, ending once more in Scandinavia.

In India yet another pathway occurs. The hot dry climate of the Indo-Pakistan plain is such that there is little or no survival of the spores of *P. graminis* or *P. recondita* from one season to another. However, both pathogens survive on out-of-season crops and volunteer plants in the Nilgiri hills of southern India and in the Himalayan foothills. From this base urediospores are dispersed to the main wheat-growing areas under the influence of tropical cyclones during the month of November. In this pathway the rust has a one way ticket and the infection on the plains is regenerated annually from the endemic rusts of the hills. There is also some evidence for the transport of urediospores of *P. striiformis* from Turkey to India by the 'western disturbances', upper air low pressure systems generated in the Caspian Sea area in the non-monsoon months. In winter they are accompanied by rain on the Indo-Pakistani Plain, which aids the deposition of the spores.

These repeated and predictable annual distributions of rust inoculum over wide geographical areas can be contrasted with the sudden spread that takes place as a result of some accidental change in circumstances, leading to a world epidemic. Such an epidemic of southern corn rust (*P. polysora*), a pathogen of maize, began in 1948. Previously confined to the southern states of the USA and the contiguous Caribbean area where it was not a serious problem, *P. polysora* suddenly turned up in West Africa and independently in the Philippines, whether by human agency or climatic accident is not known. On arrival in Africa it initially had very severe effects on the local maize varieties, and within three or four years had spread throughout the continent. Fortunately, breeding programmes to incorporate resistance from American varieties brought the situation into balance and the disease is no longer the threat it once was.

Another world traveller is the coffee rust (*Hemileia vastatrix*). This is endemic in Ethiopia, the home of arabica coffee (*Coffea arabica*), and from there has spread throughout Africa. It was first recognised, however, in Sri Lanka in 1869, where coffee had recently been introduced; the devastation it caused led to the extinction of coffee-growing and its replacement by tea, with consequent economic upheaval. The pathogen appeared in Brazil in 1970 and has become a major threat to the coffee-growing industry there.

Arabica, which accounts for 65% of coffee produced, is the best flavoured coffee and is grown for the luxury market, the finest being produced on particular soils at considerable elevation (1,850 m) where the moist climate is coincidentally favourable to the spread of the fungus. Robusta coffee (*C. canephora*), of West African origin, has much less flavour and is used in less demanding situations, including the production of instant coffees. However, it is more resistant to the coffee rust disease, and genes from this and from *C. liberica*, another West African species, are now being used in breeding programmes to produce resistant arabica cultivars.

What is interesting about both these case histories is the seriousness of the disease when it first appears in a new geographic area and meets cultivars that have not become adapted to it. Later, by selecting and breeding less susceptible varieties, sometimes aided by spray programmes, an equilibrium position may be obtained.

A variation of the pattern demonstrated by the coffee disease is found in the pine blister rust, *Cronartium ribicola*. Here intercontinental spread is involved but, since the rust is heteroecious, control is complicated by the spread of the pathogen to alternate hosts in its new homes. It is suggested that *C. ribicola* originated in Asia on relatives of the Swiss stone pine (*Pinus cembra*), a five-needled pine. It spread slowly into Europe and there found a congenial host in another five-needled pine, the imported American white pine (or Weymouth pine) – *P. strobus*. The pathogen infects the needles and from there grows into the stem, causing swelling and producing on the surface pycnia and broad aecial cups with membranous covers which tear to release the aeciospores. The pathogen continues to grow in this way for several years, eventually girdling the branch or indeed the whole tree. The aeciospores are carried several hundred kilometres to infect wild and cultivated species of the genus *Ribes*, including blackcurrant (*Ribes nigrum*), gooseberry and flowering currant (*R. sanguinem*). On these hosts urediospores and teliospores are produced, then basidia and basidiospores which reinfect the pines. Because there was no easy way the fungus could be transferred to the American continent by wind or weather systems, it should by all reasonable assumptions have stayed in Europe, where it did comparatively little damage, but that was not to be the case.

The native American white pine, which in the time of the first colonisers occupied most of eastern and some of middle America, was such a good source of timber that most of the forests had been destroyed by the end of the nineteenth century and moves were afoot to replant them during the period 1900–1910. Ironically, because there was no tradition of growing appropriate nursery stock in the USA, plants were sought from the countries in Europe to which the species had been exported. Initially a tariff added to the cost of the imported stock but, when this was removed, millions of small trees were transported back to the USA, many of which were infected by *Cronartium ribicola*, so that the disease spread throughout the country and into Canada. Once established, the pathogen began to attack the cultivated currant, and also spread to the native currants, which were particularly plentiful wherever the white pine was grown. This allowed the disease to build up still further and colossal damage was done. Furthermore, control of the epidemic was costly, for it involved the complete removal of blackcurrant plantations and the destruction of all wild *Ribes* spp. within the forests.

This is by any standards a spectacular disease. *Cronartium ribicola* still occurs in America and may occasionally be seen in Britain. In addition there are reports of swollen stems bearing *Cronartium*-like aecia on native Scots pine in the Scottish Highlands, and on the Scots pine hedges planted as windbreaks in the Breckland of eastern England. Variously called *Peridermium pini* and *Cronartium flaccidum*, (the name *Peridermium* is used for repeating aecia of *Cronartium* spp.), the identity of these rusts is still far from certain.

Plant pathology, fortunately for its practitioners, is never static. No sooner was the pine blister rust understood and the possibility of its control established in the USA than a similar native American rust began to cause significant damage in the southern states, from Maryland south to Florida and west to Texas. The fungus *Cronartium quercum* f.sp. *fusiforme* produces its pycnia and aecia on loblolly pine (*Pinus taeda*) and slash, or Cuban, pine (*P. elliottii*), important sources of timber in the southern USA, and its urediospores and teliospores on a variety of oaks, particularly black oak (*Quercus nigra*). The disease causes spindle-like swellings on the stems of young pines; these are perennial and lead to poor, irregular growth and finally the death of the host.

The fungus is endemic in the southern states, living with the natural population of pines in mixed forest. The way the forest is managed for pulpwood in a 25 year cycle provides ideal conditions for the pathogen: the short rotation combined with thorough preparation of the soil promotes the growth of young pines; while the control of forest fires and frequent cutting of the understorey encourage growth of the oak saplings and coppice shoots that form its alternate host, and from which the young pines are infected. Attack is greatest between the first and the fifth year in rapidly growing pines. It was estimated in 1981 that 30% of all loblolly pines planted developed potentially lethal stem infections, with an annual increase of 2–3%. Since then there have been advances in control, by fungicides and breeding more resistant pines, but the disease remains serious.

A mutable tribe: pathogenic variation in the rusts

As soon as the importance of rusts as pathogens of cereals was appreciated in the latter half of the nineteenth century it was established, as with *Erysiphe graminis* (see p.171), that the identifiable species such as *Puccinia graminis*, *P. striiformis* and *P. recondita* each occurred in a number of *formae speciales* which were more or less restricted to specific cereals such as wheat, oats and rye, or particular grass genera. Thus stem rust occurred as *P. graminis* f.sp. *tritici*, mostly on wheat, although it could also infect barley and rye; *P. graminis* f.sp. *secalis* mainly occurred on rye, barley and couch grass (*Agropyron repens*) but not on wheat or oats; *P. graminis* f.sp. *avenae* occurred on oats and cocksfoot grass but not on wheat or barley, and so on. By intensive work among the grasses many more *formae speciales* could be proposed so that the concept has limited value except in a local context, and in showing how the appreciation of the variability of the rusts was built up historically.

Until the nineteenth century, wheat and the other cereals existed as a number of land races. As understanding of the inherent variability of flowering plants developed, farmers began to realise the value of the selection of superior ears of wheat from the mixed genetic stock of the land races. Their work had no theoretical base, since Darwin's ideas were only promulgated after 1859. Gregor Mendel did not demonstrate the heritability of individual character until 1866, and this was not demonstrated to the world in general until much later and in Britain by publication of Bateson's exposition of *Mendel's Principles of Heredity* in 1909. By good fortune the self-fertilisation and inbreeding which we now accept as a characteristic of many cereals and grasses ensured that any superior ear selected from the crop would, in general, pass on its good qualities to its progeny, thus producing distinct varieties.

The events that followed this identification of crop varieties was most dramatically seen in the United States in the latter half of the nineteenth and the early part of the twentieth centuries, and paralleled in Europe and Australasia. One example is enough to illustrate this. The wheat 'Red Fife' was developed in the United States from a few seeds extracted from a cargo of Polish wheat, of no exact provenance or varietal status, passing through Glasgow. These seeds were sent to a farmer in Ontario called David Fife, who selected the progeny of one ear in 1847 and multiplied it up to give rise to the variety called 'Red Fife'. In 1903, when agronomists were seeking varieties adapted to the particular problems of growth on the prairies, a scientist in the Canadian Agricultural Service deliberately crossed 'Red Fife' with a nondescript form called 'Hard Red Calcutta', which was really the trade name for a particular quality of wheat imported from India. It had the virtue of early ripening and high gluten content (gluten is the protein that gives bread flour its elasticity when proved). Both varieties were common wheats (*Triticum aestivum*) with three sets of chromosomes in the haploid, making them hexaploid in the diploid. From the very varied progeny of this first cross the variety 'Marquis' was selected and multiplied on to become very widely grown in both Canada and the United States. By 1917 it was providing about 9 million tonnes of grain and is said to have contributed greatly to the food supplies of the Allies during the First World War.

'Marquis' continued to be grown into the 1930s, but together with other wheats suffered a number of epidemics of black stem rust, and a search began for resistance among wheat varieties and species. Resistance was discovered in a common wheat named 'Kota', and from this and 'Marquis' was developed the rust-resistant variety 'Ceres', distributed in 1926 and grown extensively throughout the northern Great Plains area until it became the victim of an epidemic of a new physiologic race of rust in 1935, at which point it was superseded. Resistance derived from other species of wheat, including 'Yaroslav' emmer and 'Iumillo' durum (both with two sets of chromosomes in the haploid, making them tetraploid in the diploid) were introduced into a crossing programme which was at first difficult because of the differences in ploidy. This resulted in the varieties 'Hope' (with rust-resistance genes from emmer), distributed in 1927, and 'Marquillo' (derived from a cross with 'Iumillo'), distributed in 1928. The later resistant variety 'Thatcher', with durum genes derived from the same 'Marquis' cross as 'Marquillo', was distributed in 1934 and continued to be popular until 1940. 'Thatcher' was in fact susceptible to strains of the brown rust *Puccinia recondita* and it was this susceptibility that finally caused its downfall.

The point of this rather complex story is that a situation developed where, as soon as a variety was found with resistance to stem rust, a new pathogenic 'physiologic race' of the fungus was likely to appear to attack it. These physiologic races were recognised by inoculating a range of differential wheat varieties which had emerged as resistant or susceptible over the years, and which included standard varieties such as 'Marquis' and varieties of other species of wheat such as *Triticum durum* (macaroni wheat), *T. dicoccum* (emmer wheat) and *T. compactum* (club wheat), twelve in all. In the hands of E.C. Stakman the identification of races of rust by inoculation on these differentials became something of an art. Race 15, which was first identified in 1913, developed a variant which was called 15B. This was noticed at an early stage in the populations of rust, but eventually became epidemic in 1950 when it attacked all known wheat varieties, with the exception of some from Kenya, which became the source of resistance for the next few years. Stakman referred to the situation where resistant varieties are attacked and in turn replaced by new resistant varieties as 'the plant pathologist's merry-go-round'. He might have called it 'the plant breeder's treadmill', for there was a constant struggle to identify new races of the rust, find resistance to them and incorporate this into new varieties of wheat. When we realise that the development time for a new variety is about 15 years and that the multiplication time for a new physiologic race may on occasion be less than that, we see that the plant pathologists and plant breeders have no room for complacency. These races of rust are believed to have their pathogenicity controlled by a single gene locus with genes for virulence or avirulence matched in the varietal host by a gene locus with genes for susceptibility or resistance, and are referred to as race-specific genes (see Chapter 3).

This situation, which parallels that described for potato blight in Chapter 5, only lasted a few decades until more durable forms of resistance were recognised and incorporated. The value of the Kenyan varieties has already been mentioned and the variety 'Thatcher' also showed a very stable form of resistance. Attention became focused on forms of resistance which, while not

absolute, were 'durable'. Some varieties were found which, although susceptible to black stem rust, did not develop the disease rapidly and were designated 'slow rusting varieties'. It soon became apparent that there were many
mechanisms by which varieties could escape or resist disease. 'Earliness' in
itself as a means of avoiding late build up of disease proved useful in the southern Great Plains area. 'Adult plant resistance', where seedlings were susceptible but the adult plant showed resistance when inoculated, was another useful
characteristic. Differences in the rate of multiplication of spores on different
varieties and many other small physical and physiological characteristics resulted in differences in the rate of development of epidemics and could be useful
disease controls. Sometimes these characteristics are under the control of single genes, but often the genetic situation is more complex. Gradually the concept of 'adapted crops' came into being, and of crops that combined many
characteristics of yield and hardiness with the capacity to live with or contain
disease to the extent that destruction of the crop was avoided.

In practical terms modern cereal husbandry takes a three-way approach to
the control of rust disease. In the first place, the breeder seeks to produce varieties incorporating any available resistance. This includes both single dominant genes (major genes), which can break down dramatically following a
change in the matching pathogen gene for avirulence, and race non-specific
resistance, which is generally less complete than race specific resistance. The
genetic control of the latter may be simple or complex, but because the resistance is not absolute it may not place a strong selection pressure on the
pathogen population for variants that can overcome it. It therefore does not
suffer from the dramatic breakdown of race specific resistance. These two
types of resistance are sometimes referred to as 'vertical resistance' and 'horizontal resistance' respectively.

In the second place the breeder looks for adapted varieties that, as well as
having disease resistance, show a wide range of qualities related to survival and
production. Genes incorporated from wild relatives of the cereals often confer
resistance which is effective and durable. Thus the genes incorporated into
modern American varieties from macaroni wheat and emmer wheat have
played an important part in the control of cereal rusts in that country. In
Europe it is more difficult to point to a similar example, but the variety
'Cappelle Desprez' has provided its progeny with a durable resistance to yellow
rust and is also the main source of a stable resistance to the eyespot disease of
wheat caused by *Pseudocercosporella herpotrichoides*. This suggests that in its own
parentage 'Cappelle Desprez' may have received genes from wild relatives.
Genetic engineering now makes possible the incorporation of a wide range of
genetic material from distant relatives without a long drawn out expensive
hybridisation process.

Finally, there has been the introduction of effective chemical sprays and
spraying techniques. The development of these largely parallels that described
for control of the powdery mildew diseases of cereals on pp.174–175 and will
not be repeated here.

Go out and find them

Rusts are there to be found wherever there is vegetation; in our view they are
as common or rare as their hosts. Each one is attractive, exciting and raises scientific questions, and after a little you quickly recognise those you have not

seen before. But don't get carried away; remember the old Professor of Medicine's adage about diagnosing a new disease – 'It's more likely to be a sparrow than a canary'. But sparrow or canary, it will be equally interesting. There is no group of fungi more amenable to field natural history, with or without a microscope, than the rusts. Go out and find them!

10

The Dark and Secretive
Smuts and Bunts

The smut and bunt fungi (the Ustomycetes) produce very dark powdery resting spores (ustilospores) which, en masse, are reminiscent of the black smuts that were one of the perils of travelling by rail in the days of steam. Like the rusts they are biotrophic members of the Basidiomycota, but differ in that most have a long invisible phase when they are systemic within the host. The hyphae grow between and through the living cells but do not produce haustoria. They have simpler life cycles than the rusts, with no alternate hosts and generally only two spore types. They are fairly readily grown in artificial culture and have therefore been extensively used as models for genetic research. Some are referred to as bunts; bunt is an old word of unknown etymology, sometimes used to describe pouter pigeons. Cereal grains packed with the ustilospores of smut fungi in the genus *Tilletia* are rounded, with a fat, pouter-like appearance, hence the use of the term in plant pathology, perhaps. We wonder whether Billy Bunter's name comes from the same root.

Smuts as pathogens

To the farmer smuts and bunts were at one time catastrophic diseases, especially of cereals. So effective are modern control measures, however, that they are now rarely found in agriculture, although they remain an ever-present threat. Fortunately for the naturalist, they are easily found on wild and garden plants.

In Europe most of us could walk onto any piece of rough ground in midsummer and find plants of the oat grass (*Arrhenatherum elatius*) with seeds transformed into masses of the black smut spores of a *forma specialis* of *Ustilago avenae*, the oat smut (Fig. 10.1). In infected plants every floret in a flowerhead and every flowerhead from a single plant will carry the disease. These black ustilospores are the equivalent of the teliospores of the rust fungi. They are sometimes also referred to as teliospores, brand spores or chlamydospores. In the USA, *U. bullata* is an equally common grass smut, appearing in the florets and seeds of over 50 species of wild and cultivated grasses. This species is also found in Europe, where it is confined to bromes (*Bromus* spp.).

But not all smuts form their ustilospores in the flowering parts of plants. The common stem smut of grasses (*U. hypodytes*) forms its ustilospores on the stem surface, sometimes affecting almost the whole stem. There is no protecting membrane covering the spores, but at first they are enclosed by the leaf sheaths. In Europe the disease occurs on common couch grasses (*Agropyron* spp.), some bromes and fescue, but is most spectacular on the seashore grasses, especially marram and sea lyme. The same or a closely similar fungus is found on Canadian lyme grass (*Elymus canadensis*) and ryegrasses in North America.

Fig. 10.1 Black ustilospores of *Ustilago avenae* filling the florets of oat grass (*Arrhenatherum elatius*) at Aberlady, East Lothian, Scotland. (D.S. Ingram.)

Fig. 10.2 Black ustilospores of *Ustilago tragopogonis* destroying the florets of goat's-beard (*Tragopogon pratensis*) at Gullane, East Lothian, Scotland. (Debbie White, Royal Botanic Garden Edinburgh.

Equally striking is *Ustilago longissima*, found in Europe on the leaves of reed grass (*Glyceria maxima*) and flote grass (*G. fluitans*) in slow-flowing water. It forms its ustilospore masses within the tissues as narrow stripes in the parts of the leaf not occupied by vascular bundles. They can easily be seen by holding an infected leaf up to the light or shining a torch through it. As the ustilospores mature and are blown away, narrow furrows are left in the leaves. The superficially similar but unrelated species *Urocystis agropyri* forms elongated blisters on the leaves of couch grasses in both Europe and North America. A similar pattern may be seen in the worldwide species complex *U. tritici*, the flag smut of wheat. *Urocystis* differs from *Ustilago* in having 'spore balls' made up of a small number of fused ustilospores with a covering of sterile cells.

There are over 1,100 known species of smuts and more than 60% of them occur on grasses or cereals. Another 10% or so occur on rushes and sedges and the remaining 30% are widely scattered over a number of different flowering plant families. *Ustilago vaillantii*, for example, is often found in the anthers of a number of spring-flowering liliaceous plants, including squill (*Scilla* sp.), glory of the snow (*Chionodoxa* sp.) and grape hyacinth (*Muscari* sp.). The dark violet ustilospores replace the pollen and dust and stain the mouth of the flower tube, making the infection very visible. A similar smut, *Ustilago violacea*, occurs in the anthers of members of the campion family (Caryophyllaceae). It is most easily seen in red campion (*Silene dioica*) and white campion (*S. album*), where once more the dusting of dark ustilospores readily catches the eye (Plate

7). It has also been known to cause economic loss in cultivated relatives such as glasshouse carnations (*Dianthus caryophyllus*). *Ustilago tragopogonis-pratensis* may often be seen on flowers of Jack go to bed at noon or goat's-beard, the black smut spores destroying the florets (Fig. 10.2).

There are several *Urocystis* spp. that may easily be seen on non-grass species. *U. eranthidis*, for example, infects winter aconite (*Eranthis hyemalis*) (Fig. 10.3). In old established colonies of the plant you may find on the leaves and petioles silvery swellings which open to reveal the dark spore masses of the fungus. Often the infected tissues are swollen and distorted. *U. primulicola*, a much less obvious pathogen, produces its spore balls in Europe in the ovaries of the wild primroses *Primula vulgaris* and *P. farinosa*, and is consequently not easily seen except by a special search. However, secondary spores (called sporidia) are formed in the anthers and appear as white mealy masses replacing the pollen in both thrum- and pin-eyed forms of the plant (Plate 7). They are apparently carried from plant to plant by pollinating insects. The details of the life cycle are fragmentary and need further elucidation. *Urocystis cepulae*, the onion smut, which moved from Europe to the USA in the 1880s, is a serious disease of sown onions and leeks in cool temperate climates worldwide. The ustilospore balls lie in the soil and germinate with the seed to bring about infection, resulting in swollen areas which open to release the masses of spore balls, permanently damaging the plant. The spore balls do not germinate well above 22°C, which is why the disease does not occur in warmer climates.

Fig. 10.3 Ustilospores of *Urocystis eranthidis* bursting out from a lesion on a petiole of winter aconite (*Eranthis hyemalis*) in Cambridge, England. (D.S. Ingram.)

To complete this brief overview of easily-seen smuts on wild and garden plants, mention should be made of the genus *Entyloma*. The species most frequently seen is *E. calendulae* on marigolds (*Calendula* spp.) and daisies (*Bellis* spp.). Another form of this species, *E. calendulae* f.sp. *dahliae*, which originated in South Africa, is now common on *Dahlia* spp. throughout Europe. Unusually the ustilospores are colourless or pale brown and embedded in the leaf, where they cause some discoloration of the tissues. They germinate in the lesion and sporidia appear on the surface of the leaf; the process of germination is not unlike that of *Tilletia* spp, which will be described below.

The smuts of cereals (including bunt of wheat)

Many smuts are associated with the reproductive organs of the plants they infect – sometimes they occupy the ovary, sometimes the anthers and sometimes both. Cereal smuts frequently replace the starchy endosperm of the grain with inedible spore masses, destroying a human food resource. Because of their importance they have been studied extensively and the knowledge so gained provides us with a deeper insight into the smuts in general.

There are two main families of Ustomycetes that attack cereals, the Tilletiaceae and the Ustilaginaceae, separated by the behaviour of the germinating ustilospores. In the family Tilletiaceae the ustilospores are large and frequently ornamented with a net-like pattern. They germinate to produce a non-septate mycelium (the promycelium, equivalent to a basidium), which is negatively geotrophic and therefore bends upwards. On the apex of this promycelium primary sporidia are produced (equivalent to basidiospores). These germinate, fuse in pairs and then produce dikaryotic secondary sporidia which infect the plant. The genus *Tilletia* and two genera already mentioned, *Urocystis* and *Entyloma*, are included in this family. In the family Ustilaginaceae the ustilospores are smaller than in the Tilletiaceae and the promycelium is septate, as in the rust basidium, and produces the sporidia laterally, like rust basidiospores. Most cereals are attacked by different species of the principal genus, *Ustilago*, which has already been mentioned.

Bunt of wheat (*Tilletia tritici* and *T. laevis*) continues to be potentially the most damaging disease of wheat worldwide. Significant annual losses occurred in all countries in the early years of the twentieth century, but were progressively reduced with the advent of constantly improving chemical control. Exceptions were areas such as Pacific northwest America, where soil contamination took place on a massive scale and conditions were particularly favourable to the survival of ustilospores.

The characteristic symptoms of bunt come not only from the replacement of the grain by the ustilospores, giving them in consequence a swollen appearance, but also from the associated fishy smell, caused by the presence of trimethylamine. This has given rise to the alternative common name 'stinking smut'. The yield of an infected crop is significantly reduced, and contaminated healthy grain is discounted on the market because of the odour. There is also a threat to human health from respiratory and skin allergic reactions to the ustilospores, which may on occasion form dense clouds during harvest. Moreover, these clouds of spores are easily ignited and were once the cause of spectacular explosions and fires on the prairies before steps were taken to design safer harvesting, threshing and milling machinery.

The infection cycle in bunt

As one of the most damaging diseases of cereals in the eighteenth and nineteenth centuries, bunt attracted the attention of the early plant pathologists, and became in consequence a classic disease in the development of the subject. The ustilospores, which are threshed out with the grain of bunted wheat, coat the surfaces of otherwise healthy seed and after sowing infect the young plant as it germinates. The mycelium then becomes systemic in the plant, growing between the living cells, ultimately entering the developing ovary as the flowers form. As these mature further ustilospores are formed within the grain to begin the cycle again. This sequence of events was first described in broad terms by Mathieu Tillet (hence *Tilletia*) in France in 1755, in the first attempt to demonstrate scientifically that a fungus was the cause of a plant disease. He compared the ears of plants grown from 'clean' grain with those grown from the same seed contaminated with bunt spores. The clean grain gave rise to only a few bunted ears, whereas at least half of the ears grown from contaminated seed were bunted. As a demonstration it was a success, but Tillet was lucky to have so little chance contamination by bunt in the plots grown from healthy seed. Further proof of Tillet's hypothesis came with the work of the Rev. Miles Berkeley a century later. Berkeley's first attempt to prove that fungi caused disease was made with *Phytophthora infestans*, the cause of potato blight (see p.13), but he found bunt of wheat an easier organism to deal with. His studies went some way to establish the connection between the presence of ustilospores on the seed and the development of disease. In addition he appreciated the importance of ustilospore contamination of the maturing grain, and studied ustilospore germination.

Tillet's work was also extended by another pioneer of plant pathology, Isaac Benedict Prévost from Switzerland. Working in France in 1807, Prévost not only confirmed Tillet's findings but also germinated bunt ustilospores in the laboratory and guessed that they infected the young plant. The nature of this ustilospore germination was beautifully illustrated by the brothers Tulasne in 1847. A germ tube (the promycelium) is first produced, bearing at the tip uninucleate needle-shaped structures, the primary sporidia (Fig. 10.4). These

Fig. 10.4 Tilletia caries, cause of bunt or stinking smut of cereals: (a) an ustilospore germinating to produce a promycelium bearing primary sporidia; (b) fusion of pairs of primary sporidia; (c) germination of a primary sporidium to form a secondary sporidium. (Drawings by Mary Bates, after A. De Bary and based on drawings of Tulasne.)

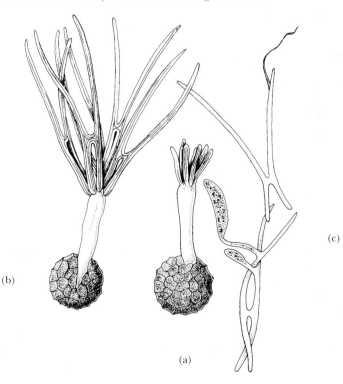

(c)

(b)

(a)

fuse and produce secondary binucleate sporidia which then infect the seedling in the region of the root node. Further knowledge of the bunt fungus, apart from a reference by Berkeley to bunt mycelium within the tissues of the leaf, came as a result of studies by Oscar Brefeld in Germany in 1888, and the subsequent development of twentieth-century methods of microscopy.

Understanding this life cycle gave the key to control. If antifungal chemicals were applied to the seeds of the host before sowing, the ustilospores would be killed at germination and the infection cycle broken. This identification of the 'Achilles heel' of the fungus led to the virtual eradication of bunt from agriculture in the twentieth century and established the principle that recognition of the weak point in a pathogen's life cycle underlies all successful control measures for plant diseases.

The cereal smuts

The members of the family Ustilaginaceae appear to occupy the same ecological niche as bunt, producing ustilospores in place of grain, and at first sight with the same infection mechanism. The most damaging pathogens are *Ustilago nuda*, the loose smut of wheat and barley and *U. hordei*, the covered smut of barley and oats. *U. avenae*, already mentioned as a pathogen of oat grass (*Arrhenatherum elatius*), has a special form that causes a loose smut of oats.

When the experimental infection of wheat by bunt had been achieved it seemed obvious that the smuts of wheat, barley and oats would have the same pattern of infection; they did not, and therein lay a mystery. Dusting the seed with ustilospores before sowing led to as many failed as successful infections. Similarly with the entirely separate corn smut (*U. maydis*) of maize, dusting the grain prior to sowing produced no infection. After much frustrating work it gradually came to be understood that there are four main types of infection pattern for smuts, depending on the time in the growth cycle of the plant at which the fungus infects.

1) SEEDLING INFECTION
The pattern described for the bunts holds only for the covered smut of barley and oats, *U. hordei*. All the grains of an ear are usually converted into ustilospores, but these are not blown by the wind because the chaffy scales, although modified, remain as a semi-transparent covering. The ustilospores are released when this covering is broken during harvest or threshing. Seed dusted with ustilospores, or clean seed sown in soil contaminated with ustilospores, gives rise to infected seedlings. The fungus enters at the coleoptile stage, before the true leaves open, and grows systemically in the developing plant, eventually forming ustilospore masses in the grain to complete the life cycle. The actual infection is probably initiated by mycelium growing from the binucleate secondary sporidia produced by fusion of the basidiospores (primary sporidia).

2) LOCAL INFECTION
The corn smut of maize (*U. maydis*), which was rare in Britain until the hot summers of 1975 and 1976, forms large galls on leaves, stems or flower spikes (Fig. 10.5) and it is in these that the ustilospores are formed. In smutted plants the male spike may bear female or hermaphrodite flowers in its lower part. Infection takes place in any growing part of the plant and can apparently arise

Fig. 10.5 Ustilospores of *Ustilago maydis* in the swollen and distorted ovaries of corn (maize) (*Zea mays*). (B. Goddard.)

from direct entry of an ustilospore germ tube (promycelium) or by the entry of a germ tube derived from a sporidium. This type of infection always remains localised and never becomes systemic.

3) SHOOT INFECTION LEADING TO SYSTEMIC GROWTH

Shoot infection was first demonstrated experimentally not with a cereal smut but with *U. violacea* on white campion. Infection takes place through the base of the emerging stem or the buds in the leaf axils and the fungus then becomes systemic and grows with the elongating shoot, eventually forming ustilospores in the flowers. There is a suggestion that ustilospores transferred, perhaps by insects, from an infected flower to a flower on an uninfected plant can initiate systemic infection in that plant. Similar shoot infection leading to the production of ustilospores in leaves and stems occurs in the sugar cane (*Saccharum officinarum*) smut (*U. scitaminea*), which periodically becomes a serious disease, mostly in response to faulty breeding programmes that lose the generalised resistance of the adapted crop.

4) FLOWER INFECTION

This type of infection occurs in loose smut of wheat and barley (*U. nuda*) and of oats (*U. avenae*). Ustilospores released from the uncovered spore masses formed in the ears at about the time of anther maturation are blown to flowers on other plants by wind, where they establish an infection. It was once thought that the ustilospores behaved like pollen grains and were trapped by the stigmas, infecting the ovary by that route. It is now clear that they lodge between the glumes (bracts enclosing the flower) around the growing ovary, germinate before the grain is ripe and put out a promycelium which forms a mycelium (without first forming sporidia) in the scutellum, the layer of secretory cells between the storage part of the grain and the embryonic plant. When the grain ripens and becomes dormant, the smut mycelium, still confined to the scutellum, also becomes dormant. At germination the mycelium comes to life again and grows systemically in the plant, keeping up with the growing apex; eventually it grows into the young developing ear. With *U. avenae* the pattern is slightly different. The fungus does not penetrate the young ovary, but instead forms a plate of resting mycelium on the surface; this becomes dormant but is reactivated when the grain starts to grow early in spring.

Control of cereal smuts

Serious attempts to control smuts and bunts of cereals date from the early experimental demonstration of the importance of ustilospores adhering to the grain in the transmission of *Tilletia caries*. Because cereal grains in the dormant state were fairly resistant to mild chemical treatment, the practice developed of 'pickling' them in various concoctions, including copper salts. Copper sulphate was first used as a steep, and a lime bath was found to reduce toxicity to the seed. Later copper carbonate was used and, to overcome the uncertainties of wet treatment, dry dusting became standard practice for a time. Spraying dilute solutions of formalin on to oat grains infected with loose smut, followed by a period under a cover, was an effective pickling method used in the early part of the twentieth century.

But all these procedures were imprecise and the parasites might escape control, or the grain might be put at risk by too great an exposure to the chemicals; moreover their effectiveness against different smuts was very variable. A breakthrough came with the development of organo-mercury materials, which were effective against a range of smuts and other seed-borne fungi such as *Pyrenophora avenae* and *P. teres*, destructive leaf diseases of oats and barley. But mercury is very poisonous and the dusts were difficult to apply without harming the operative. Slurries reduced the respiratory risk, but the overall toxicity of mercury produced pressure for its removal from the environment. Currently a variety of less damaging fungitoxic organic compounds is used and all quality seed is now supplied ready dressed at time of sowing. Many of these modern fungicides are 'systemic', being taken up by the cells of the seedling plant and translocated within it.

Because the loose smuts overwinter as mycelium in the grain a quite distinct method of control was developed for these fungi by J.L. Jensen, a self-taught Danish plant pathologist. He began life as a teacher and magazine editor, then became a seed merchant and from this base became interested in the role of hot water treatment for the control of seed-borne diseases. In 1888 he perfected a procedure for treating barley and wheat seed with hot water, to destroy the smut mycelium without killing the grain. Nowadays hot water treatment, sometimes with systemic fungicides, is used where necessary to treat nuclear stocks from which seed stocks are multiplied. Those requiring treatment can be identified by macerating grain samples in potassium hydroxide, then squashing and staining them to reveal the presence or absence of infecting mycelium in the embryo; where the percentage of infected seeds in a batch is low treatment is not necessary.

Of course seed treatment is not suitable for all plants and all situations, and resistance to smut diseases has been widely bred into cultivated cereals. Both major gene and more complex forms of resistance have been exploited. Also, there is evidence that tolerance is found naturally in 'adapted crop varieties' which have been selected, perhaps unknowingly, for survival in situations where smuts are limiting. These can coexist with the disease, resulting in only limited damage. However, such is the capacity for variation in the smuts and bunt that constant attention is required to select varieties that maintain the balance between virulence and resistance. Variation is exacerbated by the capacity of many smut species to hybridise with one another.

The secret lives of the smuts

A feature of so many of the smuts and bunts is their capacity to become systemic in their hosts without any visible antagonistic response by the tissues they are inhabiting. To the casual observer, cereal plants infected with any of the common smut or bunt diseases appear little different from their uninfected neighbours. Farmers, plant pathologists and keen-eyed naturalists, however, who know the plants well, can point to differences in stature, tillering (production of axillary grain-bearing shoots) and vigour. These are not caused by resistance reactions to exclude the fungus, but by interference with the growth control systems in the host and the diversion of nutrients to the developing pathogen. Haustoria are not formed, as they are in the rusts, but the mycelium grows between and through the living cells. When a cell is penetrated the membrane is not broken, but encloses the hypha as it would a haustorium. Layers of new material are deposited between the membrane and the hyphal wall and may have a role in communication between host and pathogen. Presumably nutrients are taken up through all parts of the mycelium, but especially through the hyphae growing within the cells. The relationship between fungus and host is a very cosy one until the moment of spore formation and the replacement of plant tissue by ustilospores.

As is often the case with biotrophs, infection sometimes causes the plant hormones controlling growth to get out of balance. Tissue overgrowth then occurs, as with the maize smut *Ustilago maydis*. In this disease, when the reproductive tissue is attacked, a whole variety of responses can be observed: sometimes the female or male flowers are replaced by leaf-like structures; sometimes the control of male and female structures appears to break down and male and female inflorescence parts follow one another on the stem axis; sometimes male inflorescences become bisexual, with the development of both male and female entities; or individual grains in the female inflorescence may become greatly enlarged.

Perhaps the most famous example of a smut affecting the sex of flowers is the change brought about in white and red campions infected with *Ustilago violacea*. When the female flowers of these normally unisexual plants are attacked they become changed, with the development of stamens (with anthers converted to smut spore masses) as well as the normal ovary.

Finally, there are bunts that induce dwarfism in their hosts. In the cereals most bunted plants are slightly smaller than their neighbours, and some strains of *Tilletia* spp. induce even smaller plants. The most extreme case has come to be recognised as a dwarf bunt, caused by *T. controversa*, which is known in both Europe and the USA. It came to be called 'stubble bunt' because it was missed by the normal settings of the combine harvester. Indeed, it has been suggested that it came into prominence as a result of the advent of new harvesting methods.

Smuts and human beings

Apart from their direct effects in reducing crop production worldwide, marring the beauty of wild and garden plants and providing excitement for plant pathologists, the smuts impinge on other aspects of human life. In a severe epidemic great quantities of spores are released by cereal crops. Breathed in by humans, they cause allergies of the respiratory tract. Even today, when present at comparatively low levels, they contribute to the population of spores over cereal fields, bringing on hay fever in susceptible people. Field and laboratory workers sorting cereal samples are particularly vulnerable, and may also develop severe dermatitis from skin contact with the spores.

Reports of humans and livestock being poisoned by eating spores of smuts have generally not been corroborated. On the contrary, some smuts are recognised as foods. *Ustilago esculenta*, which infects the wild rice plant *Zizania aquatica* in Taiwan, China and Japan, causes buds in the axils of the leaves to grow and become swollen and deformed. These swollen shoots, which contain a compact mass of white fungal mycelium, make good eating, we are told. Similarly, the young galls of *U. maydis*, gathered and eaten before the spores form, are said to be delicious. In Mexico they are given the name 'huitlacoche' and in the USA are sold as 'smoky maize mushroom'. Why not try some? But beware – being expensive they can seriously damage the bank balance.

11

Bringing Down the Trees

Introduction

As plants die and fall to the ground the remains that accumulate are mainly the cellulose cell walls of the softer tissues and the lignified cell walls of the water transport and support systems. Lignified material forms the greater proportion from plants such as forest trees, which lay down fresh wood every year. This accumulating material is available as an energy source for any organisms able to break it down. Indeed, if such organisms did not exist, enormous quantities of cellulose and lignin would accumulate in the world, a situation illustrated by the deposits in a modern peat bog or an ancient coal measure swamp, where decomposition was prevented by lack of oxygen.

The toadstools and bracket fungi (Basidiomycetes) in the Basidiomycota have developed the capacity to degrade and utilise cellulose and, unusually in the fungi, many can also degrade and utilise lignin. As a rule of thumb it is possible to picture a succession of fungi on plant remains in which the Zygomycota have enzymes capable of degrading the simpler carbohydrates (sugars), some of the Oomycota and the Ascomycota are capable of tackling longer chain carbohydrate polymers such as pectin, hemicellulose and cellulose, and the Basidiomycetes can degrade all of these, with some species also able to degrade the complex lignin molecules. The last is not an easy step because lignin has a complex structure, its breakdown products are sometimes toxic and tree wood is an inhospitable habitat for microorganisms (see p.44).

The Basidiomycetes in action

To see the Basidiomycetes in action it is sufficient to remove the surface litter and moss in a pine wood, where the active mycelium will be seen in the upper layers, sometimes as strands, spreading among the recognisable needle and twig debris. Below this will be found degraded material which is still recognisable, and again permeated by mycelium. As one goes deeper still the material becomes more and more disintegrated until eventually it merges with the humus of the soil. To complete the picture, long roots of pine may be seen with a series of short mycorrhizal branches (see p.144), the pine roots progressively more forked from tip to base and often connected by strands and cobwebs of mycelium. If the hyphae in this aromatic damp decaying world are examined with a microscope, most will be found to have the typical clamp connections of the Basidiomycetes (see below).

The Basidiomycetes (see also p.27) are unusual in their sexuality. The mycelium is divided into cells by septa with pores closed by membranes (dolipore septa). Each cell is dikaryotic, containing two separate haploid nuclei derived from the fusion of two separate monokaryotic mycelia. The paired nuclei control cell behaviour in much the same way as a single diploid nucleus does in other organisms. They divide in synchrony by an elaborate mechanism involving clamp connections, the hyphal bridges between adjacent cells that ensure that, following division, each daughter cell contains one nucleus derived from each pair in the parent cell (Fig. 1.11, p.28). Eventually, in special club-shaped dikaryotic cells called basidia, the two nuclei fuse, and then undergo meiosis to form four haploid nuclei. Finally these migrate into four (monokaryotic) basidiospores borne on short branches (sterigmata) set as a crown at the tip of the basidium (Fig. 1.12, p.29). The basidia and basidiospores are normally formed in a specialised layer, the hymenium, protected by a complex fruit body. This may be a toadstool, with the hymenium covering the surfaces of the gills (Agaricales) (Fig. 7.4a, p.145), downward-pointing tubes (Boletales) or downward-pointing fleshy spines (Hydnales). Alternatively, it may be bracket- or hoof-like (Poriales) (Plate 8) with the hymenium again lining downward-pointing tubes, or it may form a thin crust, covered with basidia and closely adhering to a host surface (Stereales). Most of the fruit bodies of the Agaricales, Boletales and Hydnales are soft and fleshy, with a relatively short life, often measured only in days. Those of the Poriales and Stereales are sometimes ephemeral, but often have a woody or leathery texture and live for long periods, sometimes years (Plate 8d). These perennial forms may produce a new zone of reproductive tissue each year.

It is well established that even in windy weather there is relatively still air close to any surface, the thickness of this laminar layer varying with the roughness of the surface but rarely being less than 1 mm. Beyond this layer the turbulence of the air varies according to wind speed. To achieve dispersal basidiospores must find their way out of the laminar layer, to be carried away rapidly by the turbulent air. In the Agaricales the basidiospores on the beautifully folded and accurately positioned gills are actively discharged by a hydrostatic mechanism that causes the sterigmata to flick them into the inter-gill space, far enough to clear the hymenial layer but not so far as to impact on the opposite gill surface. After discharge the basidiospores fall to earth from a height determined by the height of the stalk (stipe) and as they clear the body of the fungus they are caught up in moving air. The same situation obtains for the Boletales (with tubes) and the Hydnales Cantharellales (with teeth). In the Stereales the crust-like fruit body is usually formed on the lower surface of a branch or fallen trunk, and the basidiospores are then projected downwards from the smooth hymenial surface into the moving air. In the Poriales the basidiospores are projected into the tube-space with the same precision and by the same method as in the Agaricales and Boletales. Here, however, the bracket-like fruit bodies may be formed high in a tree host and the basidiospores discharged over a relatively long period.

There must be no movement of the fruit body during the time of basidiospore discharge, hence the advantage of the thick stalks and massive structures of those Basidiomycetes which are exposed to the elements, and the rigidity of the stipe construction in the more delicate species growing in sheltered areas. Moreover, the gills and tubes must be maintained in a vertical position and this is achieved by a number of responses to environmental stimuli. The developing fruit bodies seem to grow away from the soil and host surface initially by means of a type of avoidance reaction, but further growth of the stipe of most toadstools and the tissues of most brackets is normally controlled by gravity. This seems to be triggered by the movement of microscopic particles in hyphae against a network of protein fibres. The result is that the hyphae elongate differentially to bring the fruit body back to its optimum position. The response to gravity is maintained throughout the growth and maturation of the fungus and if a fruit body is moved out of position it will readjust rapidly, sometimes in a matter of hours. Light seems to be the primary trigger for fruiting body initiation in many Basidiomycetes.

The importance of gravity in controlling the growth of basidiomycete fruit bodies was recently emphasised, by implication, in some experiments in space. When colonies of the basidiomycete saprotroph *Flammulina velutipes* were carried into space by the space shuttle Colombia as part of an experiment on the D-2 Spacelab mission and grown in conditions of weightlessness for 165 hours, the toadstool fruit bodies grew in a completely random fashion, in sharp contrast to the normal vertical orientation of those grown on earth.

The development of shape and form in growing organisms is a complex matter which is only now being understood with our increasing ability to study the effects of individual genes, their interactions and the mechanisms for their control. In cellular plants and animals it is possible to visualise how genetic control of patterns of cell division in different positions leads to changes in shape. It is much more difficult to envisage the mechanism for shaping organisms such as fungi, which are made up of disparate elongated hyphae, growing only at their apices. We have a few clues, however. When an actively growing hypha is brought to a sudden stop, whether by human interference or genetic control, there are two obvious consequences. The arrested hypha swells at the soft apex and may also branch from, or below, the swelling. This, repeated throughout the tissue or in specific parts of it, can give some shape to the growing mass. In the bracket fungi, and in modified form in the toadstools, separate systems of differentiated hyphae appear to grow through the general background fungal hyphae to serve skeletal and binding functions. It is clear that the precision of gill spacing and tube size is 'written into' the development pattern of the fungus from an early stage, but how the end result is achieved is still unknown, although there is some evidence that hormones may be involved, as in plants and animals. Research on these matters is dependent on a sound foundation of observation by the natural historian. Much more needs to be known about the natural history of fruit body production in the Basidiomycetes, such as time from initiation to sporing, macroscopic structural changes and so on. This would provide useful underpinning knowledge for the laboratory experimentalist seeking a deeper understanding of the processes involved. Moreover, although we make a confident statement about the sporing behaviour of these organisms (individual sporophores of *Polyporus squamosus*, for example, are said to discharge a visible spore cloud continuously over a period of at least a week), information about the natural history of sporulation is remarkably sparse. Some comparative observations of the length of time of discharge and the areas of the spore-bearing surface that are active would be immensely useful in the study of the dispersal of pathogens.

But to return to the basidiospores. Following discharge, once they reach a suitable food source they germinate to form a monokaryotic mycelium. The life cycle begins again if monokaryotic mycelia carrying nuclei of opposite mating-type meet and fuse to re-form a dikaryotic mycelium. In many members of the Basidiomycetes there are several mating types rather than the normal two, so the chances of a mycelium encountering another of compatible mating type are maximised whilst outbreeding is maintained. The fusion of hyphae of the same mating type, which would lead to inbreeding, is prevented by sophisticated incompatibility mechanisms controlled by proteins on the hyphal surface.

Basidiomycetes as necrotrophs and biotrophs

Most of the Basidiomycetes are highly successful saprophytes, using the cellulose and often the lignin of plant and tree remains as sources of carbohydrate. Since such substrates are not available to other fungi, which lack the enzymes to degrade them, the saprophytic Basidiomycetes have a significant competitive advantage in habitats such as woodlands, forests and sometimes grasslands. Some also behave as necrotrophic parasites of trees, entering weakened or wounded individuals, often from the springboard of the large food base of another dead host. Once inside the new host they cause extensive rots as they degrade the tissue mass of the roots or trunk and from the dead tissue extend into the living tissue, killing as they go forward, until the weakened tree either dies where it stands or is blown down in a gale (Figs. 11.1 and 11.2). By this means the necrotrophic Basidiomycetes steal a march on their purely saprophytic competitors by gaining access to and colonising a potentially massive food source before its death. But more of this later. Some Basidiomycetes are undoubtedly biotrophs and hemibiotrophs and we shall deal with these first.

In the soil and litter of the forest floor some basidiomycete species form mycorrhizas of the ectotrophic type (see p.143). Insofar as they cause little disruption and provoke no marked resistance reaction in the tissue of the tree roots, these mycorrhizal Basidiomycetes qualify as sophisticated biotrophs. As yet, however, we know little of the balance between saprophytic and biotrophic

Fig. 11.1 Scots pine (*Pinus sylvestris*) and Corsican pine (*P. nigra* ssp. *corsica*) killed by *Heterobasidion annosum*, cause of butt rot, in Thetford Forest, Norfolk, England. (D.S. Ingram.)

activity in these organisms. Have they lost their capacity to produce litter-destroying enzymes in their relationship with a living plant, or are the enzymes that degrade cellulose and lignin only induced in particular situations, and if so is the anonymity this allows the reason for their parasitic success? In one study to investigate this matter seven species of the basidiomycete genus *Tricholoma* were investigated. Two were non-mycorrhizal and produced cellu-lase copiously. The others were mycorrhizal and, except for one, produced no cellulose-degrading enzymes. The exception was undoubtedly mycorrhizal but produced cellulase. So we cannot be certain that the mycorrhizal Basidiomycetes do not also digest cellulose and lignin in their scavenging phase in the soil; perhaps some do whereas others do not.

Some members of one order, the Exobasidiales in the Basidiomycetes, are undoubtedly biotrophic pathogens. These are leaf parasites that parallel in a remarkable fashion the Taphrinales (Exoascales) in the Ascomycota (see p.101), infecting by basidiospores which form short germ tubes and appresso-ria on young leaves of certain woody angiosperms, each sending an infection hypha through the immature cuticle and epidermis. The fungus then grows between the cells of the leaf tissue, which may become blistered by the over-growth of infected cells, as with *Exobasidium vexans*, the blister blight of tea (*Thea sinensis*). Here the massed basidia form a hymenial layer, generally on the lower, concave surface of the blister, but a typical basidiomycete fruit body is not produced. The basidiospores are discharged at the same time every day, usually in the four hours after midnight or after rain. The infected areas may ultimately die, but on the whole the fungus lives in balance with its host and may be described as biotrophic. Losses from this disease result from deforma-tion of the leaves and shoots, which are the source of commercial tea; the

Fig. 11.2 Paper birch (*Betula papyrifera*), its roots rotted by *Armillaria mellea*, blown down by a gale at the Royal Botanic Garden Edinburgh. (Debbie White, Royal Botanic Garden Edinburgh.)

situation is made worse by the frequent picking and pruning which are a feature of tea cultivation and which encourage vulnerable new growth.

Tea is not usually encountered in the British countryside, except in vacuum flasks, but another *Exobasidium*, *E. vaccinii*, is common on the shoots and leaves of wild bilberry (*Vaccinium myrtillus*) in both Britain and mainland Europe. Other members of the heath family (Ericaceae) attacked by *Exobasidium* include *Rhododendron* spp., marsh andromeda (*Andromeda polifolia*), crystal tea (*Ledum palustre*) and other *Vaccinium* spp. such as the commercially important blueberry (*V. corymbosum*). Some argue that all these fungi may belong to a single species, *E. vaccinii*, but others suggest that up to 18 different species may be involved. Moreover, symptoms vary from host to host. On *Rhododendron*, for example, one may see round leaf galls on alpenrose (*R. ferrugineum*) and hairy alpenrose (*R. hirsutum*), flattened marginal galls on *R. luteum*, leaf spots on *R. flavum* and witches' brooms (clusters of small branches arising from an infected bud) on *R. indicum*. The chalky white galls formed by *Exobasidium japonicum* on the leaves of greenhouse azaleas (*Rhododendron simsii* hybrids) are much prized by flower arrangers. Finally other *Exobasidium* spp. have been recorded on ornamental camellias (*Camellia japonica*), cinnamon (*Cinnamomum zeylanicum*) and other members of the laurel family (Lauraceae), citrus fruits (*Citrus* spp. hybrids) and holly (*Ilex aquifolium*).

Crinipellis perniciosus causes witches' broom disease of cacao (*Theobroma cacao*), has clear hemibiotrophic characteristics. Shoots of cacao trees become infected by the basidiospores of *C. perniciosus*, which then invades actively growing areas such as secondary shoots, buds or leaves. These proliferate, producing a bundle of shoots instead of flowers (hence the name 'witches' broom') and fail to produce fruit, resulting in economic loss. The fungus grows between the green cells of the cortex, leaf or young fruit and eventually leads to death of the tissue. In cacao the flower buds grow as cushions on the surface of the stem. The elegant toadstools of the pathogen appear on the dead tissue some time later, if the weather is moist. Witches' broom disease is currently causing major damage to cacao crops in South America and at present there is no effective control other than the removal of diseased branches. It has been suggested that the disease may lead to a steep rise in chocolate prices as supplies of cacao decline (see also p.87).

Finally, another member of the Basidiomycetes with an interesting pathology is *Chondostereum purpureum*, cause of silver leaf disease of plums, apples and their relatives. Whether it is a hemibiotroph or a necrotroph is a matter of debate. The fruiting bodies of this member of the Stereales are unusual in being like flat plates formed on the lower surface of a horizontal branch; when produced on an upright branch they may curl over to form a series of small shelves. The fruit bodies have a smooth spore-bearing surface that is mainly purple in colour, and slightly woolly; the growing edge is white. The disease gets the name 'silver leaf' from the partial separation of the surface cells of leaves of invaded plants following the secretion of a toxin by the pathogen. The exposed cell surfaces reflect the light, giving a silvery white sheen to the leaves. The toxin itself is not, however, responsible for the death of the plant.

Spores infect through wounds, being sucked into the functional vessels from a freshly cut surface. They germinate well in this protected situation and the mycelium spreads rapidly in the vessels, first in an upward or downward direction and then outwards into the younger wood and cambium, secreting enzymes and toxins as it goes and killing the living wood. Eventually a limb or even the whole tree may be killed. *Chondostereum purpureum* was a damaging disease in plum and apple orchards in Britain when fruit was grown on a larger scale than today, and is still a potential threat in Europe, North America and New Zealand.

The pathogenicity of the fungus is influenced by the season of the year, being greatest in the cooler months and lowest when the trees are actively growing. When the trees are active, gums produced in response to attempted invasion may prevent infection or, after infection and spread, may so restrict the growth of the fungus that the tree recovers. This is the basis of recommended control measures, which suggest that winter pruning should be restricted to the initial shaping of the tree, and that winter pruning wounds should be covered by an impervious protective coating. At the same time all dead wood, which might bear fruiting bodies, must be removed from the orchard before midsummer.

Here then we have a member of the Basidiomycetes which behaves like a typical wound parasite, but has extended its role to invade the vascular system of the tree and, by toxin production, cause symptoms at a distance from the infection site. It thus has some of the characteristics of a vascular wilt pathogen (see p.125).

The Basidiomycetes as wood destroyers

The rest of the pathogenic Basidiomycetes have a strong necrotrophic element in their behaviour (Table 11.1) whilst retaining many features of saprophytism. Most are members of the Poriales, and all have the capacity to degrade wood by destroying either the lignin and cellulose or the cellulose alone. The first are referred to as white rots since the rotted material and the penetrating hyphae have a whitish hue, and the second are called brown rots because the removal of the cellulose alone leaves a brownish rotted wood that still retains much of its structure, although it may show cubical fractures.

The capacity to rot wood is responsible for one of the most striking disease phenomena of the forest, the hollow tree. When a healthy tree is felled it shows a pattern of circular annual rings, and in a tree of some years the central rings tend to be darker than the peripheral rings and form the heartwood. Here the vessels are closed off by tyloses, gums and resins (Fig. 6.1, p.127) and are no longer functional, so the heartwood is of little importance for water transport. The outer, lighter rings form the sapwood that is newly laid down by the cambium. This region includes living cells which make up the medullary and xylem parenchyma. It is in the sapwood that water transport occurs.

If the tree is in the early stages of colonisation by a wood-rotting fungus when felled, the heartwood may appear to have an irregular stain superimposed on it, which does not exactly coincide with the ring structure. This denotes the presence of a wood-rotting pathogen. At a later stage this stained wood shows obvious signs of structural deterioration and the centre of the tree is gradually destroyed. As the rot progresses it may destroy the central wood from the base of the tree to high in the trunk, producing a so-called heart rot. If, as is common, the rot starts at ground level and only progresses a short distance up the central wood before the tree is felled, it is called butt rot. However, all rot does not start at the tree base and work upwards; it may also be introduced through wounds or broken branches, first invading the exposed dead tissue

Table 11.1 Common wood-rotting Basidiomycetes.

AGARICALES: Fruit body (basidiocarp) a toadstool with stalk (stipe) and cap bearing gills on its undersurface, these being lined with basidia.

Latin name	Common name	Description of Basidiocarp	Hosts	Symptoms
Armillaria mellea (but see also Table 11.2)	Honey fungus or bootlace fungus	Short-lived toadstool. Cap 3–15 cm diameter, very variable, sometimes convex, flattened or wavy; honey-coloured or darker, sometimes with olive tinge, often with darker scales near centre. Stipe 60–150 x 5–15 mm, white, tapering towards the base with a well-defined ring. Gills white, ageing to yellowish, then pinkish; basidiospores cream. Occurs in clusters around tree base or on stumps.	Almost all woody plants, with a worldwide distribution.	Soft, wettish, brown rot, eventually becoming fibrous and white, at the base of the trunk and in the roots: may be confined to the roots. White sheets of mycelium and black rhizomorphs under bark, the latter often extending into surrounding soil.
Pleurotus ostreatus	Oyster mushroom	Short-lived bracket-like toadstool. Cap 6–14 cm diameter, asymmetric, oyster-shell shaped, lobed or splitting at the margin, no clear stipe; streaky grey, brownish-white or blue-grey. Gills white, becoming yellowish; basidiospores lilac.	Mainly beech but sometimes infects other broadleaved trees.	Heart rot of standing trees, the infection being established through wounds.

PORIALES: Basidiocarp bracket-shaped or hoof-shaped; firmly attached by one edge to the tree; may be soft (ephemeral) or woody (perennial); lower surface covered with pores, these being the mouths of long tubes lined with basidia.

Latin name	Common name	Description of Basidiocarp	Hosts	Symptoms
Fistulina hepatica	Beefsteak fungus or ox tongue fungus	Large, soft, purple-red or chocolate, tongue-shaped bracket, 10–25 cm across, sometimes with short, rudimentary stipe. Pore surface off-white, bruising reddish; basidiospores pale ochre. Exudes red juice when squeezed.	Oak and sweet chestnut.	Causes brown staining of infected heartwood, with little structural damage. (Used in ornamental woodworking).

cont.

Species	Common name	Description	Host	Effect
Fomes fomentarius	Hoof fungus or tinder fungus	Hoof-shaped, woody, grey-brown to black bracket, 5–50 cm across; upper surface with horny crust, concentrically grooved or zoned. Pores greyish; basidiospores lemon-yellow.	Birch and beech, sometimes sycamore – rare in England but common in northern Europe and Scotland.	Mottled, yellowish-white rot of sapwood and heartwood, often with dark zone lines. Yellowish-white mycelium may occur in sheets in cracks and spaces. Usually infects through wounds in upper parts of tree.
Ganoderma applanatum	Artist's fungus	Flattened, hard, corky, semicircular bracket, 10–60 cm across, surface knobbly and concentrically grooved, with wrinkled, cocoa-coloured surface. Pores white, bruising brown; basidiospores brown, often dusting surface of infected trees.	Deciduous trees, especially beech.	Intensive white heart rot with dark banded margin; wood eventually becomes soft and spongy.
Heterobasidion annosum	Butt rot	Reddish-brown, knobbly perennial bracket with whitish margin, often uneven and lumpy, 5–30 cm across; formed at base of tree, sometimes enveloping twigs, litter or grass. Pores white, becoming brown; basidiospores white.	Very common on coniferous trees, but also attacks broadleaved trees.	Infected trees killed or develop an intensive white pocketed butt rot of tree base, depending on species and location.
Inonotus dryadeus	–	Thick, bumpy, annual, yellowish to brownbracket, 5–25 cm across, exuding drops of yellowish liquid; formed at soil level. Pores whitish, becoming rusty, basidiospores white to yellow.	Oak.	Soft white rot at base of tree; infection may occur through roots. Infected trees weakened and often blow over.
Inonotus hispidus	–	Annual, fan-shaped, felty, ochre-coloured bracket, becoming dark brown or blackish. Pores ochre, becoming brown; basidiospores rusty brown.	Ash, but also occurs on other broadleaved trees such as elm, apple and walnut.	Heart and stem rot associated with wounds.

cont.

Species	Common name	Description	Trees affected	Effect
Inonotus radiatus	–	Annual woody bracket, 2–9 cm across, upper surface velvety, wrinkled, apricot becoming rusty brown to blackish with yellow margin. Pores silvery, basidiospores yellowish.	Alder; occasionally on other broadleaved trees such as birch, poplar and willow.	Heart rot.
Laetiporus sulphureus	Sulphur polypore or chicken of the woods	Large, annual, thick, yellow-orange, fan-shaped to semicircular bracket, with uneven suede-like surface and margin. Yellow pores; basidiospores white.	Oak and other deciduous trees including cherry, yew, sweet chestnut, pine and willow.	Red-brown heart rot, breaking up into cube-shaped pieces, with sheets of brownish mycelium visible.
Meripilus giganteus	Giant polypore	Fan-shaped masses of annual brackets arising from a common base at foot of tree or in tufts above roots, 50–80 cm across; yellow-brown with fine scales and radial grooves, concentrically zoned dark and lighter brown. Pores whitish, bruising blackish; basidiospores translucent.	Beech and sometimes oak.	Root rot, making tree unstable. Fruit bodies are a danger sign in over-mature beech trees.
Phaeolus schweinitzii	–	Annual, bracket-like or occasionally stalked, 10–30 cm across, velvety to hairy, yellowish to rusty brown. Pores yellow to brown; basidiopores whitish to yellow.	Mature conifers, especially Douglas fir, sitka spruce, larch and pine.	Red-brown rot which breaks up into crumbly cubes with sheets of mycelium in cracks and strong smell of turpentine. Infection via roots.
Phellinus igniarius	–	Perennial, thick, hoof-shaped bracket, very hard and woody, 5–20 cm across, grey-black, cracked and ridged. Pores yellow-brown; basidiospores white.	Willow, but can affect most broadleaved trees.	Soft white heart rot with thin black zone lines; usually associated with broken branches.

Species	Common name	Description	Host	Rot
Phellinus pomaceus	–	Perennial, hoof-like or bracket, 6 cm across, dark grey; may be plate-like adhering to the bark. Pores grey to brown; basidiospores white.	*Prunus* spp., especially associated with wounds on plum trees.	Crumbly white central rot surrounded by firm, dark brown zone.
Piptoporus betulinus	Birch polypore	Annual bracket that starts as a knob and becomes smooth and kidney-shaped, 10–20 cm across, with a smooth, rounded, greyish to brown surface. Pores white; basidiospores white.	Birch.	Attacks weakened trees, producing reddish-brown heart rot that breaks into angular pieces and then becomes powdery.
Polyporus squamosus	Dryad's saddle	Annual, fan-shaped, slightly stalked bracket, 50–60 cm across, ochre-cream, overlain by concentric dark brown scales. Pores angular, ochre-cream; basidiospores white.	Elm, also beech, sycamore and other broadleaved trees.	Intense white heart rot, usually associated with branch wounds.
Rigidoporus ulmarius	–	Perennial, thick, woody bracket, 12–50 cm across, with ridged, white to ochre or dirty brown, knobbly surface. Pores reddish-orange; basidiospores yellow.	Elm.	Brown friable butt rot associated with wounds near ground level.

STEREALES: Basidiocarp forming a skin or crust adhering tightly to the host surface, or small, tiered brackets. Basidia formed on exposed surface.

Species	Common name	Description	Host	Rot
Stereum gausapatum	Pipe rot	Thin, tough and leathery, 1–4 cm across, ochre-brown to grey with white margin. Bleeds red if cut or scratched. Basidiospores white.	Oak.	Dark brown, firm heart rot with lighter to yellowish decayed streaks, or 'pipes', often associated with wounds.
Stereum sanguinolentum	–	Thin, leathery, 1–4 cm across, cream to brown. Bleeds red if cut or scratched. Basidiospores white.	Conifers.	Reddish rot becoming soft and white; associated with wounds.

and then spreading upwards and downwards through the trunk. In all these cases the infection is essentially saprophytic and the fungal growth is in dead tissue.

A hollow tree may stand for centuries, for a tubular structure has great strength, and water is conducted in the surviving sapwood. However, any internal rot or incipient decay of the trunk causes serious economic loss, since severely rotted stems cannot be used for timber. Young trees planted on a fresh site rarely develop rot, but second and later generation trees on the same site are more likely to be affected, and indeed in the worst incidences whole plantations will suffer severe loss.

Each of the wood-rotting Basidiomycetes has distinct characteristics, which the forester needs to be aware of. For example *Meripilus giganteus* (Plate 8), a wood-destroying fungus of deciduous trees in Europe, forms giant bracket-like fruit bodies around the base of over-mature beech trees, where it serves as a warning of the instability of the tree. *Phaeolus schweinitzii*, which occurs in Europe but is particularly important in natural stands of timber in the USA, causes a butt rot, and valuation of the timber has to take into account the potential loss of the first two metres or so of the most valuable part of the tree. The occurrence of the scaly, pored bracket-like fruit bodies of Dryad's saddle (*Polyporus squamosus*) on the trunks of elms, in the days when elms could easily be found in Europe, was a sure indication of discoloured wood and progressive heart rot. The irregular brackets of *Heterobasidion annosum* (Plate 8) emerging through the litter at the base of conifers is a sure sign of extensive root and butt rot, leading inevitably to death of the tree (Plate 8).

Many of the wood-rotting fungi can infect and kill a limited area of living tissue from their saprophytic base, leading to the formation of fruit bodies at the surface of the trunk. Entry to the trunk is generally through wounds that expose the dead heartwood; this occurs more readily in broadleaved trees, but in conifers an outpouring of resin forms a waterproof layer over the exposed surface and gives a measure of protection. However, stumps become infected because the resin produced in the younger tissue cannot protect the whole surface, and from these other trees may be infected by root contact. Since the Basidiomycetes produce very large numbers of spores which are efficiently dispersed into surface wind currents and distributed over a wide area, it is useful to think of the spores of most Basidiomycetes as being universally present in the air in and near woody districts, and the chances of spores being available to infect a wound or stump are relatively high. Even so, most require the presence of two mating types to produce a fruit body, and each spore has to establish itself separately, which must reduce somewhat the chances of formation of a dikaryotic, sexually complete organism.

There are, however, geographical limits to the wider spread of basidiomycete fungi, as evidenced by the local differences in the basidiomycete flora of conifer forests on the two sides of the Atlantic Ocean. Although there have been many introductions of North American coniferous trees into Europe, the toadstools and bracket fungi of the two areas are not identical. There is a rich basidiomycete flora in the Douglas fir forests of northwest America, but in Britain and as far as we know in mainland Europe the planted forests of this species are singularly lacking in basidiomycete fruiting bodies. Similarly, the trees in these forests, and the imported sitka spruce and lodgepole pine, if attacked by wood-decay fungi, are usually attacked by the common European species and not by the species common to the USA.

In addition to the production of resins, there are other inbuilt mechanisms that prevent infection by the wood-rotting Basidiomycetes. The stilbene glycosides produced by sitka spruce (see p.62), for example, present in the sapwood and bark, are toxic to a variety of potential parasitic fungi. The wood of many other trees also contains toxic chemicals. An extreme case is seen in the timber of western hemlock (*Tsuga heterophylla*) and western red cedar (*Thujopsis plicata*), which is used to make outdoor furniture, beehives, wooden shingles and so on, because it is generally resistant to decay. The resistance lies in water-soluble materials and oils which inhibit general wood-rotting fungi, even though butt rot caused by *Ganoderma applanatum* is common on the living hemlock tree in its native habitat.

Sometimes black lines may be seen which delimit various zones in the dead colonised wood, especially if the fallen tree is left where it is for a year or more. These should not be confused with resistance reactions. Similar zone lines are formed in culture as a result of an incompatibility reaction that occurs when two colonies of basidiomycete fungi meet. It is probable that the black zone lines in wood are formed in a similar way, as a result of interaction between two fungal colonies. They may also be a response to the stresses imposed on the pathogen by the environment of the woody tissue.

Heterobasidion annosum (Plate 8; Fig. 11.1) – a case history

Perhaps the best studied of all the wood-rotting fungi is *Heterobasidion annosum*, formerly *Fomes annosus*. Little was known of this member of the Poriales until the 1950s and 60s, when John Rishbeth (Plate 8) began to study the death of pines in forestry plantations on the English Breckland, an area of dry sandy soils in Cambridgeshire and Suffolk. The sand overlies chalk at various depths in spite of the low rainfall, and becomes leached because of the high percolation rate, giving rise to podzolised soils carrying a natural heathland. Scots pine and Corsican pine (*Pinus nigra* ssp. *corsica*) had been extensively planted on this infertile ground, a variety of different sites being used, including former planted woodland sites, former arable land, heathland acid soil and some areas where the chalk was near the surface and the resulting grassland neutral or slightly alkaline in nature.

When Rishbeth examined the dying trees he found that they died sooner on former woodland sites and on sites where the original trees had been destroyed by fire. Detailed examination showed that these trees had become infected by root contact with already diseased stumps and root systems from earlier plantings. In the plantations on the heathland and former arable sites, where there was little possibility of infection from such sources, infection only took place after the first thinning. Thinning is a normal forestry operation where the young trees, having been planted close together to minimise the effect of any failure and suppress branching on the lower trunk, are reduced in number once the plantations have become crowded by cutting out sufficient individuals to leave room for the remaining trees to grow strongly. Such felling takes place at different times depending on variable growth rates, the trees removed being sold for pulp or pit props. Rishbeth observed that in thinned plantations the group of trees surrounding the stump of a culled tree soon became infected by the butt rot fungus *Heterobasidion annosum*, and that its characteristic bracket-shaped fruit bodies soon became visible on the central stump. Experiments showed that freshly cut stumps of pine could readily be infected by the basidiospores of *H. annosum*, which then spread to the roots.

The pieces of the jigsaw puzzle were beginning to come together, but several problems remained. One was that there were many more infected pines on some sites than on others. Detailed observation of the growth of *H. annosum* on the roots of trees from alkaline and from acid sites showed that the growth rate was slower on the roots from acid sites. This was associated with the presence of the common soil mould *Trichoderma viride*, which produces antifungal materials, including the antibiotic griseofulvin; *T. viride* was generally absent at alkaline sites. There was thus an indication of the possible reason why the disease flourished more at alkaline sites than acid sites.

Where the fungus was infecting roots it spread superficially as a fan of mycelium ahead of the already infected area, presumably carrying nutrients forward from this established supply. The bark and sapwood were then penetrated at intervals as the mycelium advanced along the root. At these infection points the fungus first grew in the living tissue and then spread saprophytically in the woody cylinder. This behaviour is paralleled by the infection behaviour of *Rigidoporus* (formerly *Fomes*) *lignosus*, *Phellinus noxius* and *Ganoderma philippii*, which attack tropical plantation crops such as rubber after the felling of the natural forest.

During his studies Rishbeth had identified the Achilles heel of *Heterobasidion annosum*; the cut stump through which the disease first became established. If stump treatments could be devised to prevent the ingress of the fungus, he reasoned, the disease could be controlled. First various paints were tried: a bituminous sealant excluded spores of the pathogen but later cracked, so that late infection was possible; creosote gave good control but again when its effects wore off infection of the stump still occurred. Other substances were tried, including ammonium sulphamate (commonly used in stump eradication) urea and sodium octoborate; all had advantages. Each chemical treatment encouraged the growth of particular saprophytes on the stump surface, thus keeping out *H. annosum*. With the sulphamate, growth of moulds like *Trichoderma viride*, an antagonist of Heterobasidion, was encouraged; with the borate, *Botryotinia fuckeliana*, a fungus which one would not expect to find on stumps, and various wood-rotting fungi which were natural components of the stump flora proliferated. An intensive study of the ecology of the fungi of the stump surface eventually showed that one saprophyte, a bracket-forming member of the Poriales, *Peniophora* (*Phlebiopsis*) *gigantea*, was an early natural colonist and that when this fungus entered the stump ahead of *Heterobasidion annosum* it rapidly colonised the tissue and successfully prevented the entry of the pathogen, destroying the stump so rapidly that no further infection by *Heterobasidion* was possible. The advantage of *P. gigantea* compared with other saprophytes was that it would be incapable of attacking a healthy tree but would be able to colonise the still living stump immediately after felling, along with *H. annosum* but ahead of other saprophytes. Being a more effective saprophyte than *H. annosum*, however, it would then race ahead, colonising the stump tissue so rapidly that *H. annosum* could not compete with it.

Artificial inoculation with *P. gigantea*, Rishbeth decided, could be used as a means of controlling infection by *H. annosum*; a 'biological control' in which an antagonistic organism rather than a chemical or resistance gene would be deployed against the pathogen. Unusually in the Poriales, *P. gigantea* forms

asexual spores (oidia) very readily in culture. A method was devised for producing these in large quantities, drying them off and incorporating them into
pills with sugar, talc and sodium carbomethylcellulose. The pills were suspended in water by the forester, with the addition of a harmless dye as an indi-
cator, and painted on the stump surfaces of newly felled trees, with outstanding success in controlling the pathogen. This was among the first successful
biological controls of a fungal pathogen, and it is salutary to note that its invention depended on an intimate knowledge of the biology of the pathogen. More
recently, tests in which inoculum of *P. gigantea* is added to the oil on chain saws
or mechanical harvesters have given promising results.

Heterobasidion annosum is a necrotroph but behaves differently in different situations and on different hosts. On Scots pine (especially on alkaline soils) it
colonises fresh (living) stumps through the cut surface and proceeds to rot the
woody part of the stump. It may occupy the stump for some years and fruit on
it, producing both conidia (its anamorphic phase) and basidia from the bracket-like fruit body. It also proceeds along the main roots of the stump in their
dying phase, and from this base attacks living roots of neighbouring trees. In
pine it proceeds to invade and kill the whole root system and the tree. This frequently shows in young plantations as the death of a group of trees. In sitka
spruce on the other hand, although the main features of infection through the
stump are the same, the trees are generally not killed, and the fungus establishes itself in the central wood at the base of the trunk and grows upward to
cause (frequently) several metres of butt rot, often accompanied by copious
external resin. Since the lower portion of the tree is the most valuable, losses
can be considerable. Other tree species are affected in different ways although
the elements of the attack are similar.

How far other wood-rotting fungi behave in a similar way is still being established. Many of them are only visible as the tree becomes over-mature, and little or nothing is known of the early stages of their parasitism. Thus stands of
American aspen (*Populus* spp.) develop serious decay when they are between
50 and 100 years old, depending on location, while Douglas fir and sitka
spruce in natural forest in northwest America may apparently remain free
from rots for 200 or more years, despite the presence of pathogens nearby. Is
this because it takes a long time to accumulate sufficient wounds and entry
points for the wood-rotting fungi to get established, or is there a relationship
between potential life span and response to infection, with older trees becoming more susceptible, or does the rot simply develop very slowly with fruiting
occurring only when most of the tree has been colonised?

Heterobasidion annosum can spread for very considerable distances. In woods
where it is present, spores are released at most times of the year, and there is a
clear relationship between actively sporing fruit bodies and the infection of
neighbouring stumps. But just how widely does infection spread from a
sporing fruit body? There are a number of methods of catching and counting
spores to attempt to answer this question: they can be caught on sticky slides
over which air is passed at a constant rate, but spores trapped in this way need
to be easily recognisable; alternatively they can be passed across and deposited
on Petri plates containing an agar gel with appropriate added nutrients, but
here the choice of nutrients influences which fungi will grow. To overcome
these difficulties Rishbeth developed a method that was at once breathtaking-
ly effective and very simple. He used muslin cloths which he sterilised in steam

and stretched on a metal frame mounted on the roof of his car, which was then
driven over a prescribed route at a standard speed. (Rishbeth was as careful a
driver as he was an experimentalist!) The cloths, or rather standard areas cut
from them, were washed in standard quantities of sterile water and then a portion of the resulting suspension spread on the surface of slices of freshly cut
stumps of pine trees. Individual colonies of *H. annosum* that grew from any
spores collected could easily be counted. Also, as an insurance against misiden-
tification, he spread samples of the suspensions on Petri dishes containing
malt agar, a medium on which it was known that *H. annosum* grew well.

Using this method it was possible to say that approximately 30 spores could
be recovered at one mile from an actively sporing *Heterobasidion* fruit body,
three spores at five miles from source and one or no spores at 80 miles. Of
course, as Rishbeth recognised, there are many ifs and buts about these figures, but the main pattern of effective spread over distance is secure. Indeed
spores were recovered from traps exposed at a distance of 30 miles from land,
in the middle of the Irish Sea!

Troublesome saprotrophs

Many of the wood-rotting Poriales are efficient saprophytes, but there is often
an inverse relationship between pathogenic and saprophytic ability. A most
spectacular example is the dry rot fungus (*Serpula lacrymans*). There is evidence that this is an occasional cause of decay in living trees in Himalayan
forests, but otherwise it can be looked on as a thoroughly domesticated fungus
that destroys worked timber in buildings. It mounts its attack by invading timber that has become damp because of roof leaks, dampcourse failure or poor
ventilation. Sometimes in these situations another related basidiomycete,
Coniophora puteana, is the first colonist, soon followed by *Serpula*.

Once established, the dry rot fungus spreads superficially over the wood and
forms thick strands of wide hyphae bound into a 'rope' by thinner branches.
These are credited with the ability to carry moisture forward to sustain the
growth of the fungus into dry woodwork; hence the name dry rot. *Serpula* also
has the capacity to spread through brick and plasterwork to colonise woodwork at a distance. When fruit bodies are formed on wall surfaces or on woodwork they are flattened, or more rarely bracket-like, and coloured yellow to
rust depending on age. (To our eyes young fruit bodies are reminiscent of
poached eggs.) The surface of the fruit body is smooth to wrinkled with broad
irregular pores; the basidiospores are rust-coloured. Sometimes piles of these
characteristic spores are blown under skirting boards or through cracks in
floorboards, providing hapless householders with the first indication of an
alien presence under their roof. Further investigation usually reveals severely
rotted timber with typical cuboidal cracks. The enemies of dry rot are good
ventilation and warmth, and a series of chemical treatments has been devised
that protects timber from further attack. In cases where conditions are right
for the fungus, as in the cellars and roofs of large old houses and churches, failure to eradicate an attack can lead to spectacular collapse of the structure.

In the United States *Poria incrassata* replaces *Serpula lacrymans*. It has a similar ecology and also transports water in very large strands to several storeys
high. It has a fruit body with pores, initially orange in colour then maturing
through a series of darker shades to sinister deep purplish-black, at the same
time becoming leathery in texture. The occurrence of different genera of dry

rot fungi in different areas of the world is probably partly due to the different ecological environments provided by differences in climate, domestic heating arrangements and house construction, and partly to the geographic isolation of the fungi. Another *Poria* species, *P. vaporaria*, occurs in centrally heated buildings in mainland Europe, in association with leaks of water, with catastrophic results. The difficult taxonomy of these fungi leaves their true identity in doubt.

The puzzle of the honey fungus

Apart from *Crinipellis perniciosus* (see p.216) there are few pathogens among the gill-forming members of the Basidiomycetes. A number of gilled fungi of the genera *Flammulina, Lentinus, Pholiota, Pleurotus* and *Schizophyllum* occasionally cause rots in living trees, but their pathogenic status is largely uninvestigated. There is, however, one devastating pathogen in this group, the honey fungus *Armillaria mellea* (Figs. 11.2 and 11.3). It has been known for many years as an apparent cause of death of individual trees in gardens and as a sporadic cause of death in forests. The shiny honey-coloured toadstools, each with a long stout stem and a marked ring below the cap, are characteristic features of stumps and dead trees in old hardwood forests. A particular feature of *A. mellea* is its capacity to produce rhizomorphs (see p.55) (Fig. 11.4). A mature rhizomorph, often referred to as a bootlace, is capable of apical growth and spreads over roots and out through the litter and the soil to make contact with neighbouring trees. In old established gardens it is quite common to observe the death of a small ornamental tree and so be able to trace bootlace connections coming to it from neighbouring living, though possibly moribund, trees. *Armillaria* is also commonly present on the roots of aged trees that have been exposed by felling, stump-extraction or wind blow. Whether in these cases the fungus is acting as a primary parasite or as a saprophyte on the dead tissues is not clear (but see taxonomy below). It is also capable of causing the death of planted forest trees, both conifers and hardwoods.

The wide ecological range of this pathogen is very confusing. Some clarification comes from an understanding of the behaviour of biologically distinct forms within the species *A. mellea*. A virulent pathogenic form, apparently only able to attack coniferous trees, was shown to gain initial entry through stumps

Fig. 11.3
Toadstools (basidiocarps) of *Armillaria mellea* (honey fungus) on a dead tree in Madingley Wood, Cambridge, England. (D.S. Ingram.)

Fig. 11.4 Rhizomorphs of *Armillaria mellea*, revealed by removing the bark from a killed tree. (D.S. Ingram.)

in a fashion similar to *Heterobasidion annosum*; this could be controlled by stump treatment with ammonium sulphamate, which encouraged rapid rotting of the stump by saprophytic competitors of *Armillaria*. This contrasted with an apparently saprophytic type found on dead roots and an equally distinct form that attacked living plants from a food base in a dead tree. The problem was illuminated by the discovery of a corresponding taxonomic heterogeneity in what, in the past, had been regarded as a single species. At least nine species have now been identified, worldwide, to replace the former single species *Armillaria mellea*. Those most likely to be encountered by the naturalist are listed in Table 11.2.

Table 11.2 Some commonly found *Armillaria* species.

Name	Pathogenicity	Fruit body
Armillaria mellea	On a range of broadleaved trees and conifers.	Cap honey-brown and reddish-brown; ring white; yellow flecks.
A. ostoyae	Pathogenic only on conifers.	Cap pinkish; ring white; brown scales.
A. bulbosa	Much less pathogenic than the other species; grows on roots of stunted and shaded trees, occasionally causing butt rot.	Swollen stem, cap honey-brown, ring white.
A. tabescens	Virtually non-pathogenic; grows saprotrophically on broadleaved trees.	Stem not swollen, no ring, cap honey-brown.

Armillaria spp. are interesting in a number of ways. The mycelium and infected wood may be luminous, but whether one or all of the now recognised species behave in this way remains to be determined. Their capacity to infect over a distance by means of rhizomorphs was used by Denis Garrett (see p.111) in developing the concept of inoculum potential (the energy of growth of a parasite available for infection of a plant at the surface of the organ to be infected). He mentioned rhizomorphs 10–20 metres in length and at the same time pointed to evidence that the success of infection by rhizomorphs was closely related to the proximity of the food base. In Tanzania, where the fungus was troublesome in coffee and tea plantations, observations established that infection was generally from infected material lying close to the new victim. Moreover, in an experimental situation, Garrett found that the speed of infection declined with increase in the length of the rhizomorph. The concept of inoculum potential, even though it cannot be narrowly quantified, has been invaluable to generations of plant pathologists seeking to understand infection phenomena in plant disease.

Whatever the apparent limitations of long rhizomorphs in Garrett's experiments, they are probably more efficient in a field situation, provided they can leapfrog from root to root gathering food as they go. And they are formidable weapons for any fungus, for once formed they are protected by their black rind and are able to grow at the tip by means of closely-packed rows of advancing hyphae, which over a short distance have an enormous advantage over a single hypha or spores. They resemble military columns on the march, but are similarly dependent on the efficiency of their supply lines.

To end with a world record: *Armillaria* has been shown to be among the largest and oldest of living organisms. A colony of *A. bulbosa* growing in woodland in California, USA, occupied an area of at least 15 hectares. Sampling for DNA analysis revealed that the colony was homogeneous, and therefore probably originated from a single dikaryon rather than having been formed by the fusion of several colonies. Calculations suggested that it weighed over 10,000 kilograms, whilst from a knowledge of its rate of progression through soil in a woodland situation its age was estimated to be in excess of 1,500 years. As with an iceberg, there may be far more to a toadstool if you look beneath the surface!

12

The Smallest and the Largest Pathogens

Whereas large pathogens, the plants that parasitise other plants, are measured in metres and millimetres, smaller pathogens, the bacteria and viruses, are measured in micrometres (= μm or 10^{-6} m) and nanometres (10^{-9} m), each measurement being a thousand times smaller than the last. Fungi, which have been our major concern so far, fall in the middle of this range, being measured in μm; thus the oospores of *Phytophthora* are 30–60 μm in diameter and the ascospores of *Sclerotinia* are approximately 10 x 6 μm. Bacteria are about ten times smaller than the fungi: the cells of *Erwinia carotovora,* for example, are about 1 μm in length, this being about ten times their breadth. And the viruses are at least 1,000–10,000 times smaller than the bacteria!

Bacteria, like the fungi, have a cell wall composed of complex carbohydrates and a protoplast made up of membranes, cytoplasm and organelles. The nuclear apparatus is less precisely delineated than in the fungi, with no enclosing membrane, but the presence of deoxyribonucleic acid (DNA) and genetic evidence for a linear arrangement of genes suggest a certain similarity to the nucleus of fungi and higher plants. The viruses, by contrast, have no cellular structure and have a central core of ribonucleic acid (RNA) or, in a limited number of cases, DNA; genes can be identified in both the RNA and DNA. The nucleic acid is contained within a protein coat. There are some organisms intermediate between bacteria and viruses, the mycoplasmas, and some 'viroids' which consist of circular low molecular weight RNA without a protein coat. These are probably the smallest pathogens of all, where we include the prions, the proteins thought to be the causal agents of some encephalopathies.

The flowering plant parasites of other flowering plants (there is one known conifer parasite of another conifer) can be very large. Members of the mistletoe family, Loranthaceae, can reach the size of a large bush in the crowns of tropical trees, and *Rafflesia* spp., which live in the roots of tropical vines, have a vegetative body impossible to measure, but it must be of considerable size to nourish the periodic solitary flower, one of which was found to weigh 5 kg. That said, the flowering plant parasites, if they are to survive, must not be larger or more voracious than their hosts can tolerate.

At the other extreme the viruses and viroids have constraints at the molecular level: the infectious virus particle must contain sufficient nucleic acid to programme the continued multiplication of the nucleic acid and the production of the protein coat; also, these organisms are completely dependent on a living host in order to reproduce, for although all the genetic information to reproduce a new virus or viroid particle is contained within the DNA, the biosynthetic machinery of the host must be harnessed to synthesise the new nucleic acid and protein molecules. Despite this, it is not unknown for viruses to cause a slow decline of a host, but this usually allows plenty of time for fur-

Fig. 12.1
Variegated
Abutilon striatum
at Downing
College,
Cambridge,
England. The
variegation is
caused by a viral
infection. (D.S.
Ingram.)

ther spread to another host before death finally occurs.

Being non-mycelial and composed of single particles or cells, the viruses and bacteria have evolved very different mechanisms from the fungi for infecting their hosts. Viruses sometimes enter through transient wounds but are more often carried by a vector such as an insect that feeds on plants or a fungus that infects plants. Bacteria are occasionally carried by vectors but more usually enter a host through a wound or via the stomatal pores.

The virus diseases of plants*

From the earliest historical times there are written accounts that probably refer to virus diseases of animals and man such as those subsequently identified as smallpox, foot-and-mouth disease and rabies. But it was not until the early nineteenth century that Louis Pasteur reported that no microscopically identifiable organism could be found in infectious fluid causing rabies, and suggested that pathogens might exist that were too small to be seen. The effect of viruses on plants was first recognised in the sixteenth century, when the 'breaking' of flower colour in tulips became a common subject for early flower painters, especially in Holland. This condition is now known to be of viral origin, the single base colour of the tulip flower giving way to complex patterns of white or yellow stripes where it has been eliminated from the epidermal cells. By the eighteenth century various degenerative diseases of potatoes were recognised but not categorised, and many of these were probably of viral origin. 'The curl', a disease of potatoes now variously ascribed to potato leafroll virus and/or a mixture of the viruses producing severe mosaic symptoms in potatoes, was a cause of crop failure at the time of the potato blight epidemics in the nineteenth century, and in 1869 the variegation of *Abutilon striatum* (Fig. 12.1), a decorative plant used extensively in Victorian conservatories, was found to be infectious when variegated scions were grafted on to green stocks in the course of propagation.

---

*The nomenclature of viruses is constantly changing. We have used, in general, the traditional names that describe the symptoms caused. Modern names are given in the reference texts included in the Bibliography.

This was all circumstantial evidence for the existence of a 'new' infective agent until Mayer, in 1886, transmitted tobacco mosaic disease to healthy tobacco (*Nicotiana tabacum*) plants using juice from infected plants, and Ivanowski, in 1892, showed that the juice containing this tobacco mosaic-inducing agent remained infective after passage through bacteria-proof porcelain filters. Nobody paid much attention until in 1898 the great microbiologist Beijerinck independently confirmed these results and referred to the filter-passing material as a *contagium vivum fluidum*. Also in that year it was shown by Loeffler & Frosch that foot-and-mouth disease of cloven-hoofed animals was caused by an agent that passed the bacterial filters. At about the same time, Erwin Smith (who later went on to study plant bacteria – hence *Erwinia*) suggested that peach yellows disease in the United States, for which no pathogen was known, might be transmitted to healthy plants by leafhoppers. In fact it was later shown to be caused by very small xylem-inhabiting bacteria near the limit of light microscope resolution, but the research was important because it focused attention on the possibility that viruses could be transmitted by insects.

By the beginning of the twentieth century, viruses were being recognised and described, the mechanisms for their transmission from host to host were becoming apparent and methods for infecting plants experimentally were being worked out. This provided enough information for a measure of practical control of the diseases, but was not intellectually satisfying for it did not solve the mystery of the *contagium vivum fluidum*. Then in 1935 W. M. Stanley at Princeton University in the USA isolated and purified a protein from tobacco mosaic which he obtained as crystals. This substance retained its infectivity even after successive recrystallisations and after considerable dilution; moreover it was powerfully antigenic and when injected into rabbits yielded a serum that reacted with the sap of diseased plants. Stanley was careful to point out, however, that purification of proteins was difficult and he anticipated that his claim would require modification, as indeed it did. In 1937 Bawden & Pirie showed that the 'protein' from diseased tobacco was in fact a nucleoprotein appearing as liquid crystals comprising 5% RNA and an associated protein, and in 1938 the tomato bushy stunt virus was shown to contain 15% RNA in addition to a viral protein.

Once methods of transmission by grafting, sap inoculation or insects had been worked out it became possible to give some identity to viruses in terms of their transmission characteristics and the range of hosts they attacked, but their taxonomy still remained difficult. With the establishment of the chemical nature of viruses, however, the way was opened for their study using a variety of chemical and physical techniques. Thus it was found that some were rod-like (tobacco mosaic virus), consisting of a coiled chain of RNA enclosed within regularly positioned protein molecules, which formed a protective coat (Fig. 1.15a, p.33). This protein coat, in different forms, was characteristic of particular viruses and a consistent accompaniment of the virus particles, but was not necessary for infection. In other viruses, tomato bushy stunt and turnip yellow mosaic for example, the virus particles were arranged in a somewhat spherical structure with the protein sub-units arranged in a lattice to form the surface of an icosahedron (Fig. 1.15b, p.34). The nucleic acid was closely associated with the inner portion of this protein shell in a regular pattern. Crick & Watson, who with Wilkins & Franklin first solved the puzzle of the arrangement of the nucleic acids in chromosomes, described the structure of small viruses in 1956, and there is no doubt that the X-ray crystallographic study of tobacco mosaic virus was an important precursor of their later work on the helical structure and base pairing in DNA.

But before chemistry took over, two important techniques were used in the quantitative study of viruses, based on their ability to form local lesions and on their antigenic properties. The first technique resulted from the discovery by Francis Holmes in 1928 that tobacco mosaic virus produced systemic mottle symptoms in the 'Burley' variety of tobacco (*Nicotiana tabacum*), but when inoculated in precisely the same way onto the leaves of *N. glutinosa* produced a dark spot at each infection point – a 'local lesion'. It was thus possible to establish a relationship between the concentration of virus particles in a sample and the number of local lesions produced on a reactive leaf. Similarly, viruses could at lower dilutions be very precisely identified and quantified using serum precipitation tests and other technically more complex serological methods.

Since then, more refined techniques have been developed, such as the enzyme-linked immunosorbent assay (ELISA), where an enzyme-labelled antibody binding to a virus sample precipitates and digests an added enzyme substrate, with consequent colour change; the intensity of the colour change gives a measure of the virus present in the sample. Immunological techniques are not the normal tools of the naturalist, and indeed because of the stringent regulations surrounding their production and use are only likely to be available to research professionals, but it is useful to be aware of their contribution.

However, a great deal of useful field work on the viruses of wild plants can still be carried out using routine tools of the virologist such as grafting and sap inoculation. Successful horticultural grafting for long-term permanent growth is a difficult technological operation but such refinement is not needed for virus transmission; it is sufficient for a temporary union to take place. This may be achieved with simple wedge grafts on herbaceous as well as woody material; with patch grafts on woody material, where pieces of bark cut with a standard cutter are exchanged with similar sized pieces removed from the potential host; with standard budding techniques; and in some instances (e.g. *Citrus* spp.) with a cut piece of leaf or strip of other tissue imprisoned under a flap of bark on the potential host, where it makes a temporary union, allowing transmission to occur.

For all virus work a source of virus-free seedlings is also necessary. While more complex work requires such seedlings to be grown in an insect-free glasshouse, simple diagnostic work can be carried out at cooler times of year with seedlings grown and maintained in a normal glasshouse, without special exclusion of insects, provided routine spraying for insect control is maintained and sufficient numbers of uninoculated plants are used as 'controls'.

Sap transmission remains a convenient method for comparative work on the viruses of a number of annual plants. The method used is to grind up symptom-bearing leaves of the donor plant and to rub the juice, roughly filtered through muslin, onto the surface of the tester plant. It is usual to add a small quantity of a fine abrasive such as carborundum powder to the juice to create small wounds on the leaf surface. After this inoculation procedure the juice is washed off the leaves and the plants returned to the glasshouse. With highly infectious viruses like tobacco mosaic virus or cucumber mosaic a high level of transmission will be obtained. Some plants have high levels of inhibitory substances such as tannins in their leaves and it is necessary in these cases to use phosphate buffers or additional reducing agents such as sodium sulphite to achieve adequate transmission. With new and previously unknown viruses it is a matter of trial and error to find the right conditions for transmission, and sometimes transmission by sap inoculation proves difficult or erratic until the virus has been obtained from a more suitable host.

Viruses can be transmitted in nature by other biological means, such as insects with sucking mouthparts, as with aphids and potato virus Y, mealy bugs and cacao swollen shoot virus, leafhoppers and rice dwarf and maize mosaic viruses, thrips and tomato spotted wilt, or white flies and cotton leaf curl. Others may be transmitted by insects with biting mouthparts, for example flea beetles and turnip yellow mosaic. The transmission pattern shown by these insect-virus combinations varies. Many of the aphid-borne viruses are transmitted to the new host immediately after feeding and the insect does not remain infective. Here the mouth parts of the insect appear to act like a dirty syringe which carries some virus with it into the tissues of the new host. Other viruses transmitted by aphids, leafhoppers and beetles are ingested and the insect does not immediately transmit the virus, a latent period of 12 hours or more being required to allow the insect to become infective. Once infective it remains so for at least a week. There is evidence that these 'persistent' viruses pass from the gut of the insect and circulate in the body fluid, eventually reaching the salivary glands. From the salivary glands they are injected together with the saliva into the vascular elements of the plant when the insect next feeds, and thus reinfect. In some virus-insect combinations there is evidence that the virus multiplies within the body of the insect and may even be included in the eggs, thereby being passed to a subsequent generation.

Insect transmission may be routinely used in the study of viruses, but is probably not the vehicle of choice for the amateur seeking to make sense of a virus problem. The same is likely to be true of the soil fungi or nematodes which transmit soil-borne viruses, but they too are of considerable intrinsic interest. In some fungi, such as *Olpidium brassicae* (see p.83), the zoospores pick up the virus from the soil solution around the roots of infected plants and, swimming to a new host, carry the virus into the roots; there is no long term association between fungus and virus. In *Spongospora subterranea* (see p.78) the virus is incorporated into the fungal tissue during parasitism of a virus-infested plant, and the particles enter the resting spores, where they are maintained for a long period before a new host is infected by the fungus and the virus transferred at the same time.

Eelworm transmission is usually by members of the free-living genera *Trichodorus* and *Paratrichodorus*. These carry the Tobra viruses responsible for tobacco rattle disease, which affects potatoes and bulbs and other hosts as well as tobacco. Similarly, free-living eelworms of the genera *Longidorus* and *Xiphinema* carry the nepoviruses, including arabis mosaic, grapevine fan leaf, raspberry ringspot and many others. In the transmission of viruses by eelworms the infected sap, including the virus particles, is sucked from the plant root and the virus particles are adsorbed onto the walls of the pharynx and oesophagus. They are in part released when the salivary fluid is injected into a new host during feeding. There is no evidence of multiplication of the virus within the eelworm vector.

Viruses are also sometimes passed from plant to plant by way of the seed. This is not a common happening, but in lettuce mosaic, for example, identification of mosaic infection of mother plants (which leads to seed infection) is one of the important steps in controlling the disease. Transmission through vegetative propagation is often the reason for the deterioration of stocks of vegetatively propagated crops such as potatoes, raspberries, strawberries, fruit and ornamental trees and bulbs. Once the damaging effects of the resulting diseases were recognised a number of methods of obtaining disease-free stocks were developed. Thus it was found in the 1940s that if the apical meristem was surgically excised from an infected plant and then grown on in culture, a virus-free plant might result which could then serve as the basis for the production of new virus-free clones. Viruses are often absent from the apical meristem of plants and this explains their elimination during meristem culture; sometimes too they seem to die out during the culture process itself, perhaps because of destabilisation of the delicate balance between host cells and pathogen. Virus elimination may be aided by keeping the infected plants at high temperatures; this appears to reduce the relative amount of virus in the tissues and makes it easier to excise virus-free apical meristems.

Other methods of dealing with viruses in crop plants include the use of resistance genes, or control of the vector using appropriate insecticides or fungicides.

Finally it should be mentioned that both fungi and bacteria, as well as plants, may be attacked by viruses. Evidence for their existence can be found in many fungi, but particularly in the watery stipe disease of cultivated mushrooms, where infection leads to the collapse of individual fruit bodies. Virus infections are also responsible for the reduced parasitic vigour that can occur in the fungi responsible for chestnut blight and Dutch elm disease (Chapter 6). In bacteria the virus parasites are called phages and may be revealed experimentally as plaques (circular translucent areas of dead bacteria) on agar plates seeded with the host bacterium. These viruses, as in the T2 phage of *Escherichia coli*, may have a protein coat, and under the electron microscope can be seen to have a distinct head and a tail. The virus becomes attached by the tip of the tail to the bacterial surface and the nucleic acid contained within the protein coat finds its way through the modified tail tip into the bacterial cell, where it multiplies.

The biology, particularly the ecology, of viruses is full of interest. Some are true latent viruses carried in the host with no visible symptoms (see p.152). Others may produce severe initial symptoms and then settle down to a more or less symptomless phase. Thus the cacao swollen shoot disease occurs in forest trees related to cacao without apparent symptoms unless the tree is coppiced, when infection may show in the coppice shoots. Inoculation of seedlings of the natural host with the virus leads to initial symptom production, which gradually becomes less apparent. However, inoculation of cacao leads to marked symptoms and a slow decline of the tree, the rate of decline differing with viruses of different origin.

Other viruses are dependent on the ecology of their insect vectors. For example, potato virus Y is carried by the aphid *Myzus persicae*; the eggs of this insect overwinter on the shoots of trees of the genus *Prunus* (e.g. cherry and plum) and on hatching pass through several wingless generations before giving rise to a winged generation which moves to surrounding herbaceous plants, including the potato. In the potato field the aphids multiply in discrete areas of the field and when these become crowded migrate, by crawling or by producing further winged forms, to spread through the whole crop, taking any virus infection with them. In cool conditions the build-up is less and in cool windy conditions the distant spread of the aphids by flying is inhibited; for this reason the production of virus-free seed potatoes is restricted to cooler and hillier regions such as Scotland.

The eelworm vectors of the viruses of potatoes, bulbs and soft fruits are less susceptible to environmental changes above ground. *Longidorus* and *Xiphinema*, for example, live mainly on the roots of hedgerow trees and shrubs, from whence they spread inwards to the area of cultivation.

The naturalist is unlikely to become acquainted with viral diseases in the same way as with fungal diseases. However, there are some that are worth looking for. Many viruses occur on raspberries and related plants both in the wild and in gardens, producing symptoms such as mosaic, yellowing of the veins, banding of the veins (yellow and brown) and changes in the morphology of the plant such as curled leaves and reduced side shoots. Potatoes are no longer a source of readily available virus symptoms, except where gardeners save their own seed, since commercial seed potatoes (tubers) now have very high health standards and stocks are renewed from healthy mother stocks on a recurring basis. There are a number of rose viruses in the USA, Australia and Europe, however, which may sometimes be found in collections of older varieties. A range of symptoms may be observed including mosaic, yellow leaf flecks and stunting, probably caused by rose mosaic virus, strawberry latent ringspot virus and rose wilt virus respectively. Cucumber mosaic virus is capable of attacking a wide range of hosts in the garden and glasshouse, being found not only in marrows (*Cucurbita pepo*), courgettes, cucumbers, tomatoes and melons (*Cucumis melo*) but also in a wide range of ornamental plants such as aquilegia (*Aquilegia* spp.), delphinium (*Delphinium* spp.), lupin, geranium (*Pelargonium* spp.) and aster. We see it commonly in Christmas roses such as *Helleborus niger* and *H. orientalis*. In the former it appears to be responsible for the rapid decline in vigour that affects some stocks; in the latter it produces mosaic, vein banding and occasionally ring spot symptoms on the maturing leaves but does not obviously hasten the plants' decline.

Any unusual morphology or discoloration in wild plants that cannot be attributed to known fungal diseases may be investigated for possible viral origin. But be warned, this is not a road one can go down lightly; the study of viruses needs organisation and resources that can be provided only by the most dedicated amateur worker.

The bacterial diseases of plants

The morphological variety of bacteria is not great: they may be rod-shaped (sometimes with clubbed ends), spherical (cocci), comma-shaped or spiral. These variously-shaped cells may appear as single entities or grouped in twos (*Diplococcus*), in chains (*Streptococcus*), in groups like bunches of grapes (*Staphylococcus*) or in packets (*Sarcina*). The individual cells may or may not have flagella, and these appear in different positions and in different numbers on the cells with such regularity that they can be used as criteria for classification, as can the enclosing cell wall, which may appear as a structure that only reveals its presence when stained or may be surrounded by a well-marked mucilaginous wall (largely polysaccharide), the capsule. This last is sometimes only formed if the bacteria are grown in culture. In non-encapsulated bacteria a particular staining technique has proved useful in differentiating two groups, Gram-positive and Gram-negative bacteria. The procedure uses the crystal violet stain followed by a mordant such as iodine to stain the cells, already dried and fixed on a slide; washing with dilute ethyl alcohol or acetone removes the stain in a number of instances but leaves it completely fixed in others. This procedure, discovered empirically by a Danish bacteriologist called Gram, appears to distinguish some fundamental property of the wall and of the cell, and is used as a criterion in the classification of bacteria, together with their limited morphology.

By contrast the metabolic activities of the bacteria are very varied. They actively metabolise components of substrates presented to them and secrete into their environment a wide variety of other materials, particularly enzymes. These may be coded not only by the bacterial chromosome but also, in some cases, by a piece of extra-nuclear DNA called a plasmid. Plasmids are infective and may be passed from one bacterial cell to another. Whether they may be regarded as sub-viral parasites of bacteria is a moot point. The metabolic properties of bacteria, along with the morphological features already mentioned, are used in the detailed identification and classification of the organisms, and are of great importance in determining their pathogenic behaviour: most are necrotrophs although some have an interesting biotrophic behaviour. Bacteria also have a rapid multiplication rate, especially in higher temperatures; they are in general, therefore, pathogens of the tropics although some can multiply and spread under temperate conditions.

A few plant pathogenic bacteria are Gram-positive. The genus *Clavibacter*, for example, with rod-shaped cells lacking flagella, is responsible for a large number of vascular diseases, especially wilts. The genus contains a number of well-marked species and also, within some species, subspecies which are confined to separate hosts but are otherwise not separable. Thus the tomato canker and vascular wilt originally attributed to *C. michiganensis* is now thought to be caused by a subspecies *C. michiganensis* ssp. *michiganensis*. Other forms, such as *C. insidiosus*, which causes wilt in lucerne and *C. sepedonicus*, which causes ring rot and wilt in potatoes, are now included in *C. michiganensis* as ssp. *insidiosus* and ssp. *sepedonicus* respectively. This taxonomic complexity is mentioned here because it reflects the pathogenic behaviour of the bacteria, where one species is capable of forming distinct, specific, pathogenic relationships with different plant genera whilst retaining the same morphology and basic biochemical equipment.

Ring rot of potatoes is a particularly damaging disease worldwide. It has caused much loss in the USA and Canada and is prevalent in northern Europe, although currently it is not present in Britain on any permanent basis. In Britain quarantine is maintained with careful scrutiny of samples of all imported potatoes. This is particularly important because field symptoms tend to appear late and may not reveal the existence of the disease before the crop foliage has been burned down and the tubers dug at an early date (a good practice for the control of other diseases). Apparently healthy tubers may thus contain the organism, and it is the planting of infected tubers that spreads the disease; the vascular system of emerging shoots is invaded, as is that of the stolons on which the new tubers will be borne. Of these tubers some may be lightly and some heavily infected; the former give rise to infected plants in their turn while the latter rot in the soil. The bacteria from these rotting tubers do not infect other tubers or succeeding crops, for they do not persist in the soil from season to season. The spread and intensity of the disease is greatest where potato tubers are cut up to supply seed for the next crop, since cross contamination occurs when unsterilised knives carry the bacteria to freshly cut xylem vessels.

A disease caused by *Clavibacter* spp. that the naturalist could look out for is that caused by *C. rathayi* on cocksfoot. Here the inflorescence is covered with a yellow slime and the flowering structure dwarfed. It is occasionally seen in Britain and is sometimes a serious problem in mainland Europe. Possibly the same or very similar organisms attack bent grasses (*Agrostis* spp.) in the USA and wheat in a number of subtropical countries. In these hosts the bacteria are spread by the eelworms *Anguina tritici* and *A. graminis*. On their own, *Anguina* species cause individual wheat grains to swell, forming 'corn cockles'. The biology and taxonomy of *C. rathayi* and its allies need to be examined further, particularly in wild grasses, and this could be achieved by means of straightforward field studies to determine host ranges.

The genus *Pseudomonas* (Gram-negative organisms with rod-shaped cells bearing single or multiple polar flagella) contains a large number of species responsible for plant diseases, including *P. solanacearum* and *P. syringae*. *P. solanacearum* causes the brown rot of potatoes and is, like *Clavibacter*, a vascular disease, but is not confined to the vascular tissue and breaks out to invade and rot the cortex of the potato tuber. Furthermore, it can contaminate and invade tubers in the soil. It occurs in at least three races, each of which is further divided into pathotypes (fine divisions of a species classified in terms of pathogenicity). The races here attack (1) hosts in the potato family (Solanaceae) but not potato itself (and in culture produce a dark colour because of the action of the enzyme tyrosinase on phenolic chemicals); (2) potatoes (isolates show a weak tyrosinase reaction); (3) the banana and its relatives (isolates of this race show no evidence of tyrosinase production).

In potatoes the disease is widespread in warmer parts of the world, is present in mainland Europe and has recently been identified in imported seed in Britain. It is not restricted to seed transmission, in which it differs from ring rot. Rotting tubers release bacteria which can gain entry to a new host through injured roots or wounds on the stem and also via the stomata, where they enter the substomatal chamber, multiply and destroy the surrounding tissue by means of enzymes.

Pseudomonas syringae is a name used for a number of disease organisms once separately identified as distinct species; now they are thought to be pathogenic forms of the single species, and about 50 pathotypes have been identified. These include the pathotypes causing halo blight of oats, halo blight of beans and celery leaf spot, which contrast with pathotypes causing wilt and canker on the twigs of woody plants with associated leaf infection as in, for example, lilac blight, *Prunus* stem canker (especially of cherry and plum), poplar canker and the olive knot and ash canker. Infection takes place through stomata, lenticels or wounds. Invaded areas may exude bacterial slime which is further spread by rain splash, and reservoirs of infection in the tissues may give rise to later infection when the organisms are in turn released. Infected leaves often remain attached to the stem into the autumn and winter. In cherry and plum the bacterium invades the cortex of the young shoot through a leaf base or lenticel, and killing the tissue causes splits in the bark. Depending on the success of infection the bacterium at the edge of these splits may advance into neighbouring healthy tissues to form a canker, and may even expand sufficiently to ring the stem. Control in all these forms of *P. syringae* depends on an intimate knowledge of the individual pathogen and the introduction of sprays or pruning to remove infection foci at appropriate times.

One other important group of bacterial plant pathogens is found in the genus *Erwinia*, which is characterised by Gram-negative rod-shaped cells with flagella produced generally over the whole surface of the cell (Fig. 1.14, p.31). *E. amylovora* is responsible for the notorious fireblight disease of members of the rose family (Rosaceae) such as apple, cotoneaster (*Cotoneaster* spp.), hawthorn and pyracantha (*Pyracantha coccinea*). About a hundred hosts have been recorded. The bacteria overwinter in cankers which begin to ooze in the spring, whence bacterial cells are carried to the flowers by insects and spiders. The invaded flowers become water-soaked and then darken and dry to give a burnt appearance, hence the name fireblight, and as the disease spreads to the leaves these too take on a burnt appearance. The bacteria produce a non-specific polysaccharide toxin called amylovorin which plays a significant role in symptom development. Some host varieties are able to resist attack by the toxin, but in general the disease must be controlled with chemical sprays and by pruning.

Altogether there are more than twenty taxonomic entities in the genus *Erwinia* that are recognised as producing disease. *E. carotovora*, in various pathotypes and subspecies, is responsible for a range of soft rots of carrot (*Daucus carota*), beet, rhubarb (*Rheum* spp.), rhizomatous iris (*Iris* spp. hybrids) and so on. The subspecies *E.c. carotovora*, which grows at higher temperatures in the range 36–7°C, and rapidly rots the host tissues, is responsible for the soft rot of vegetables in store or in transit. *E.c.* ssp. *atroseptica* causes the blackleg disease of potato, which extends from the infected tuber as a vascular wilt and then rots the stem bases; tubers for the next season may be infected and rot in store. However, all rotting of potatoes cannot be attributed to this organism and the complex of closely related organisms attacking potato needs expert assessment. What is clear is that a variety of rotting organisms allied to this bacterium can be found in the soil, in association with roots and in watercourses, and in favourable moist, warm conditions these will rot potatoes. In the days before temperature-controlled stores were common, if potatoes slightly infected by *Phytophthora infestans* found their way into a store the disease would spread within the individual tubers, and although it would not spread from tuber to tuber the moisture given off in the course of infection would, under particular conditions, condense on the surrounding tubers and activate various organisms related to *E. carotovora* to rot the tubers. If this went on unnoticed large tonnages could be lost.

A distinct species of *Erwinia*, *E. salicis*, causes watermark disease of the cricket bat willow (*Salix alba* var. *coerulea*) and other willows, in the eastern counties of England. This bacterium occurs in the xylem vessels, where it spreads and causes a brown staining and watermark which spoils the wood for cricket bat production. The bacterium enters the tree through small insect wounds and is spread when larger insect wounds allow the bacteria to exude and be picked up in turn by biting and sucking insects. The leaves of recently infected branches turn brown and these brown 'flags' at midsummer are characteristic indications of the disease, easily seen on railway journeys through East Anglia.

Fig. 12.2 Nitrogen-fixing nodules on the roots of broad bean (*Vicia faba*). (Debbie White, Royal Botanic Garden Edinburgh.)

All these bacteria share a common property: they can only enter the plant by accident or by force: by accident when a stem breaks or a knife cuts or an insect bites, introducing the bacteria into the xylem; and by force when the bacteria spread via water droplets to the plant surface and find their way into the substomatal chamber or lenticel, where they multiply and secrete enzymes that destroy the cells ahead of them and allow invasion. They are in both these situations behaving as very successful necrotrophs.

The relationship between bacterium and plant may, however, be much more intimate. For example, there are two Gram-negative species which are not destructive of the host tissue in the normal sense and which depend on a close integration with the metabolism of the host for their survival and multiplication. They are difficult to place taxonomically. The first is *Rhizobium leguminosarum*, which in a mutualistic relationship causes nodules to form on the roots of plants in the pea family (Leguminoseae); these can easily be seen if pea or bean seedlings are gently dug up and their roots washed free of soil (Fig. 12.2). In the nodules gaseous nitrogen from the atmosphere is fixed into soluble nitrogenous salts. In simple terms the green plant supplies carbohydrate to the bacterium and in turn receives a supply of nitrogen in excess of that present in the soil, hence the ecological role of many legumes as pioneer colonisers of infertile ground. The organism has a wide range within the Leguminoseae, but particular legumes such as clover and lucerne (alfalfa) are infected by specialised strains of the organism. The bacteria do not produce spores but seem capable of surviving for long periods in the soil in the absence of a host. There is a gradual linear decline in their presence in soil over several years but figures of ten years or more are quoted for their survival in air-dried soil.

The bacterial cells multiply in the soil around the seedling root to be infected, causing curling and branching of the root hairs, a symptom which may be induced by the bacterial conversion of tryptophan, released by the roots, to the plant hormone indoleacetic acid. The bacterial cells then become attached to and penetrate the root hairs, forming an infection thread that grows along the root hair into the body of the root; this penetration process may be aided by the release of cell-softening enzymes such as polygalacturonases. Changes in plant hormone levels as a result of bacterial activity cause the cells to divide and a nodule to form, and *Rhizobium* cells are released from the infection thread into these living nodule cells, where they are enclosed by an extension of the plasmalemma. The bacteria finally enlarge and differentiate into spherical or branched nitrogen-fixing cells called 'bacteroids'. The mature nodule is a highly organised structure with an inner core containing the bacteroids and a growing outer region. Xylem and phloem vessels connected to the host's plumbing system maintain a supply of water, mineral ions and carbohydrates to nurture the bacteroids and carry nitrogenous salts to the rest of the plant.

The nodule is also highly organised biochemically to support the nitrogen-fixing activity of the bacteria while preventing damage to the host cells. This is well illustrated by the processes which protect the nitrogenase enzyme responsible for fixation from being deactivated by oxygen, to which it is sensitive. Oxygen is essential to the metabolism of the host cells, so a supply must be maintained. Root nodule cells contain the reddish pigment haemoglobin which binds oxygen in a similar way to the pigment of animal blood cells, and this oxygen can then be released into the plant's respiratory pathways direct from the haemoglobin, without reaching the bacteroids. Such is the level of integration between the metabolism of the plant and bacterium that the haemoglobin is probably synthesised jointly, with the bacterium producing the oxygen-binding 'haem' portion of the molecule and the plant producing the colourless protein to which it is attached.

Agriculturally, the nodule-forming bacteria are of great importance. There appears to be a reservoir of symbionts in the soil for indigenous crops such as the clovers in Europe, but exotic crops such as lucerne in the UK, soya (*Glycine max*) in the USA and subterranean clover in Australia require initial inoculation of the seed with a compatible strain of the bacterium. Estimates of the world production of nitrogen by *Rhizobium* (in wild and cultivated plants) suggest an annual yield of 1,000 million tonnes. In conventional temperate agriculture it is reckoned that a hectare of clover/rye grass sward will fix about 200 kg of nitrogen annually to yield over 7,500 kg of dry matter, compared with a potential yield of 11,000 kg of dry matter from an application of 400 kg of fertiliser nitrogen. Clearly, so long as fertiliser nitrogen is cheap there is no commercial advantage in relying on legume nitrogen, although in environmental terms the reverse may be the case.

Nitrogen-fixing nodules are also formed on the roots of some shrubs and trees, but here the cause is not a bacterium but a Streptomycete, an organism intermediate between a bacterium and a fungus. For example, species of alder form nodules in association with members of the Streptomycete genus *Frankia* and these can be seen if young alder roots are carefully excavated. A related Streptomycete species, *Streptomyces scabies*, causes the common scab disease which mars the skin of so many potatoes.

The second bacterium with a close biotrophic relationship with its host is *Agrobacterium tumefaciens*, which causes crown gall tumours in a wide range of dicotyledonous plants (Figs. 12.3 and 12.4), but is not known to attack monocotyledons. The bacterium, which can survive in soil, gains entry to its host through wounds and is particularly troublesome in nurseries, where the stem and root pruning of plants for sale leads to the later development of tumorous growths, particularly at soil level. Tumours may also form at graft unions, especially on fruit trees. The most surprising thing about *A. tumefaciens* is that the tumours it induces continue to grow in a disorganised way, even when the

Fig. 12.3 (left) A crown gall tumour induced by artificial inoculation of a sunflower plant with *Agrobacterium tumefaciens.* (Dennis Butcher.)

Fig. 12.4 (below) A tumour, probably induced by *Agrobacterium tumefaciens*, on a stem of *Corylus* sp. from Thetford Forest, Norfolk, England. (B. Golding.)

bacterium has died out in the cells or has been eliminated experimentally by raising the temperature or exposing it to antibiotics. Similarly, pieces of bacterium-free tumour are capable of inducing a new tumour if grafted onto a healthy host. The bacterium, in producing a tumour, has transformed otherwise healthy tissue into tumorous tissue which then remains tumorous, irrespective of the presence or absence of the infecting bacterium.

At first this seemed to be beyond all reasonable biological explanation, but in the 1970s the presence of a plasmid was discovered in *Agrobacterium* and found to be passed to the potential tumour cells of the host. *Agrobacterium* cells without the plasmid were non-pathogenic, but could receive the plasmid from a plasmid-carrying pathogenic strain of the bacterium to become pathogenic themselves. But more was to come. The pathogenic bacteria (but not non-pathogenic forms) were found to induce the production in the attacked host tissue of opines (carboxy ethyl and dicarboxy propyl derivatives of amino acids such as arginine, lysine and histine) and to utilise these opines as a source of carbohydrate and nitrogen. Moreover, strains that induced and utilised octopine, for example, did not induce or utilise another opine such as nopaline and *vice versa*. So here we have a pathogen inducing the production of its preferred substrate in the host, thus favouring its pathogenicity.

The story became even more interesting when it was discovered that the tumour-inducing plasmid (or part of it), following transfer to the host cells, became closely associated with the DNA of the nucleus. There it coded for the production of messenger RNA which affected the metabolism of the cell, causing it to produce excessive quantities of hormones and to become tumorous. Opine production was induced in the same way. This is a truly remarkable three way plant-pathogen relationship of bacterium, plasmid and plant. The only other well-documented example we know of involves the related bacterial species *Agrobacterium rhizogenes*, which induces root proliferation in its hosts, leading to the so-called 'hairy root' disease.

Finally, the tumour-inducing properties of *A. tumefaciens* and its plasmid have been harnessed by the genetic engineer to carry genes from the nucleus of one plant to the nucleus of another in order to introduce new characteristics into a crop or ornamental plant in a breeding programme. In commercial genetic engineering new ways have now been found to replace the living bacteria with microprojectiles, but there is no doubt that work on *A. tumefaciens* laid the foundation for modern biotechnology.

Other bacteria such as *Corynebacterium facians* cause distortion of host tissues and the formation of 'fasciations' – flattened, fused masses of stems in plants as diverse as aster, wallflower, sweet pea and forsythia (*Forsythia* spp.) – but the cells are not transformed as they are with *Agrobacterium*.

The parasitic flowering plants

Having looked at the viruses, the smallest plant pathogens, and the bacteria and their allies, the smallest 'cellular' pathogens, we now turn to the largest pathogens of all, the flowering plants (although the estimates for the great size and age of the fungus *Armillaria mellea* – see p.229 – might be at odds with that statement). It is estimated that about 1%, or about 3,000 species, of the known flowering plants have evolved a parasitic lifestyle, far more than the casual observer might suspect.

Parasitic plants attack in one of two ways. Firstly they may invade the roots of their host, a mode of parasitism adopted by more than half the known parasitic species. Such plants may have a complete above-ground structure which is green and superficially indistinguishable from that of their non-parasitic neighbours. Some of these are in fact facultative parasites (Fig. 12.5) and may live without attachment to a host plant for some weeks or months (e.g. eyebright (*Euphrasia officinalis*) and yellow rattle (*Rhinanthus minor*). Others may invade the host at an early stage and disappear from superficial view until the flowering shoot, without chlorophyll, emerges to flower and set seed (e.g. the broomrapes (*Orobanche* spp.) and toothworts (*Lathraea clandestina* and *L. squamaria*), and the tropical *Rafflesia* spp.) (Fig. 12.6).

Alternatively parasitic plants may invade the stems of their host, competing for

Fig. 12.5 Yellow rattle (*Rhinanthus minor*), a facultative parasitic flowering plant. (Royal Botanic Garden Edinburgh.)

Fig. 12.6 (above)   A single flower (approximately 60 cm diameter) of *Rafflesia pricei* in Mount Kinabalu National Park, Sabah, Malaysia. (Colin Pendry, Royal Botanic Garden Edinburgh.)

Fig. 12.7 (below)   Mistletoe (*Viscum album*) growing far to the north of its normal range, on hawthorn (*Crataegus* sp.) at the Royal Botanic Garden Edinburgh. (Debbie White, Royal Botanic Garden Edinburgh.)

space with the branches of the host and deriving nutrients from the vascular system which they tap. These forms (e.g. the mistletoes (Loranthaceae) (Fig. 12.7)) are often much branched and green, although the balance of the photosynthetic pigments may be different from normal, and the stems may become filament-like and pale in colour, as in the dodders (Cuscutaceae) (Fig. 12.8), with a major reduction in the chlorophyll pigments.

Fig. 12.8 Dodder (*Cuscuta* sp.) parasitising plants in the Rio de Janeiro Botanic Garden, Brazil. (D.S. Ingram.)

The species with chlorophyll are referred to as hemiparasites and, if they have a free-living phase, as facultative hemiparasites. Those without chlorophyll, which make up some 20% of all known parasitic plants, are known as holoparasites. Most of these are obligate holoparasites since they are totally dependent on the host for a supply of carbohydrates.

Many of the parasitic flowering plants are notable features of particular ecosystems while others are troublesome as parasitic weeds. In temperate regions the hemiparasitic members of the figwort family (Scrophulariaceae), for example, are important constituents of natural, mature grassland communities, but also occur as cornfield weeds. In Britain rattles (*Rhinanthus* spp.) have a damaging effect, reducing the number of species in a community over time. In the warmer parts of the USA and South Africa their obligately parasitic relatives are found in communities inhabiting open sunny sites or areas kept open by periodic fires.

Members of the sandalwood and dwarf mistletoe family (Santalaceae), which inhabit similar ecological niches in nature, are now mainly recognised as troublesome parasitic weeds of cultivated plants. An exception is bastard toadflax (*Thesium humifusum*), which is found in chalk pastures in Britain and also in grassland in Europe and Russia. The dwarf mistletoes (*Striga* spp.) commonly grow on the roots of various legumes and grasses, attacking particularly cereals in semi-arid regions with poor soil. They also occur in agricultural systems in moist savannah areas of Africa and extend into Asia, Australia and the USA, where the accidental introduction of one species, *Striga asiatica*, led to a serious infestation of parts of North and South Carolina. According to current estimates over 40% of arable land in sub-Saharan Africa is infested with *S. hermantica*, the main cereal root parasite, and well over 60% of the 73 million hectares in cereal production in savannah regions worldwide is contaminated with this or other *Striga* spp. Since the sorghums and millets of these dry regions are the main staples of the people who live there and are hosts to the *Striga* spp., the economic importance of parasitic plants is obvious.

The broomrapes (Orobanchaceae), relatives of the Scrophulariaceae, are particularly damaging on the Leguminoseae (hence broomrape) and Solanaceae. They can be found in Britain and mainland Europe attacking

Galium spp. (bedstraws), *Achillea* spp. (yarrow etc.) and other wild hosts. The destructive pathogen *Orobanche crenata* is confined to the Mediterranean seaboard, but other species have spread north and south from there, and also eastwards to India and China. They have also been introduced into the USA, South America and South Africa.

The most striking of all the root holoparasites, although seen by few naturalists, are members of the Rafflesiaceae. *Rafflesia arnoldii* was first discovered in Sumatra in 1818, when a male flower was observed and measured at nearly 1 m in diameter. A female flower nearly as big was found a few years later and given the name *R. titan*, although how one names a species which is only found occasionally and which has flowers difficult to preserve is far from clear. These flowers are said to weigh about 5 kg. No specimens which can be referred to either of these two species have been found again. Most modern sightings are of *R. tuan-mudae*, which is still spectacular but has smaller flowers. The plants, which are strictly limited in their hosts to members of the vine family (Vitaceae) of the genus *Tetrastigma*, live within the roots of their host. The only visible evidence of their existence is the appearance of the flower at ground level, where it forms a flat basin surrounded by four to six petaloid sepals and with a central column bearing either anthers or a stigmatic disk and surrounded by a diaphragm. The heavy stench, similar to rotting meat, attracts pollinating insects. The ovary lies below the petaloid flower parts and ripens to become a berry containing many small seeds which are disseminated by fruit-eating animals. The cycle from seed to seed appears to be about five years and the flower remains visible for about five days. It is not surprising therefore that there are gaps in our knowledge of these plants.

The stem parasites are epitomised by the mistletoes (Loranthaceae). *Viscum album* has a long folk history in Europe, where it has been a significant element in winter solstice celebrations, in druidical ceremonies and more recently in Christmas decorations. It is temperature-sensitive, being most common in the south of Britain and virtually absent from the north (but see Fig. 12.7). In the USA the role of *Viscum* at Christmas is taken by mistletoes of the native *Phoradendron* spp. There too the endemic dwarf mistletoe *Arceuthobium* spp. is a major pathogen of native conifers. In the tropics mistletoes of the genus *Loranthus* are conspicuous in the crowns of forest trees and can be economically important pathogens of fruit and beverage crops.

On the whole, because of their size and life span, the flowering plant parasites are not easy experimental material and our knowledge of them is fragmentary, but occasionally revealing. One of the best researched genera is *Striga*, the obligate parasite described above. The seeds of this plant germinate only in response to root exudates of its preferred host, an obvious device to prevent fruitless germination. Thus *S. hermantica* seeds from millet parasites respond to millet root exudate, while those from sorghum do not, and *vice versa*. While there is a series of conditioning factors that affect germination, the overriding stimulus seems to come from specific compounds in the root exudate. These have been difficult to characterise because of their instability (the stimulating effect of an extract disappears at room temperature after about 2 hours) but there is now good evidence that a complex hydroquinone is the stimulant present in the root exudate of sorghum.

In facultative root parasites the seeds can apparently germinate without the presence of a root stimulus and can live autotrophically (without dependence on the host) for some time. Not so *Striga* spp.; they produce tiny seeds about 100 μm in diameter which need to make contact with the host root at an early stage. Clearly, in a relationship depending on a highly labile root stimulant and a small seed without reserves, the position of the seed in relation to the root is critical for infection to take place. Indeed there is a considerable amount of experimental evidence on this point and on the mechanism of induction by the hydroquinone in *S. hermantica*, together with the production also of a material by the *Striga* root which is thought to act as a stabiliser of the stimulatory material.

Once seed germination in flowering plant parasites has taken place, either close to the root or on the shoot surface, multicellular haustoria are produced to link the parasite to the host's root or shoot. There is an underlying theme to all such structures, whether on root or shoot, but because of their economic importance most is probably known about *Striga* spp.. In this genus the first response as the emerging root of the parasite approaches the host root is enlargement of the cortex of the parasitic root, generating a lateral or terminal protuberance. This is followed by the division of the cells surrounding the vascular tissue and the production of epidermal hairs, which appear to become attached to the surface of the host root, as does the surface of the swollen end of the parasite root. From this surface a peg-like protuberance grows forward and penetrates the host tissue, where it is called the 'endophyte'. The mature haustorium that develops from the enlarging penetration peg forms a vascular extension which grows towards the vascular system of the host. The area inbetween, known as the hyaline body, contains only a few strands of rather simply differentiated xylem and no clearly differentiated phloem. There is thus a complete, though sparse, connection between the xylem tissue of host and parasite but no clearly visible phloem connections. This is intriguing and suggests that the hyaline body, in spite of its simple appearance, may substitute for phloem and have complex functions in the storage and transport of carbohydrates and other substances from host to parasite in a manner yet to be understood.

Once germination has been induced there is ample evidence that attachment to the plant is a physical response to a surface, and a great variety of inert substances are capable of bringing about attachment with more or less equal facility, although there is some conflicting evidence that epidermal secretions may be involved. In *Striga* spp. attachment occurs within 24 hours of the seed receiving the chemical signal for germination; and growth and penetration to link with the host's vascular system follow in another 24 hours. Where parasite and host are incompatible the host rapidly develops responses such as localised cell necrosis (hypersensitivity), production of corky layers, callose deposition and lignification of the cell walls, very much as in response to incompatible fungi and bacteria (see p.65).

And so we come to the end of our exploration of the world of plant parasitism. We have seen that this is a niche occupied by viruses, bacteria, fungi and higher plants themselves, and that there is a remarkable degree of parallelism in the strategies that have evolved. The relationships range from necrotrophy through hemibiotrophy and biotrophy to mutualism. Sometimes the pathogen or parasite has evolved an obligate relationship with its host and in others it has retained the flexibility to adopt an independent existence. It would be wrong, however, to think of any one of these relationships or strategies as being more successful, more advanced or more specialised than another; each enables the parasite or pathogen to invade and succeed in one of a range of potential niches not available to its competitors. In the case of mutualistic relationships, which we have touched on only occasionally, the ecological range of both partners is extended by their shared existence.

We find plant diseases fascinating subjects for study at every level, whether it be their natural history, their taxonomy, the complexities of the structural and biochemical relationships between host and pathogen, their importance in agriculture, their control or their effects on history. And again, it would be wrong to think of any one of these approaches as more sophisticated or more important than another; to a scholar all knowledge is important in achieving a measure of understanding of the world we inhabit. We hope that in writing this book we may have communicated some of our all embracing interest and enthusiasm for pathogens to fellow natural historians.

We end with a final plea: enjoy your pathogens and don't let them intimidate you (see Appendix, p.249).

Postscript

Those who study plant diseases must 'get their eye in'. It is easy to walk past vegetation without seeing the infected plants. One day, however, the difference between infected and healthy individuals will leap to the eye, and once recognised will never be forgotten. While this revelation can take place anywhere to reward the careful observer, it happens particularly and effortlessly, in our experience, at picnics or lunch breaks in the countryside, especially if wine or beer rather than tea provides the liquid refreshment. Not when lying flat looking up at the sky, but when replete and resting with the back against a rock or tree; at that time the eye roves in a relaxed way over the nearby vegetation and soon distinguishes anything strange. Perhaps the most spectacular picnic find of our experience was the discovery of *Schizonella melanogramma* by a student during a lunch break in the Cairngorm mountains in Scotland. This is an unusual smut with its spores in attached pairs, occurring on the leaves of a sedge, and until that time it had not been recognised in Britain. If you are not so lucky, don't worry; admire the view and try again another day.

Appendix: Practical Matters

Signs and symptoms

In human medicine the patient presents two sets of information to the doctor: the 'signs' and the 'symptoms' of the illness. Some of the signs are obvious when the patient walks over the threshold limping, for example, or clutching an arm or a leg. Other signs are revealed when specific tests are carried out with the stethoscope or by chemical and microbiological techniques. The symptoms are revealed when the patient describes, for example, feeling nauseous, breathless or lethargic. The diagnosis of the illness is made by a careful analysis of this information, checking for particular headmarks and rigorously excluding solutions which do not fit the expected range of signs and symptoms.

Examining the plant

Plants are much less complex than humans. They cannot speak, however, so the task of diagnosing plant disease is harder (the cynic might say easier). The 'symptoms', here defined as the host response, such as yellowing, wilting or root death, are valuable pointers to the nature of the disease. Time should, therefore, always be allowed for a considered appraisal of the diseased plant and a comparison with healthy neighbouring plants of the same and of a different species: is the condition confined to one species, is there evidence of weed-killer damage, a previous fire, flooding, or a shortage or over-supply of specific nutrient chemicals? Once it is established from the symptoms that there is a disease condition, by far the greatest attention is paid to the signs: the presence of a specific organism, its morphology and its identity. Plant pathologists use the terms 'signs' and 'symptoms' very loosely and most often refer to symptoms to embrace both sets of information.

Getting started

To begin the study of plant diseases, a keen eye is required, to spot the signs and symptoms of disease (see Waller *et al*, 1998) and a sharp knife to cut off small pieces of diseased tissue for closer examination or a small trowel to expose the roots, if infection below ground is suspected. Since good records are essential in plant pathology, also needed are a camera (with a close-up lens) and pencil and notebook (or an electronic equivalent), to record the observations. A basic library of reference books is necessary to help identify the diseases and their causal pathogens (see Bibliography). Since, as already suggested, lesions and blemishes on plants are not always caused by a pathogen but may result from nutrient deficiency or toxicity, an appropriate reference book (Bennett, 1993) is included in the Bibliography.

Examination with the naked eye and hand lens

The first examination is with the naked eye. Indeed, the naturalist trains his or her eye to recognise unusual colours, shapes or architecture in the natural plant community, and distortions or spots on individual leaves. When something unusual is found the next step is to examine it more closely with a hand lens. This may be purchased in most photographic or optical shops. Useful magnifications are x10, x15 and x20 and the lens should be one which minimises distortions and colour changes. More detail is revealed by the highest magnification, but the greater the magnification the more difficult it is to position the eye, the lens and the specimen to achieve best focus. The x15 magnification is a convenient compromise. The lens should be positioned so that sufficient light reaches the surface of the specimen, unshaded by the observer's head or hand. Naturalists who wear glasses may find it useful to attach them to a halter so that they can be removed and replaced easily and safely. It is not impossible to observe through a magnifying glass while wearing glasses and this is necessary for those with particular eye problems, but the additional lens makes manipulation and focusing more difficult. It is also convenient to attach the magnifier to a cord and carry it at all times.

Most plant diseases likely to be encountered by the field naturalist are caused by fungi and their spores and surface mycelium can usually be easily seen with the naked eye or hand lens. Bacteria and viruses are too small to be seen in this way, but much can still be learned by close examination of the diseased specimen. Further guidance for the study of bacterial diseases is given by Bradbury (2000) and Lelliot and Stead (1987), and for virus diseases by Hill (1984), Walkey (1990) and Matthews (1991, 1992).

Examination with a microscope

More detailed examination may be carried out by using a stereoscopic (low power: x20 – x200) dissecting microscope or a compound (high power: x100 – x1000+) microscope. The naturalist can do quite a lot without a microscope, but there is no doubt that possession of such a piece of equipment adds greatly to the enjoyment, precision of examination and understanding of the material. Stereoscopic microscopes have a considerable depth of focus and are relatively easy to use. However, the magnification is insufficient for the identification of smaller fungal spores and structures or bacteria and viruses. Compound microscopes give excellent magnification, but ideally the specimen must be mounted on a slide and specially prepared for examination by, for example, clearing, squashing or sectioning, and often staining. Appropriately prepared bacteria are visible at higher magnifications, but viruses are beyond the reach of even the best light microscopes. We suggest that you consult your local optical equipment supplier for advice on the different kinds of microscope available, and the range of prices. Do not rush into the purchase of such an expensive piece of equipment without careful consideration. Microscopes can usually be fitted with a camera to record selected images, but purchase costs are relatively high. Excellent advice on the use of microscopes and the preparation of specimens for examination is given by Waller et al (1998), Hawksworth *et al* (1995) and Johnston and Booth (1983). Most suppliers also provide instruction booklets.

Simple microscopic examinations can be made with the minimum of knowledge and a little trial and error. (**Hazard:** artificial light must not be too powerful and should be diffused through a filter to protect the eye. Similarly, the very greatest care should be taken if daylight is used as the light source.) A device to allow the substitution of filters is useful to bring out contrasts in stained material. From the point of view of safety, it is essential to study the instructions supplied with the microscope before embarking on an examination.

The identification of fungal and bacterial pathogens frequently requires that the cells or spores are carefully measured. This is achieved using a graduated eyepiece micrometer on the microscope, calibrated using a micrometer slide. These may be purchased at the same time as the microscope.

Handling the material

Material collected and brought back for examination should be processed as soon as possible. Good preparations can usually be made by removing a small portion of the infected material, transferring it to a drop of tap water on a microscope slide and repeatedly teasing it into tiny fragments with a pair of dissecting needles. These can be purchased from laboratory suppliers but adequate tools can easily be made by cutting off portions of sycamore or hazel twigs (10–15 cm long) of convenient thickness and pushing the base of a large pointed needle into the centre of the twig with a pair of pliers. The grip of wood on metal increases as the twig dries out. With a packet of mixed sewing needles a range of dissecting needle sizes may be manufactured.

When the material has been well teased out a few of the smaller portions with fungal attachments should be transferred to a fresh drop of water using a glass pipette, covered with a cover glass and gently tapped for about 30 seconds with the handle of the needle, a pencil or other suitable object. The scattered fragments should then be scanned under the low power of the microscope. When suitable spores, conidiophores or portions of fruiting bodies are found they may be examined in more detail under the higher powers of the microscope. The thickness of the coverglass is important. Too thin and it will be delicate to work with, too thick and it will impede the search at higher magnifications (size No 1.5, 22 x 22 mm square is a useful compromise).

A great deal can be discovered using simple water mounts, but for more sophisticated analysis materials need to be chemically fixed and stained with, for example, lactophenol containing 0.01% cotton or trypan blue. Details of the procedures involved for this and other fixatives and stains are given by Waller *et al* (1998) and Johnston and Booth (1983). (**Hazard:** take great care in using fixatives and stains, for many of the reagents are toxic or carcinogenic to humans.)

Sectioning Material

While satisfactory examinations can be made by teasing out infected material, a better view of the relationship between a fungus and the host tissues, and often of the structure and morphology of the fungus, may be obtained in a transverse section of the infected stem, root or leaf. An older generation of botanists wielded 'cut throat' razors to produce sections of material held in the hand, often with spectacular results. Nowadays such razors are hard to obtain and recourse is made to safety razor blades backed with a stiff metal strip, sold for cosmetic purposes, or sharp disposable scalpels obtained from medical instrument or graphic materials suppliers.

If the material to be sectioned is delicate it should be placed in a cleft cut in a piece of carrot tissue, or better in a piece of pith of the elder (*Sambucus nigra*, not the red-berried elder *Sambucus racemosa*). The pith is gathered from recently suppressed twigs and obtained by stripping away the thin bark and wood. The cylinders of pith are stored in dilute alcohol (25–50% surgical or methylated spirit). Sectioning involves cutting a cleft in the pith, trimming the piece of stem or leaf to an appropriate size for incorporation therein and then slicing off thin sections of pith and specimen. The top of the pith and the enclosed material should be moistened with a drop of water, and the blade dipped in water. Then, with the pith and specimen grasped firmly between thumb and curled index finger of the left hand, and the blade grasped in the right hand (or vice versa) a series of horizontal cuts is made into the pith and enclosed specimen, striving all the time to cut as thinly as possible. (**Hazard:** take great care in carrying out such operations, for it is easy to slip and cause injury). It is not necessary, indeed it is counter productive, to try to produce a complete transverse section; a series of thin fragments is much more useful. The fragments should be removed from the blade using a small moistened paintbrush and mounted in a drop of water on a slide, for preliminary assessment. Fragments showing interesting features may then be transferred to a separate slide for closer examination. If it is useful to substitute a staining material for the water mount without disturbing the fragment of infected plant material, a drop of the stain (see Waller *et al*, 1998) may be placed on one side of the coverslip in contact with the water in the mount and drawn under the coverslip by applying a piece of blotting or filter paper to the opposite side.

Preserving material

The foundation of the study of natural history is recording. As Captain Cuttle said in Dickens' *Dombey and Son*, "When found, make a note of!" With plant diseases, not only is it necessary to write down details of the specimen or phenomenon observed, it is also especially helpful to keep specimens of the diseases and pathogens for later examination and comparison.

After preliminary examination and approximate identification of the fresh material, herbaceous specimens and tree leaves may be dried in an herbarium press. This may be purchased or easily manufactured from two wooden frames formed of upright and cross pieces of lath to form a lattice of right-angled squares, the whole frame approximately 45 x 60 cm. Between these two frames pieces of dry newspaper cut to the appropriate size are stacked and the whole held together with a webbing strap. Specimens for drying are placed between the papers along with a note of the place and time of collection and any preliminary identification or notes on appearance, and so on. Drying is faster under warm conditions and the appearance of the specimens is enhanced if pressure is applied to the press using a heavy weight.

Specimens of wood or bark should not be pressed but dried directly in a warm atmosphere.

Smaller specimens (the majority unless one is making a special study of tree diseases) may be stored in paper envelopes folded to a standard pattern from sheets of A4 paper. These store easily in card index boxes alphabetically, or they may be numbered. It is useful to make special classifications in pursuit of a research project. Thus rusts on thistles, for example, could be classified by host, by location or by altitude or latitude to unravel their complex taxonomy. Larger specimens may be stored in A4 envelopes and kept together in a box file, and the very large woody specimens may be kept in cardboard boxes.

Specimens may deteriorate in storage from attacks of herbarium beetle. Another small beetle commonly destroys smuts and leaves behind packets of digested spores (to the confusion of the novice). To overcome such problems various strategies may be used. For example, the dried specimens may be placed in a deep freeze at -25°C before being stored. If the problem persists resort must be made to insecticides. (**Hazard:** most insecticides are toxic to humans and must therefore be used only in accordance with the manufacturer's instructions.) In the case of a large collection it is worth consulting the nearest herbarium (such as at the Royal Botanic Gardens, Kew, the Royal Botanic Garden Edinburgh or the International Mycological Institute, Egham, UK) for details of the insect control protocols which they observe.

Important Conservation Note

The collection of large samples or whole plants from the field, whether diseased or not, must be carried out with due regard for the conservation status of the host, the pathogen and the habitat (see Ingram, 1999). Plants should not, therefore, be collected from the field without the advice of a national herbarium such as one of those listed above, a local natural history society or museum, or an appropriate national society (see below).

Handling the dried material

When dried material is removed from the collection for examination it is dry and full of air sucked into the drying tissues. It is also extremely precious because the chances are that there will not be the opportunity to collect exactly the same material in the same place again. Therefore only very small portions of the material should be removed for examination. These are placed on a slide in water or glycerine or lactophenol (see Waller *et al*, 1998) and allowed to swell. The swelling and the dispersal of air bubbles in water mounts may be hastened by gentle heating over a flame, but care must be taken not to distort the material.

Inoculating Living Plants

The inoculation of plants with a suspected pathogen cannot be done on the spur of the moment, for the seed of the appropriate host must be collected or purchased, stored and grown under correct conditions. Nature is a good guide here, and seed of wild plants (but see Conservation Note above) collected in the autumn can often be sown direct and exposed to winter temperatures. Germinated seedlings are pricked out into pots and grown on in a cool greenhouse or frame.

Inoculation may be achieved in the case of some fungi and bacteria by suspending or laying the infected material on top of the seedlings. With some pathogens spores can be shaken off the infected specimen and with some rusts and powdery mildews the spores may be removed with a knife or brush and placed on the experimental seedlings. Once inoculated the seedlings must be kept free from further contamination and kept in a moist atmosphere (they may be lightly sprayed before or after inoculation, although the presence of free water can inhibit infection by fungi such as powdery mildews). Good conditions for infection may be achieved using bell-jars made from plastic soft drinks bottles, with the bases cut off. The cap may be left in place or substituted by a wad of cotton wool. The soil should be moist and the surrounding staging should be kept well watered. It is essential also to keep an equal number of healthy, uninoculated seedlings treated in exactly the same way as the experimental plants to act as controls.

Inoculation of plants with viruses requires a large number of healthy seedlings grown in a glasshouse under insect-free conditions, to prevent contamination from outside sources. The simplest procedure with the commoner sap-transmissible viruses is to grind up the suspected diseased plant and to rub the juice onto the leaves of the plants to be inoculated. The efficiency of the process is aided by the use of a chemical buffer and the addition of an inert abrasive such as fine carborundum powder to the juice and, after inoculation, by washing the surface of the host leaf with fresh water. Refinements of this technique and methods for the inoculation of plants using insect and other vectors are dealt with by Walkey (1990) and Hill (1984).

More advanced studies

More advanced studies of fungal diseases and identification of bacteria require access to a simple laboratory in which culture media may be made up and the pathogens isolated from diseased tissue and grown on in aseptic conditions (i.e. in the absence of other contaminating organisms), at a constant temperature. Excellent advice on setting up such a laboratory, making media and culturing pathogenic fungi is given by Waller et al (1998) and Smith and Onions (1994). The recipes for many culture media for fungi and bacteria are also listed by Johnston and Booth (1983). Methods for use with bacteria are also discussed by Waller et al (1998) and Bradbury (2000), and are described in detail by Lelliot and Stead (1987).

Viruses are normally identified using immunological or molecular biological techniques and powerful electron microscopes. The use of these normally requires access to a professional virus research laboratory, although immunological tests for some viruses of crop plants are now available in kit form (Hill, 1984; Walkey, 1990; Matthews, 1992).

Modern assays for the identification of fungi may also rely on immunological or molecular biological techniques (e.g. Schots, Dewey and Oliver, 1994) and again access to a professional laboratory is essential for their use.

Koch's Postulates

If a specific pathogen is to be unequivocally identified as the cause of a particular disease, Koch's postulates must be fulfilled. These may be summarised as follows (after Waller et al, 1998).
i) The pathogen must be consistently associated with the disease.
ii) The pathogen must be isolated and grown in pure culture, and its characteristics described. In the case of obligate parasites, which are incapable of growth in culture, the pathogen must be grown in isolation on a susceptible host plant and its appearance and effects noted.
iii) The pathogen from pure culture must be inoculated onto healthy plants of the same species or variety as those on which the disease was first noted, and it must produce the same disease on the inoculated plants.

iv) The pathogen must be re-isolated into pure culture and its characteristics must be exactly the same as those observed in postulate (ii).

For most routine purposes it is necessary only to note that a putative pathogen is consistently associated with the signs and symptoms of a particular disease and to identify it in an appropriate reference book. However, if a new pathogen is to be described or if a new host for an existing pathogen is suspected, all of Koch's postulates must be satisfied before the discovery can be verified, recorded officially in an appropriate journal and a specimen deposited in an internationally recognised reference collection (see Waller *et al*, 1998; Hall and Winter, 1994).

Joining a Society

The enjoyment of studying plant diseases is much enhanced by joining a local natural history society or a specialist national society such as the British Society for Plant Pathology or the British Mycological Society. It then becomes possible to share one's enthusiasm and benefit from the knowledge of others. Moreover, membership of a society usually provides access to a journal or bulletin for reading about plant diseases or for publishing one's own observations.

Valediction

The simple techniques described above will carry the reader forward to seek out and develop further ways of studying plant diseases. Techniques are important, but so too are observation and thought. The question "Why?" is always with the naturalist, and observation will often give an answer, or at least allow the development of an hypothesis for experimental testing.

Glossary

Acervulus (plural **-i**; from Latin *acervus* = heap). A layer of closely packed hyphae giving rise to short conidiophores bearing conidia.

Aeciospores (from Greek *aikia* = injury + *sporos* = seed). Binucleate spores formed from a dikaryotic mycelium following fusion of pycniospores with receptive hyphae during the life cycles of rust fungi (Uredinales).

Amino acids. Water soluble, organic nitrogenous compounds which may join together to form short chains (peptides) or long chains (polypeptides). Proteins are made up of various proportions of approximately 20 different amino acids.

Amorphous (from Greek *a* = without + *morphe* = form). Lacking shape, form or organised internal structure.

Anamorph (from Greek *anamorphosis* = forming anew). The asexual (non-sexual, imperfect) reproductive state of a fungus.

Antheridium (plural **-ia**; from Greek *antheros* = flowery). The male gametangium (sexual structure) in members of the Oomycota.

Apothecium (plural **-ia**; from Greek *apotheke* = storehouse). A saucer- or cup-shaped ascocarp (sexual fruiting body of the Ascomycota) bearing asci on the open surface.

Appressorium (plural **-ia**; from Latin *apprimere* = to press against). A swollen or flattened structure, formed at the end of the germ tube of a fungal pathogen and attached to the surface of the host immediately before penetration.

Arbuscule (from Latin *arbor* = tree). A finely branched haustorium of a vesicular-arbuscular (VA) mycorrhizal fungus.

Ascocarp (from Greek *askos* = skin sack + *karpos* = fruit). A fruiting body of a member of the Ascomycota.

Ascus (from Greek *askos* = skin sack). A sack-like structure in which the sexual spores of members of the Ascomycota are formed.

Ascospores (from Greek *askos* = skin sack + *sporos* = seed). Haploid, sexual spores of members of the Ascomycota.

Asexual (from Greek *a* = without + Latin *sexus* = sex). Non-sexual, as of a spore or phase of a life cycle (see also anamorph).

Aseptic (from Greek *a* = without + *septikos* = putrefy). Uncontaminated by microorganisms.

Autoecious (from Greek *autos* = self + *oikos* = home). Of a fungus (usually a rust – Uredinales) which completes its life cycle on a single host species.

Auxins. A group of plant growth regulators (hormones) which interact with other growth regulators to control such processes as cell division and enlargement, cell differentiation and apical dominance in higher plants.

Avirulent (from Greek *a* = without + Latin *virulentus* = poisonous). Used of a strain of a pathogen unable to cause disease on a resistant host.

Bacteriophage (from Greek *bacterion* = little stick + *phagein* = devour). A virus that infects prokaryotes, especially bacteria.

Bacterium (plural **-ia**; from Greek *bacterion* = little stick). Prokaryotic, usually unicellular microorganism.

Basidium (plural **-ia**; from Greek *baseidion* = small base). A cell in a member of the Basidiomycota in which nuclear fusion and meiosis occur, and on which the basidiospores (usually 4) are formed.

Basidiocarp (from Greek *baseidion* = small base + *karpos* = fruit). A fruiting body (such as a mushroom, toadstool or bracket) in the Basidiomycota in which basidia bearing basidiospores are formed.

Basidiospores (from Greek *baseidion* = small base + *sporos* = seed). Haploid spores of members of the Basidiomycota, formed on a basidium.

Callose. A structural polysaccharide made up of long chains (polymers) of glucan units, formed by plants and fungi, often in response to injury.

Cellulase. An enzyme involved in the breakdown of cellulose.

Cellulose. A structural polysaccharide made up of long chains (polymers) of glucan units; a major constituent of the cell walls of plants and members of the Oomycota.

Chemotaxis (from Greek *taxis* = arrangement). A response of an organism to a chemical stimulus; used to describe the swimming of zoospores or growth of hyphae towards chemicals such as sugars secreted by a plant.

Chitin (from Greek *chiton* = tunic). A polysaccharide containing nitrogen, with long, fibrous molecules; a major constituent of the hyphal walls of members of the Kingdom FUNGI and of the exoskeleton of insects.

Chlamydospore (from Greek *chlamys* = mantle + *sporos* = seed). A resting spore formed by thickening of the wall of a hyphal cell.

Chlorosis (from Greek *chloros* = green). Yellowing of green tissue, by breakdown of the green pigment chlorophyll.

Chromosozmes (from Greek *chroma* = colour + *soma* = body). Thread-like structures which occur in all living cells and carry genetic information in the form of deoxyribonucleic acid (DNA).

Cleistothecium (plural **-ia**; from Greek *kleistos* = closed + *theke* = case). An ascocarp in which the asci are completely enclosed by the outer wall.

Clamp connection. A specialised hyphal bridge involved in maintaining the dikaryotic condition during growth of hyphae of the Basidiomycetes.

Conidiophore (from Greek *konis* = dust + *phorein* = to bear). A specialised hypha, sometimes branched, which bears conidia.

Conidium (plural **-ia**; from Greek *konis* = dust). A non-motile, asexual spore formed at the tip of a specialised hypha.

Cultivar. A variety, produced by selection or breeding, of a cultivated plant.

Cuticle (from Latin *cutis* = skin). The largely 'waterproof' layer, mainly composed of the fatty, hydrophobic (water-repellent) polymer cutin, covering the above ground surfaces of a plant.

Cutin (from Latin *cutis* = skin). A fatty, hydrophobic (water-repellent) polymer that forms the major component of the plant cuticle.

Cutinase (from Latin *cutis* = skin). An enzyme involved in the breakdown of cutin.

Cytokinins (from Greek *kutos* = vessel + *kinein* = to move). Plant growth regulators (hormones) which, amongst other things, control cell division and differentiation in the presence of auxins, and retard senescence in higher plants.

Cytoplasm (from Greek *kutos* = vessel + *plassein* = to mould). The living contents of a plant cell, bounded by a membrane, the plasmalemma.

Digitate (from Latin *digitatus* = divided like fingers and toes). Divided like fingers or toes.

Dikaryon (from Greek *dis* = double + *karyon* = nut [nucleus]). A fungal cell containing two complementary haploid nuclei, usually of opposite mating type.

Diploid (from Greek *diploos* = double). Used of a cell or organism containing two complementary sets of chromosomes.

Dolipore-septum (plural -a; from Latin *dolium* = large jar + Greek *poros* = pore). A type of septum found in the hyphae of the Basidiomycota in which a central pore is extended into a barrel-shaped structure with open ends surrounded by membranes.

Dioecious (from Greek *dis* = double + *oikos* = home). Used of a species in which the male and female reproductive structures are formed on separate individuals.

Echinulate (from Greek *echinos* = hedgehog). Spiny.

Ectomycorrhiza (from Greek *ektos* = outside + *mykes* = mushroom + *rhiza* = root). A mutualistic symbiosis involving a fungus and the roots of a plant in which the fungus is restricted to a sheath around the outside of the root and the spaces between the surface cells.

Endomycorrhiza (from Greek *endon* = inside + *mykes* = mushroom + *rhiza* = root). A mutualistic symbiosis involving a fungus and the roots of a plant in which the fungus grows within the tissues of the root and may penetrate the cells.

Endophyte (from Greek *endon* = inside + *phyton* = plant). A fungus that grows within the tissues of a plant host without causing any noticeable symptoms of disease.

Enzyme. A protein that functions as a catalyst in biochemical reactions.

Epidermis (from Greek *epi* = upon + *derma* = skin). The outermost layer of cells of a plant.

Epinasty. (Of a leaf); to grow (bend) downwards from the base in response to altered levels of plant growth regulators (hormones).

Eukaryote (from Greek *eu* = well + *karyon* = nut). An organism in which the genetic material (DNA) of the cells is contained within a nucleus bounded by a membrane.

Facultative. Used of a parasite that may also grow as a saprophyte.

Flagella (sing. **-um**; from Latin *flagellum* = whip). Long, thin, whip-like structures that serve to propel motile cells such as zoospores.

Forma specialis (plural **-ae**; **-es**; abbreviated to **f.sp.**). A physiological form of a fungal species with a specific host range different from the host ranges of other *formae speciales* of the same species.

Fungistasis (from Latin *fungus* = mushroom + Greek *stasis* = standing). Prevention of germination of a fungal spore or of hyphal growth, usually by chemicals secreted by other microorganisms; if the inhibitor is removed, growth is resumed.

Fungitoxin. A chemical toxic to fungi.

Fungus (plural **-i**; from Latin *fungus* = mushroom). Used in the book to describe eukaryotic microorganisms that lack chlorophyll, usually grow as filaments (hyphae) and reproduce by means of asexual or sexual spores. See also the Kingdom FUNGI (Chapter 1).

Gametangium (plural **-ia**; from Greek *gametes* = husband). A structure which contains gametes or which functions as a gamete.

Gamete (from Greek *gamete* = wife, *gametes* = husband, *gamein* = to marry). A haploid cell that fuses with another complementary haploid cell during sexual reproduction to produce a diploid zygote.

Genome. All the genes contained in a single set of chromosomes of an organism.

Genotype. The genetic make-up of an organism.

Gibberellins. A group of plant growth regulators (hormones) involved in, for example, the regulation of cell expansion, stem elongation and seed germination in higher plants. Gibberellins were originally studied as products of the fungus *Gibberella fujikuroi*, hence the name.

Glucan. A short polymer made up of glucose molecules.

Glycoprotein. A carbohydrate linked to a protein.

Haploid (from Greek *haploos* = single). Used of a cell or organism containing a single set of chromosomes.

Haulm. The stems of a potato plant.

Haustorium (plural **-ia**; from Latin *haurire* = to drink dry). A specialised hyphal branch of a biotroph or hemibiotroph that penetrates the host cell wall and enters into an intimate association with the living cytoplasm of the cell. Also, a multicellular structure produced by higher plant (angiosperm) parasites that establishes an interface with the tissues of the host.

Hemibiotroph (from Greek *hemi* = half + *bios* = life + *trophein* = to feed). A plant pathogen exhibiting features of both biotrophy and necrotrophy, in sequence.

Hemicellulase. An enzyme involved in the breakdown of hemicelluloses.

Hemicelluloses. Complex polymers (chains) of various sugars, mainly linear but sometimes branched; major constituents of plant cell walls.

Heteroecious (from Greek *heteros* = other + *oikos* = home). Used to describe a fungus, usually a rust, that must infect two host species in order to complete its life cycle.

Heterokaryon (from Greek *heteros* = other + *karyon* = nut). A fungal mycelium containing a mixture of genetically different nuclei.

Heterothallic (from Greek *heteros* = other + *thallos* = shoot). Used of a fungus in which sexual reproduction can only occur between mycelia and nuclei that carry different, complementary genes controlling mating type.

Heterozygous (from Greek *heteros* = other + *zygon* = yoke). Having different genetic information at a particular point (locus) on the complementary chromosomes of a diploid cell.

Homothallic (from Greek *homo* = same + *thallos* = shoot). Used of a fungus which is self-fertile, without distinct mating types.

Hydrolysis. Reaction of a chemical compound with water.

Hymenium (from Hymen, Greek and Roman god of marriage). The fertile layer in an ascocarp or basidiocarp where the asci or basidia are produced.

Hypersensitive (from Greek *hyper* = over). Increased sensitivity of a cell to infection by a parasite, leading to rapid death.

Hyphae (sing. **-a**; from Greek *hyphe* = web). The tube-like filaments that are the basic body form of most fungi.

Hypovirulence (from Greek *hupo* = under + Latin *virulentus* = poisonous).

Lower than normal virulence resulting in reduced pathogenicity.

Infection peg. An infection structure produced by the appressorium of a fungus to penetrate the host cell wall.

Imperfect. See Anamorph.

Latent. Used of a pathogen that lives within a host without causing visible symptoms of disease but may do so when conditions change.

Lesion (from Latin *laesio* = injury). Site of damage or injury caused by a pathogen.

Lignin (from Latin *lignum* = wood). A tough, three dimensional, amorphous polymer of various aromatic molecules deposited in the walls of the water-conducting and strengthening cells of plants. Wood is composed of mainly lignified cells.

Lignituber (from Latin *lignum* = wood + *tuber* = bump or tumour). A proliferation of a host cell wall that ensheaths a penetrating hypha of a fungal pathogen, thus restricting its growth.

Lipids. Fatty substances.

Macroconidium (plural **-ia**; from Greek *makros* = long + *konis* = dust). A large, as distinct from a small (micro), conidium.

Meiosis (from Greek *meiosis* = reduction). Two nuclear divisions, one after the other, resulting in the reduction of the chromosome complement of the nucleus from diploid to haploid. In meiosis one diploid nucleus gives rise to four haploid nuclei, each in a new daughter cell.

Microconidia (from Greek *mikros* = small + *konis* = dust). See Macroconidia.

Microorganism (from Greek *mikros* = small). An organism that is so small as to be invisible or virtually invisible to the naked eye.

Middle lamella (from Latin *lamella* = a thin plate). The amorphous, glue-like layer joining adjacent plant cells, dominated by pectic substances.

Mitochondria (from Greek *mitos* = thread). Membrane-bound, sub-cellular bodies in which respiration occurs in living cells.

Mitosis (from Greek *mitos* = thread). Nuclear division followed by cell division in non-sexual cells of a living organism, resulting in two identical daughter nuclei, each containing the same number of chromosomes, identical to those of the mother nucleus.

Monokaryotic (from Greek *monon* = alone + *karyon* = nut). Containing only one nucleus; used of a spore or fungal cell.

Mushroom (possibly from old French *mousseron* = moss). An edible toadstool.

Mutualism (from Latin *mutuus* = mutual or reciprocal). An interaction between two species from which both benefit.

Mycelium (from Greek *mykes* = mushroom). A mass of hyphae of a fungus.

Mycorrhiza (from Greek *mykes* = mushroom + *rhiza* = root). A mutually beneficial association between a fungus and the roots of a plant.

Mycoplasmas (from Greek *mykes* = mushroom + *plassein* = to mould). A group of Prokaryotes, significantly smaller than bacteria, that lack a cell wall and are thought to be the smallest living cellular organisms. Mycoplasmas that infect plants are often called phytoplasmas.

Mycostasis. Prevention of germination of a fungal spore or hyphal growth, usually by chemicals secreted by other microorganisms; if the inhibitor is removed, growth is resumed.

Necrosis (from Greek *nekrosis* = death). Cell or tissue death.

Necrotroph (from Greek *nekros* = corpse + *trophe* = food). A pathogen that derives its nutrients from the dead cells of its host.

Nucleic acid. A complex organic substance in living cells that consists of a chain of nucleotides. There are two types: deoxyribonucleic acid (DNA), which is the genetic material of most living organisms; and ribonucleic acid (RNA), which is involved in protein synthesis and may also contain genetic information.

Nucleus (from Latin *nucleus* = kernel). A membrane-bound body within a living cell that contains the chromosomes.

Obligate. Used of a parasite unable to grow saprophytically.

Oogonium (plural **-ia**; from Greek *oon* = egg + *gonos* = child or product). The female gametangium in members of the Oomycota.

Oospores (from Greek *oon* = egg + *sporos* = seed). The thick-walled, sexual resting spores of members of the Oomycota which develop in the oogonia following fertilisation.

Paraphyses (sing. **-is**; from Greek *paraphysis* = monstrous growth). Sterile, elongate cells or hairs within a hymenium.

Parasexual (from Greek *para* = beside). A process in which nuclear fusion and genetic recombination occur within the vegetative hyphae of a fungus, quite separate from the sexual cycle.

Parasite (from Greek *parasitos* = one who dines at another's table). An organism that invades another (the host) and derives nutrients from it, often causing disease or debilitation.

Pathogen (from Greek *pathos* = suffering). A parasite that causes disease in its host.

Pathogenesis-related protein (PR-protein). An antifungal protein, usually an enzyme, produced by a host in response to invasion by a fungus.

Pathogenicity. The capacity to cause disease.

Pathology (from Greek *pathos* = suffering + *logos* = word/account/reason). The study of disease (hence 'Plant Pathology').

Pathotoxin (from Greek *pathos* = suffering). A toxin produced by a pathogen and involved in causing disease.

Pathovar (abbreviated to **pv.**). A form of a plant pathogen (usually a bacterium) having a different host from other morphologically identical pathovars of the same pathogen.

Pectate (from Greek *pektos* = congealed). An insoluble gel formed when a pectic acid combines chemically with calcium ions (Ca^{++}); found in the plant cell wall and middle lamella.

Pectic acid (from Greek *pektos* = congealed). Soluble gel composed of long chains (polymers) of acidic sugar molecules; found in the plant cell wall and middle lamella.

Pectin (from Greek *pektos* = congealed). A gel formed by the chemical reaction of a pectic acid with methanol; found in the plant cell wall and middle lamella.

Pectinase. An enzyme involved in the breakdown of pectic substances.

Perfect. See Teleomorph.

Perithecium (plural **-ia**; from Greek *peri* = around + *theke* = case). An enclosed ascocarp with a neck and pore at the top, through which the ascospores are released.

Phage. See Bacteriophage.

Phenols. Complex organic compounds containing a hydroxyl group (-OH) attached to a carbon atom in a benzene ring. Many phenolic substances are toxic to fungi and may play a role in resistance to disease.

Phloem. The tissue in which sugars are transported in a plant.

Phylloplane (from Greek *phyllon* = leaf + Latin *planus* = flat). The surface of a leaf.

Phytoalexin (from Greek *phyton* = plant + *alexein* = to ward off). A toxic chemical, produced metabolically as a response to infection by a potential pathogen and involved in resistance to that and other pathogens.

Phytotoxin (from Greek *phyton* = plant + *toxicon* = poison). A poisonous chemical, toxic to plants.

Plasmadesmata (sing. -a; from Greek *plasma* = moulded). Threads of cytoplasm which pass through pores in plant cell walls and connect the living contents of one cell with those of another.

Plasmalemma (from Greek *plasma* = moulded). The membrane enclosing the (living) cytoplasm of a plant cell.

Plasmid. A small circular piece of parasitic DNA containing genetic information, but separate from the chromosome(s) of the cell.

Plasmodium (from Greek *plasma* = moulded). A fungal body lacking a cell wall.

Polymer (from Greek *polloi* = many). A chemical molecule made up of many repeated smaller molecules.

Polyol (Polyhydric alcohol). A fungal 'sugar'.

Polysaccharide. A polymer composed of repeated sugar molecules.

Prokaryota (from Greek *pro* = before + *karyon* = nut). A superkingdom of organisms in which the genetic material consists of a single strand of DNA not separated from the rest of the cell by a nuclear membrane.

Propagule (from Latin *propagare* = to spread). A part of an organism such as a spore, cell or hyphal fragment involved in dispersal and reproduction.

Protocorm. A minute cellular structure, without chlorophyll, which is the first stage in the development of an orchid from a seed.

Pseudoplasmodium (plural -ia; from Greek *pseudos* = falsehood + *plasma* = moulded). An aggregation of uninucleate fungal bodies lacking cell walls.

Punctate. Marked or studded with points or dots.

Pycnia (sing. -ium; from Greek *pyknos* = close-packed). Structures formed by rust fungi (Uredinales) that give rise to pycniospores and receptive hyphae.

Pycnidia (sing. -ium; from Greek *pyknos* = close-packed). Minute, hollow, asexual fruiting bodies, lined with conidiophores bearing conidia.

Pycniospore (from Greek *pyknos* = close-packed + *sporos* = seed). Also called a spermatium. A haploid (monokaryotic) spore of a rust fungus (Uredinales) produced in a pycnium.

Receptive hyphae (also called flexuous hyphae). Haploid, monokaryotic hyphae produced by pycnia, which fuse with pycniospores from other pycnia to restore the dikaryotic phase in the rust life cycle.

Recombination. The reassortment of genes that occurs during meiosis.

Resting spore (from Greek *sporos* = seed). A dormant spore capable of surviving adverse environmental conditions such as drought or lack of a host.

Reticulate (from Latin *reticulum* = a net). Net-like.

Rhizoid (from Greek *rhiza* = root). A small root-like structure.

Rhizomorph (from Greek *rhiza* = root + *morphe* = form). A mass of hyphae bound together to form an elongate, root-like structure capable of transporting water and nutrients and able to grow at the tip.

Ribosome. A small, roughly spherical body, within a cell, which is the site of protein synthesis.

Rugose (from Latin *rugosus* = wrinkle). Corrugated or wrinkled.

Saprophyte (from Greek *sapros* = rotten). An organism that obtains its nutrients by breaking down the dead tissues of other organisms.

Sclerotium (plural **-ia**; from Greek *skleron* = hard). A mass of hyphae aggregated together to form a hard structure capable of remaining dormant for long periods before germinating to form a fungal growth.

Senescent (from Latin *senescere* = to grow old). Old and dying.

Septa (sing. **-um**; from Latin *saepes* = fence). Cross walls dividing hyphae into compartments.

Shikimic acid pathway. A complex biochemical pathway in plants leading to the synthesis of phenols and phenolic polymers, including lignin.

Slime mould. A multinuclear saprophytic fungus lacking a cell wall.

Spermatia (sing. **-ium**; from Greek *sperma* = seed). See pycniospore.

Spinose (from Latin *spinosus* = thorny). With long spines.

Spiroplasmas. Helically coiled mycoplasmas.

Sporangium (plural **-ia**; from Greek *sporos* = seed + *aggeion* = vessel). A walled, sack-like structure, the contents of which divide to form spores.

Sporangiophore (from Greek *sporos* = seed + *aggeion* = vessel + *phorein* = to bear). A hypha on which a sporangium or sporangia are formed.

Spore (from Greek *sporos* = seed). A minute propagule of a fungus or bacterium, functioning as a seed but without a pre-formed embryo.

Sporulate. To produce spores.

Sterigma (plural **-ata**; from Greek *sterigma* = support). A small, specialised branch that supports a sporangium or spore.

Stipe (from Latin *stipes* = stem). The stalk of a basidiocarp such as a toadstool.

Stoma (plural **-ata**). A pore, which can be opened or closed, in the epidermis (surface layer) of a leaf, through which water vapour, carbon dioxide, oxygen and other gases may pass.

Striate (from Latin *stria* = linear marking). Furrowed or streaked.

Stroma (plural **-ata**; from Greek *stroma* = bed). A mass of hyphae on or in which spores or fruit bodies are formed.

Suberin (from Latin *suber* = cork). A complex, fatty, hydrophobic polymer which is the major constituent of cork.

Symbiotic (from Greek *syn* = together with + *bios* = life). Two organisms living together either in a mutualistic or parasitic relationship.

Teleomorph (from Greek *telos* = end + *morph* = form). The sexual (perfect) state of a fungus.

Teliospore (from Greek *telos* = end + *sporos* = seed). A thick-walled, diploid resting spore of a rust fungus (Uredinales) formed following fusion of haploid nuclei of a dikaryon. (Synonymous with Teleutospore).

Toadstool. A non-edible (sometimes poisonous), mushroom-shaped fruiting body (basidiocarp) of a member of the Basidiomycetes.

Tomentose (from Latin *tomentum* = interwoven hair covering). Covered with matted hairs.

Urediospore (from Greek *urere* = to burn + *sporos* = seed). A dikaryotic (binucleate) non-sexual dispersal spore produced by a rust fungus (Uredinales). (Synonymous with Uredospore).

Ustilospore. A spore of a smut fungus (Ustilaginales) from which the basidiospores are formed.

Vacuole (from Latin *vacuus* = empty). Large membrane-bound, fluid-filled space within the cytoplasm of a cell.

Vegetative (from late Latin *vegetabilis* = animating). The growing, non-sporing hyphae and mycelium of a fungus.

Verrucose (from Latin *verruca* = wart). Warty.

Vesicle. Small membrane-bound, fluid-filled space within the cytoplasm of a cell; or a swollen hypha of a vesicular-arbuscular mycorrhizal fungus.

Vesicular-arbuscular (VA) mycorrhiza. A form of endomycorrhiza in which the fungus forms specialised feeding branches called arbuscules and swollen storage hyphae called vesicles.

Viroid. A simple form of a virus which lacks a protein coat.

Virulent (from Latin *virulentus* = poisonous). Used to describe a strain of a pathogen capable of causing disease on a host.

Virus (from Latin *virus* = poison). A microorganism capable of replication, but only in association with another, more complex organism, and having no cellular structure, consisting only of nucleic acid (RNA or DNA) with a protein coat.

Xerophyte (from Greek *xeros* = dry + *phyton* = plant). An organism adapted to live and reproduce in dry conditions.

Xylem. Lignified, water-conducting and supporting tissue in plants, forming the major component of wood.

Zoosporangium (plural **-ia**; from Greek *zoion* = animal + *sporos* = seed). A sporangium in which zoospores are formed.

Zoospore (from Greek *zoion* = animal + *sporos* = seed). A motile, asexual spore which swims by means of a flagellum or flagella.

Zygospore (from Greek *zygon* = yoke + *sporos* = seed). A diploid, thick-walled sexual resting spore formed by the fusion of two gametangia.

Zygote (from Greek *zygon* = yoke). The result of the fusion of two gametes, and containing a diploid nucleus formed by the fusion of two haploid nuclei, one from each gamete.

Bibliography

Further Reading: General Reference
*Agrios, G.N.** (1998). *Plant Pathology.*
4th Edn. Academic Press, Inc., San
Diego, USA.
Alexopoulos, C.J., Mims, C.W. &
Blackwell, M. (1996). *Introductory*
Mycology. 4th Edn. John Wiley & sons,
New York, USA.
*Bowes, B.G.** (1996). *A Colour Atlas of*
Plant Structure. Manson Publishing
Ltd., London, UK.
Brooks, A. & Halstead, A. (1999).
Garden Pests and Diseases, 3rd Edn.
Mitchell Beazley, London, UK.
*Buczacki, S. & Harris, K.** (1998).
Pests, Diseases and Disorders of Garden
Plants. 2nd Edn. HarperCollins,
London, UK.
Butler, E.J. & Jones, S.G. (1961).
Plant Pathology. MacMillan & Co Ltd,
London, UK.
Carlile, M.J. & Watkinson, S.C.
(1994). *The Fungi.* Academic Press,
London, UK.
Clapham, A.R., Tutin, T.G. & Moore,
D.M. (1987). *Flora of the British Isles.*
3rd Edn. Cambridge University Press,
Cambridge, UK.
Cooke, R. (1977). *The Biology of*
Symbiotic Fungi. John Wiley & Sons,
London.
Deacon, J.W. (1997). *Modern Mycology*
3rd Edn. Blackwell Science Ltd,
Oxford, UK.
Dixon, G.R. (1981). *Vegetable Crop*
Diseases. Macmillan, London, UK.
*Ellis, M.B. & Ellis, J.P.** (1997).
Microfungi on Land Plants. 2nd Edn.
Richmond Publishing Co, Slough, UK.
Fincham, J.R.S., Day, P.R. & Radford
A. (1979). *Fungal Genetics.* Blackwell
Scientific Publications, Oxford, UK.
Gow, N.A.R. & Gadd, G.M. (1995).
The Growing Fungus. Chapman & Hall,
London, UK.
Greenwood, P. & Halstead, A. (1997).
Plant Pests and Diseases. Dorling
Kindersley, London, UK.

Grove, W.B. (1935–7). *British Stem-*
and Leaf-Fungi (2 Vols). Cambridge
University Press, Cambridge, UK.
*Hawksworth, D.L., Kirk, P.M.,**
Sutton, B.G. & Pegler, D.N. (1995).
Ainsworth and Bisby's Dictionary of the
Fungi, 8th Edn. CAB International,
Wallingford, UK.
*Holliday, P.** (1998). *A Dictionary of*
Plant Pathology. 2nd Edn. Cambridge
University Press, Cambridge, UK.
Hudler, G.W. (1998). *Magical*
Mushrooms, Mischievous Molds.
Princeton University Press, Princeton,
USA.
Hudson, H.J. (1986). *Fungal Biology.*
Edward Arnold, London, UK.
Ingold, C.T. & Hudson, H.J. (1993).
The Biology of the Fungi. Chapman &
Hall, London, UK.
Isaac, S. (1991). *Fungal-Plant*
Interactions. Chapman & Hall, London,
UK.
Jones, D.G. (1987). *Plant Pathology:*
Principles and Practice. Open University
Press, Milton Keynes, UK.
*Lucas, J.A.** (1998). *Plant Pathology*
and Plant Pathogens. 3rd Edn.
Blackwell Scientific Publications,
Oxford, UK.
*Mabberley, D.J.** (1997). *The Plant*
Book. 2nd Edn. Cambridge University
Press, Cambridge, UK.
Manners, J.G. (1982). *Principles of*
Plant Pathology. Cambridge University
Press, Cambridge, UK.
Ramsbottom, J. (1953). *Mushrooms*
and Toadstools. Collins, London, UK.
Scheffer, R.P. (1997). *The Nature of*
Disease in Plants. Cambridge University
Press, Cambridge, UK.
Sidhu, G.S. (1988). *Genetics of Plant*
Pathogenic Fungi (Advances in Plant
Pathology Vol.6). Academic Press,
London, UK.
Smith, I.M., Dunez, J., Lelliott, R.A.,
Phillips, D.H. & Archer, S.A. (1988).
European Handbook of Plant Diseases.

Blackwell Scientific Publications, Oxford, UK.

*__Stace, C.__ (1997). *The New Flora of the British Isles.* 2nd Edn. Cambridge University Press, Cambridge, UK.

+__Various editors and authors__ (1963–1999). *Annual Review of Phytopathology* Vols 1–38 *et seq.* Annual Reviews, Palo Alto, USA.

*__Webster, J.__ (1980). *Introduction to Fungi.* 3rd Edn. Cambridge University Press, Cambridge, UK.

__Wheeler, H.__ (1975). *Plant Pathogenesis.* Springer-Verlag, Berlin, Germany.

* These titles are highly recommended and together constitute an excellent basic reference library for the amateur plant pathologist.
+ This series contains a very wide range of in-depth reviews of plant pathological topics.

__Further Reading: Chapter 1__

*__Agrios, G.N.__ (1998). See General Reference, above.

__Ainsworth, G.C.__ (1976). *Introduction to the History of Mycology.* Cambridge University Press, Cambridge, UK.

__Alexopoulos, C.J.__ *et al* (1996). See General Reference, above.

__Andrews, J.H.__ (1991). *Comparative Ecology of Microorganisms and Macroorganisms.* Springer-Verlag, New York, USA.

__Berkeley, M.J.__ (1846). Observations botanical and physiological on the potato murrain. *Journal of the Horticultural Society of London* 1, 9–34.

*__Burdon, J.J.__ (1993). The structure of pathogen populations in natural plant communities. *Annual Review of Phytopathology* 31, 305–23.

__Clarke, D.D.__ (1996). Coevolution between plants and pathogens of their aerial tissues. In *Aerial Plant Surface Microbiology*, Ed by Morris, C.E., Nicot, P.C. & Nguyen-The, C. Plenum Press, New York.

__Cooke, R.C.__ (1977). *Fungi, Man and His Environment.* Longman, London, UK.

__De Bary, A.__ (1887). *Comparative Morphology and Biology of the Fungi.* Trans by Garney, H.E.F. & Balfour, I.B. Oxford University Press, Oxford, UK.

*__Hawksworth, D.L.__ *et al* (1995). See General Reference, above.

__Ingram, D.S.__ (1999). Biodiversity, Plant Pathogens and Conservation, *Plant Pathology* (in press).

*__Jones, D.G.__ (1998). *The Epidemiology of Plant Diseases.* Kluwer, Dordrecht, Netherlands.

*__Large, E.C.__ (1940). *The Advance of the Fungi.* Jonathan Cape, London, UK.

*__Lucas, J.A.__ (1998). See General Reference, above.

__Orlob, G.B.__ (1971). History of Plant Pathology in the Middle Ages. *Annual Review of Phytopathology* 9, 7–20.

__Pirozynski, K.A. & Hawksworth, D.L.__ (1988). *Co-evolution of Fungi with Plants and Animals.* Academic Press, London.

__Rodriguez, R.J. & Redman, R.S.__ (1997). Fungal life-styles and ecosystem dynamics: biological aspects of plant pathogens, plant endophytes and saprophytes. *Advances in Botanical Research* 24, 169–93.

__Tinline, R.D. & MacNeill, B.H.__ (1969). Parasexuality in Plant Pathogenic fungi. *Annual Review of Phytopathology* 7, 147–70.

*__Webster, J.C.__ (1980). See General Reference, above.

*__Woodham-Smith, C.__ (1962). *The Great Hunger, Ireland 1845–1849.* Hamish Hamilton, London, UK.

* Key references

__Further Reading: Chapter 2__

__Agrios, G.N.__ (1998). See General Reference, above.

__Ayres, P.G.__ (ed) (1981). *Effects of disease on the Physiology of the Growing Plant.* Cambridge University Press, Cambridge, UK.

__Ayres, P.G.__ (ed) (1992). *Pests and Pathogens – Plant Responses to Foliar Attack.* Bios Scientific Publications Ltd,

Oxford, UK.

*Brian, P.W. (1967). Obligate parasitism in fungi. *Proceedings of the Royal Society* Ser.B **168**, 101–118.

*Callow, J.A. (1983). *Biochemical Plant Pathology.* John Wiley & Sons, Chichester, UK.

Cooke, R (1977). See General Reference, above.

*Cooke, R. & Whipps, J. (1980). The evolution of modes of nutrition in fungi parasitic on terrestrial plants. *Biological Reviews* **55**, 341–62.

*Daniels, M.J., Downie, J.A. & Osbourn, A.E. (1994). *Advances in Molecular Genetics of Plant-Microbe Interactions, Vol.3.* Kluwer Academic Publishers, Dordrecht, Netherlands.

Garrett, S.D. (1956). *Biology of Root Infecting Fungi.* Cambridge University Press, Cambridge, UK.

Garrett, S.D. (1970). *Pathogenic Root Infecting Fungi.* Cambridge University Press, Cambridge, UK.

*Heath, M.C. & Skalamera, D. (1997). Cellular interactions between plants and biotrophic fungal parasites. *Advances in Botanical Research* **24**, 194–225.

Heitefuss, R. & Williams, P.H. (1976). *Physiological Plant Pathology.* Springer-Verlag, Berlin, Germany.

Ingram, D.S., Sargent, J.A. & Tommerup, I.C. (1976). Structural aspects of infection by biotrophic fungi. In: *Biochemical Aspects of Plant-Parasite Relationships.* Ed by Friend, J. & Threlfall, D.R. Academic Press, London, pp.43–78.

Kolattukudy, P.E. (1985). Enzymatic penetration of the plant cuticle by fungal pathogens. *Annual Review of Phytopathology* **23**, 223–50.

*Lewis, D.H. (1973). Concepts in fungal nutrition and the origin of biotrophy. *Biological Reviews* **48**, 261–78.

*Lucas, J.A. (1998). See General Reference, above.

*Mendgen, K. & Deising, H. (1993). Infection structures of fungal plant pathogens – a cytological and physiological evaluation. *New Phytologist* **24**,
193–213.

*Mendgen, K., Hahn, M. and Deising, H. (1996). Morphogenesis and mechanisms of penetration by plant pathogenic fungi. *Annual Review of Phytopathology* **34**, 367–86.

*Parbery, D.G. (1996). Trophism and the ecology of fungi associated with plants. *Biological Reviews* **71**, 473–527.

Pyrozynski, K.A. & Hawksworth, D.L. (1988). See Chapter 1, above.

Scheffer, R.P. (1997). See General Reference, above.

*Spencer-Phillips, P.T.N. (1997). Function of fungal haustoria in epiphytic and endophytic infections. *Advances in Botanical Research* **24**, 309–33.

White, N.H. (1957). Host-parasite relations in plants. *The Journal of the Australian Institute of Agricultural Science* **23**, 129–136.

White, N.H. (1992). A case for the antiquity of fungal parasitism of plants. *Advances in Plant Pathology* **8**, 31–7.

Wood, R.K.S. (1967). *Physiological Plant Pathology.* Blackwell Scientific Publications, Oxford, UK.

* Key references

Further Reading: Chapter 3

+**Bateson, W.** (1909). *Mendel's Principles of Heredity.* Cambridge University Press, London, UK; MacMillan, New York, USA.

+**Biffen, R.H.** (1905). Mendel's laws of inheritance and wheat breeding. *Journal of Agricultural Science* **1**, 4–48.

+**Biffen, R.** (1907). Studies in the inheritance of disease resistance. *Journal of Agricultural Science* **2**, 102–28.

+**Biffen, R.** (1912). Studies in the inheritance of disease resistance II. *Journal of Agricultural Science* **4**, 421–29.

Callow, J.A. (1983). See Chapter 2, above.

Cruickshank, I.A.M. (1963). Phytoalexins. *Annual Review of Phytopathology* **1**, 351–74.

*Crute, I.R., Holub, E.B. & Burdon, J.J. (1997). *The Gene for Gene Relationship in Plant Parasite Interactions.* CAB International, Wallingford, UK.

*Daniels, M.J., Downie, J.A. & Osbourn, A.E. (1994). See Chapter 2, above.

Day, P.R. (1974). *Genetics of Host-Parasite Interaction.* Freeman, San Francisco, USA.

*de Wit, P.J.G.M. (1995). Fungal avirulence genes and plant resistance genes – unravelling the molecular basis of gene-for-gene interactions. *Advances in Botanical Research* 21, 147–85.

de Wit, P.J.G.M. (1997). Pathogen avirulence and plant resistance: a key role for recognition. *Trends in Plant Science* 2, 452–58.

Dixon, R.A., Harrison, M.J. & Lamb, C.J. (1994). Early events in the activation of plant defense responses. *Annual Review of Phytopathology* 32, 479–501.

Draper, J. (1997). Salicylate, superoxide synthesis and cell suicide in plant defence. *Trends in Plant Science* 2, 162–65.

Ellingboe, A.H. (1981). Changing concepts of host-pathogen genetics. *Annual Review of Phytopathology* 19, 125–43.

*Ellingboe, A.H. (1984). Genetics of host-parasite relations: an essay. *Advances in Plant Pathology* 2, 131–51.

[+]Flor, H.H. (1946). Genetics of pathogenicity in *Melampsora lini. Journal of Agricultural Research* 73, 335–57.

[+]Flor, H.H. (1955). Host-parasite interaction in flax rust – its genetics and other implications. *Phytopathology* 45, 680–85.

*Flor, H.H. (1971). Current status of the gene-for-gene concept. *Annual Review of Phytopathology* 9, 275–96.

Hahlbrock, K. & Somssich, I.E. (1998). Pathogen defence in plants – a paradigm of biological complexity. *Trends in Plant Science* 3, 86–90.

Hammond-Kosack, K. & Jones, J.D.G.

(1995). Plant disease resistance genes: unravelling how they work. *Canadian Journal of Botany* 73 (Suppl. 1), 5495–505.

**Hartleb, H., Heitefuss, R. & Hoppe, H.H. (1997). *Resistance of Crop Plants Against Fungi.* Gustav Fischer, Jena, Germany.

Heitefuss, R. & Williams, P.H. (1976). See Chapter 2, above.

Hutchen, S.W. (1998). Current concepts of active defense in plants. *Annual Review of Phytopathology* 36, 59–90.

*Johal, G.S., Gray, J., Gruis, D. & Briggs, S.P. (1995). Convergent insights into mechanisms determining disease and resistance responses in plant-fungal interactions. *Canadian Journal of Botany* 73 (Suppl. 1), 5468–74.

Jones, J.D.G. (1997). A kinase with keen eyes. *Nature* 385, 397–98.

*Jones, D.A. & Jones, J.D.G. (1997). The role of leucine-rich repeat proteins in plant defences. *Advances on Botanical Research* 24, 90–167.

Kessman, H., Staub, T., Hofmann, C., Maetzke, T. & Herzog, J. (1994). Induction of systemic acquired disease resistance in plants by chemicals. *Annual Review of Phytopathology* 32, 439–59.

Kuc, J. (1972). Phytoalexins. *Annual Review of Phytopathology* 10, 207–32.

Kuc, J. (1995). Phytoalexins, stress metabolism, and disease resistance in plants. *Annual Review of Phytopathology* 33, 275–97.

Laug, Ç.R. & de Wit, J.G.M. (1998). Fungal avirulence genes: structure and possible functions. *Fungal Genetics and Biology* 24, 285–97.

Lindgren, P.B. (1997). The role of *hrp* genes during plant-bacterial interactions. *Annual Review of Phytopathology* 35, 129–52.

Lucas, J.A. (1998). See General Reference, above.

Pieterse, C.M.J. & van Loan, L.C. (1999). Salicylic acid-independent plant defence pathways. *Trends in*

Plant Science **4**, 52–9.
Royle, D.J. (1976). Structural features of resistance to plant disease. In: *Biochemical Aspects of Plant-Parasite Relationships*, ed. by Firend, J. & Threlfall, D.R. Academic Press, London, UK, pp.161–93.
Sidhu, G.S. (1988). See General Reference, above.
Sticher, L., Mauch-Mani, B. & Métraux, J.P. (1997). Systemic acquired resistance. *Annual Review of Phytopathology* **35**, 235–70.
Wood, R.K.S. (1967). See Chapter 2, above.

** The most recent comprehensive treatment of modern research on all aspects of plant disease resistance; a key reference.
* Valuable reviews of different aspects of plant defence.
+ Seminal papers of great historical significance.

Further Reading: Chapter 4
Andrews, J.H. (1976). The pathology of marine algae. *Biological Reviews* **51**, 211–53.
Brasier, C.M. (1992). Evolutionary biology of *Phytophthora*. I. Genetic system, sexuality and the generation of variation. *Annual Review of Phytopathology* **30**, 153–72.
Brasier, C.M. & Hansen, E.M. (1992). Evolutionary biology of *Phytophthora*. II. Phylogeny, speciation and population structure. *Annual Review of Phytopathology* **30**, 173–200.
Buczacki, S.T. (ed) (1983). *Zoosporic Plant Pathogens – A Modern Perspective*. Academic Press, London, UK.
Canter-Lund, H. & Lund, J.W.G. (1995). *Freshwater Algae – Their Microscopic World Explored*. Biopress Ltd, Bristol, UK.
Crute, I.R. (1992). From breeding to cloning (and back again?): a case study with lettuce downy mildew. *Annual Review of Phytopathology* **30**, 485–506.

Ellis, M.B. & Ellis, J.P. (1997). See General Reference, above.
Erwin, D.C., Bartnicki-Garcia, S. and Tsao, P.H. (eds) (1983). *Phytophthora: Its Biology, Taxonomy, Ecology and Pathology*. American Phytopathological Society, St Paul, USA.
Erwin, D.C. & Ribeira, O.K. (1996). *Phytophthora diseases worldwide*. American Phytopathological Society, St Paul, USA.
Gregory, P.H. (1961). *The Microbiology of the Atmosphere*. Leonard Hill (Books) Ltd, London, UK; Interscience Publishers, New York, USA.
Hendrix, F.F. & Campbell, W.A. (1973). Pythiums as plant pathogens. *Annual Review of Phytopathology* **11**, 77–98.
Holliday, P. (1995). *Fungus Diseases of Tropical Crops*. 2nd Edn. Dover Publications Inc, New York, USA.
Ingram, D.S. & Williams, P.H. (1991). *Advances in Plant Pathology Vol 7: Phytophthora infestans, the Cause of Late Blight of Potatoes*. Academic Press, London.
Karling, J.S. (1964). *Synchytrium*. Academic Press, New York, USA.
Karling, J.S. (1968). *The Plasmodiophorales*. 2nd Edn. Hafner Publishing Co, New York, USA.
Newhook, F.J. & Podger, F.D. (1972). The role of *Phytophthora cinnamomi* in Australian and New Zealand forests. *Annual Review of Phytopathology* **10**, 292–326.
Sparrow, F.K. (1960). *The Aquatic Phycomycetes*. 2nd Edn. Revised. University of Michigan Press, Ann Arbor, USA.
Spencer, D.M. (1981). *The Downy Mildews*. Academic Press, London, UK.
Weste, G. & Marks, C. C. (1987). The biology of *Phytophthora cinnamomi* in Australian forests. *Annual Review of Phytopathology* **25**, 207–29.
Williams, R.J. (1984). Downy mildews of tropical cereals. *Advances in Plant Pathology* **2**, 1–103.

* Key references

Further Reading: Chapter 5
Agrios, G.N. (1998). See General
Reference, above.
Barger, G. (1931). *Ergot and Ergotism.*
Gurney and Jackson, London, UK.
Boone, D.M. (1971). Genetics of
*Venturia inaequalis. Annual Review of
Phytopathology* **9**, 297–318.
**Cannon, P.F., Hawksworth, D.L. &
Sherwood-Pike, M.A.** (1985). *The
British Ascomycotina: an annotated check-
list.* CAB International, Wallingford,
UK.
**Carmichael, J.W., Kendrick, W.B.,
Conners, I.L. & Sigler, L.** (1980).
Genera of Hyphomycetes. University of
Alberta, Edmonton, Canada.
Coley-Smith, J.R. & Cooks, R.C.
(1971). Survival and germination of
fungal sclerotia. *Annual Review of
Phytopathology* **9**, 65–92.
Cooke, R.C. (1977). See Chapter 1,
above.
Cooke, R. & Whipps, J. (1980). See
Chapter 2, above.
Deacon, J.W. (1992). Stephen Denis
Garrett: Pioneer leader in plant
pathology. *Annual Review of
Phytopathology* **30**, 27–36.
Dennis, R.W.G. (1960). *British Cup
Fungi.* Ray Society, London, UK.
Dennis, R.W.G. (1978). *British
Ascomycetes.* 3rd Edn. J. Cramer, Vaduz,
Liechtenstein.
Ellis, M.B. (1971). *Dematiaceous
Hyphomycetes.* CAB International,
Wallingford, UK.
Ellis, M.B. (1976). *More Dematiaceous
Hyphomycetes.* CAB International,
Wallingford, UK.
Ellis, M.B & Ellis, P. (1997). See
General Reference, above.
Garrett, S.D. (1956 and 1970). See
Chapter 2, above.
Grove, W.B. See General Reference,
above.
***Hawksworth *et al* (1995), pp.3–34.
See General Reference, above.
Parbery, D.G. (1956). See Chapter 1,
above.
Samuels, G.S. & Seifert, K.A. (1995).
The impact of molecular characters

on systematics of filamentous
Ascomycetes. *Annual Review of
Phytopathology* **33**, 37–67.
Sidhu, G.S. (1988). See General
Reference, above.
Sutton, B.C. (1980). *The Coelomycetes.*
CAB International, Wallingford, UK.

** This reference lays out clearly the
problems of classification represented
by the diversity of members of the
Ascomycota.
* Key references

Further Reading: Chapter 6
***Beckman, C.H.** (1987). *The Nature
of Wilt Disease in Plants.* American
Phytopathological Society Press, St
Paul, USA.
Bowes, B.G. See General Reference,
above.
Brasier, C.M. (1986). The population
biology of Dutch elm disease: its prin-
cipal features and some implications
for other host-pathogen systems.
Advances in Plant Pathology **5**, 53–118.
Brasier, C.M. (1996). New horizons
in Dutch elm disease control. *Report
on Forestry Research 1996,* Forestry
Commission, Edinburgh, UK, pp.20–28.
Carefoot, G.L. & Sprott, E.R. (1967).
Famine on the Wind. Angus Robertson,
London, UK.
Gordon, T.R. & Martyn, R.D. (1997).
The evolutionary biology of *Fusarium
oxysporum. Annual Review of
Phytopathology* **35**, 111–28.
Richens, R.H. (1983). *Elm.* Cambridge
University Press, Cambridge, UK.
Scheffer, R.P. (1997). See General
Reference, above.
Smalley, E.B. & Guries, R.P. (1993).
Breeding elms for resistance to Dutch
elm disease. *Annual Review of
Phytopathology* **31**, 325–52.

** An excellent synthesis; a key
reference
* Key references
Further Reading: Chapter 7
Bacon, C.W. & Hill, N.S. (1997).

Neotyphodium/Grass Interactions.
Plenum Press, New York, USA.
Cooke, R. (1977). See General
Reference, above.
Harley, J.L. & Smith, S.E. (1983).
Mycorrhizal Symbiosis. Academic Press,
London, UK.
*****Helgason, T., Daniell, T.J., Husband,
R., Fiter, A. & Young, J.P.W.** (1998).
Ploughing up the wood-wide web?
Nature **394**, 431.
*****Hetherington, A.M.** (1998). *The
Tansley Review Collections. 2: Mycorrhizas
– Structure and Function.* New
Phytologist Trust/Cambridge
University Press, Cambridge, UK.
Hudson, H.J. (1986). See General
Reference, above.
*****Parbery, D.G.** (1966). See Chapter 2,
above.
Pirozynski, K.A. & Hawksworth, D.L.
(1988). See Chapter 1, above.
*****Read, D.J.** (1993). Mycorrhiza in
plant communities. *Advances in Plant
Pathology* **9**, 1–31.
Schardl, C.L. (1996). *Epichloâ* species:
fungal symbionts of grasses. *Annual
Review of Phytopathology* **34**, 109–30.
**Siegel, M.R., Latch, G.C.M. &
Johnson, M.C.** (1987). Fungal endo-
phytes of grasses. *Annual Review of
Phytopathology* **25**, 293–315.
**Simard, S.W., Perry, D.A., Jones,
M.D., Myrolds, D.D., Durall, D.M.,
Molina, E.** (1997). Net transfer of car-
bon between ectomycorrhizal tree
species in the field. *Nature* **388**,
579–82.
Smith, D.C. & Douglas, A.E. (1987).
The Biology of Symbiosis. Edward Arnold,
London, UK.
******Smith, S.E. & Read, D.J.** (1997).
Mycorrhizal Symbiosis, 2nd Edn.
Academic Press, San Diego, USA.
*****Tommerup, I.C.** (1993). *Advances in
Plant Pathology Vol. 9: Mycorrhiza
Synthesis.* Academic Press, London,
UK.
Verhoeff, K. (1974). Latent infections
by fungi. *Annual Review of
Phytopathology* **12**, 99–110.

** An excellent synthesis of research
to date; a key reference
* Key references

Further Reading: Chapter 8
Aust, H-J. & Hoyningen-Huene, J. v.
(1986). Microclimate in relation to
epidemics of powdery mildew. *Annual
Review of Phytopathology* **24**, 491–510.
Blumer, S. (1933). *Erysiphaceen
Mitteleuropas.* Fretz, Zurich,
Switzerland.
Braun, U. (1987). *A monograph of
Erysiphales (Powdery Mildews).* Nova
Hedwigia supplement **89**. J Cramer,
Berlin, Germany.
Callow, J. (1983). See Chapter 2,
above.
Clarke, D.D. (1996). See Chapter 1,
above.
*****Ellis, M.B. & Ellis, J.P.** (1997). See
General Reference, above.
Sidhu, G.S. (1988). See General
Reference, above.
*****Spencer, D.M.** (1978). *The Powdery
Mildews.* Academic Press, London, UK.
*****Spencer-Phillips, P.T.N.** (1997). See
Chapter 2, above.
*****Wolfe, M.S. & McDermott, J.M.**
(1994). Population genetics of plant-
pathogen interactions: the example of
the *Erysiphe graminis - Hordeum vulgare*
pathosystem. *Annual Review of
Phytopathology* **32**, 89–113.

* Key references

Further Reading: Chapter 9
Arthur, J.C. (1929). *The Plant Rusts
(Uredinales).* John Wiley & Sons, New
York, USA.
Arthur, J.C. & Cummins, G.B. (1962).
*Manual of the Rusts in the United States
and Canada.* Hafner, New York, USA.
Cummins, G.B. (1959). *Illustrated
Genera of Rust Fungi.* Burgess,
Minneapolis, USA.
Cummins, G.B. (1971). *The Rust Fungi
of Cereals, Grasses and Bamboos.*
Springer Verlag, Berlin, Germany &

New York, USA.
*Cummins, G.B. & Hiratsuka, Y.
(1983). *Illustrated Genera of Rust Fungi*,
Revised Edn. American
Phytopathological Society, St Paul,
USA.
*Ellis, M.B. & Ellis, J.P. (1997). See
General Reference, above.
Gaeumann, E. (1959). *Die Rostpilze
Mitteleuropas.* BÅcher & Co, Berne,
Switzerland.
Grove, W.B. (1913). *The British Rust
Fungi (Uredinales).* Cambridge
University Press, Cambridge, UK.
Guyot, A.L. (1935–67). *Uredineana:
recueil d'études systematiques et biologiques
sur les urédines du Globe.* Encyclopédie
mycologique VIII–XXXIV. Le
Chevalier, Paris, France.
Hawksworth *et al* (1995). See General
Reference, above.
Klebahn, H. (1904). *Die
Wirtswechselnden Rostpilze.* Borntrager,
Berlin, Germany.
*Littlefield, L.J. & Heath, M.C.
(1979). *Ultrastructure of Rust Fungi.*
Academic Press, New York, USA.
McAlpine, D. (1901). *The Rusts of
Australia.* Department of Agriculture,
Victoria, Melbourn, Australia.
Nagarajan, S. & Singh, D.V. (1990).
Long distance dispersal of rust
pathogens. *Annual Review of
Phytopathology* **28**, 139–53.
*Scott, K.J. & Chakravorty, A.K.
(1982). *The Rust Fungi.* Academic
Press, New York, USA.
Sidhu, G.S. (1988). See above
Reference, above.
Sydow, P. & Sydow, H. (1904–24).
Monographia Uredinarum, Vols 1–4.
Fratres Borntrager, Leipzig, Germany.
*Wilson, M. & Henderson, D.M.
(1966). *British Rust Fungi.* Cambridge
University Press, Cambridge, UK.

* Key References

Further Reading: Chapter 10
Ainsworth, G.C. & Sampson, K.
(1950). *The British Smut Fungi.* CAB,
Kew, UK.
*Banuett, F. (1992). *Ustilago maydis,*
the delightful blight. *Trends in Genetics*
8, 174–80
Burnett, J.H. (1968). *Fundamentals of
Mycology.* Edward Arnold, London,
UK.
*Ellis, M.B. & Ellis, J.P. (1997). See
General Reference, above.
Fischer, G.W. (1953). *Manual of the
North American Smut Fungi.* Ronald
Press, New York, USA.
Fischer, G.W. & Holton, C.S. (1957).
Biology and Control of the Smut Fungi.
Ingold, C.T. (1983). A view of the
basidium. *Bulletin of the British
Mycological Society* **17**, 82–93
Liro, J. (1924). *Die Ustilagineen
Finnlands,* SANA, Helsinki, Finland.
*Mordue, J.E.M. & Ainsworth, G.C.
(1984). *Ustilaginales of the British Isles.*
CAB International, Wallingford, UK.
*Sidhu, G.S. (1998). See General
Reference, above.

* Key references

Further Reading: Chapter 11
Blanchette, R.A. (1991).
Delignification by wood-decay fungi.
Annual Review of Phytopathology **29**,
381–98.
Bowes, B.G. (1996). See General
Reference, above.
Boyce, J.S. (1948). *Forest Pathology.*
McGraw-Hill, New York, USA &
London, UK.
*Buczacki, S. (1992). *Collins Guide:
Mushrooms and Toadstools of Britain and
Europe.* HarperCollins, London, UK.
*Buczacki, S. *et al* (1998). See General
Reference, above.
***Greig, B.J.W., Gregory, S.C. &
Strouts, R.G.** (1991). *Honey Fungus,*
8th Edn. HMSO, London, UK.
Holden, M. (1982). *Guide To The
Literature for the Identification of British
Fungi.* British Mycological Society,
London, UK.
Holliday, P. (1980). See Chapter 4,
above.

Moore, D. (1998). *Fungal Morphogenesis.* Cambridge University Press, Cambridge, UK.

*****Peace, T.R.** (1962). *Pathology of Trees and Shrubs.* Oxford University Press, Oxford, UK.

*****Pegler, D.** (1990). *Field Guide to the Mushrooms and Toadstools of Britain and Northern Europe.* Kingfisher Books, London, UK.

*****Phillips, D.D. & Burdekin, D.A.** (1992). *Diseases of Forest Trees and Ornamental Trees,* 2nd Edn. Macmillan, London, UK.

Rishbeth, J. (1951). Observations on the biology of *Fomes annosus. Annals of Botany* **14**, 365–83; **15**, 1–21.

Rishbeth, J. (1964). Stump infection by basidiospores of *Armillaria mellea. Transactions of the British Mycological Society* **47**, 460.

Rishbeth, J. (1981). Species of *Armillaria* in southern England. *Plant Pathology* **31**, 9–17.

Ryvarden, L. (1976–78). *The Polyporaceae of Northern Europe, Vols 1 and 2.* Fungiflora, Oslo, Norway.

*****Strouts, R.G. & Winter, T.G.** (1994). *Diagnosis of Ill Health in Trees.* Forestry Commission/HMSO, London, UK.

* Key references

Further Reading: Chapter 12
Anon (1998). *Association of Applied Biologists' Descriptions of Plant Viruses* (CD-ROM). Association of Applied Biologists, Wellesbourne, UK.

Bové, J.M. (1984). Wall-less prokaryotes of plants. *Annual Review of Phytopathology* **22**, 361–96.

*****Bradbury, J.F.** (1999). *Guide to Plant Pathogenic Bacteria,* 2nd Edn. CAB International, Wallingford, UK.

*****Brunt, A., Crabtree, K., Dallwitz, M., Gibbs, A. & Watson, L.** (1996). *Viruses of Plants.* CAB International, Wallingford, UK.

Fahy, C.P. & Persley, G.J. (1983). *Plant Bacterial Diseases – A Diagnostic Guide.* Academic Press, Sydney, Australia.

Gibbs, A. & Harrison, B. (1979). *Plant Virology: The Principles.* Edward Arnold, London, UK.

Goto, M. (1990). *Fundamentals of Bacterial Plant Pathology.* Academic Press, San Diego, USA.

*****Kujit, J.** (1969). *The Biology of Parasitic Flowering Plants.* University of California Press, Berkeley, USA.

*****Matthews, R.E.F.** (1991). *Plant Virology,* 3rd Edn. Academic Press, San Diego, USA.

*****Matthews, R.E.F.** (1992). *Fundamentals in Plant Virology.* Academic Press, San Diego, USA.

Merlo, D. (1982). Crown gall: a multipotential disease. *Advances in Plant Pathology* **1**, 139–79.

Parker, C. & Riches, C.R. (1993). *Parasitic Weeds of the World: Biology and Control.* CAB International, Wallingford, UK.

*****Press, M.C. & Graves, J.D.** (1995). *Parasitic Plants.* Chapman & Hall, London, UK.

Smith, K.M. (1972). *A Textbook of Plant Virus Diseases,* 3rd Edn. Longman, London, UK.

Smith, K.M. (1977). *Plant Viruses,* 6th Edn. Chapman & Hall, London, UK.

Walkey, D.G.A. (1991). *Applied Plant Virology,* 2nd Edn. Heinemann, London, UK.

* Key references

Further Reading: Appendix
Bennett, W.F. (1993). Nutrient Deficiencies and Toxicities in Plants. APS Press, St Paul, USA.

Dijkstra, J. & de Jager, C.P. (1998). *Practical Plant Virology: Protocols and Exercises.* Springer Laboratory Manual, Berlin, Germany.

Fox, R.T.V. (1993). *Principles of Diagnosis of Plant Diseases.* CAB International, Wallingford, UK.

Hall, G. & Minter, D.W. (1994). *International Mycological Directory,* 3rd Edn. International Mycological

Association and International Mycological Institute, Egham, UK.

Hawksworth, D.L., Kirk, P.M., Sutton, B.G. & Pegler, D.N. (1995). *Ainsworth and Bisby's Dictionary of the Fungi*, 8th Edn. CAB International, Wallingford, UK.

Hill, S.A. (1984). *Methods in Plant Virology*. Blackwell Scientific Publications, Oxford, UK.

Ingram, D.S. (1999). Biodiversity, plant pathogens and conservation. *Plant Pathology*. In press.

Johnston, A. & Booth, C. (1983). *Plant Pathologist's Pocketbook*, 2nd Edn. Commonwealth Agricultural Bureau, Slough, UK.

Lelliot, R.A. & Stead, D.E. (1987). *Methods for the Diagnosis of Bacterial Diseases of Plants*. Blackwell Scientific Publications, Oxford, UK.

Lucas, G.B., Campbell, C.L. & Lucas, L.T. (1985). *Introduction to Plant Diseases – Identification and Management*. AVI Publishing Co., Connecticut, USA.

Schots, A., Dewey, F.M. & Oliver, R. (1994). *Modern Assays for Plant Pathogenic Fungi: Identification, Detection and Quantification*. CAB International, Wallingford, UK.

Smith, D. & Onions, A.H.S. (1994). *The Preservation and Maintenance of Living Fungi*, 2nd Edn. CAB International, Wallingford, UK.

Waller, J.M., Ritchie, B.J. & Holderness, M. (1998). *Plant Clinic Handbook*. CAB International, Wallingford, UK.

Index